GLENCOE

GEOMETRY
INTERACTIVE STUDENT GUIDE

Cover: Seifert Surface. By Paul Nylander (bugman123.com)

mheducation.com/prek-12

Send all inquiries to:
McGraw-Hill Education
8787 Orion Place
Columbus, OH 43240

ISBN: 978-0-07-906177-5
MHID: 0-07-906177-X

Printed in the United States of America.

4 5 6 7 8 9 QVS 22 21 20 19 18

Contents

Each chapter of the *Interactive Student Guide* has features to help you succeed.

Chapter Overview

Each **Chapter Focus** section gives you an overview of what you will learn. The key standard from each lesson is listed along with preview questions. As you work through each chapter, revisit this section to complete each question.

> ### Quadrilaterals
>
> **CHAPTER FOCUS** Learn about some of the objectives that you will explore in this chapter. Answer the preview question. As you complete each lesson, return to these pages to check your work.
>
> | What You Will Learn | Preview Question |
> |---|---|
> | **Parallelograms** | |
> | • Prove theorems about parallelograms using two-column and paragraph proofs.
• Use coordinates to prove theorems about parallelograms. | **SMP 2** Three vertices of a parallelogram are (0, 4), (5, 0), and (10, 4). List all possible locations of the fourth vertex. |
> | **Proving...** | |
> | ...making ...about | **SMP 3** Carly drew the following figure to prove that if the diagonals of a quadrilateral are congruent, then the quadrilateral is a parallelogram. Draw a counterexample to show that Carly is incorrect. What mistake did Carly make? |
> | | **SMP 2** The vertices of quadrilateral ABCD are A(−2, 3), B(1, 6), C(7, 3), and D(5, 1). Find the slope of each side and determine if ABCD is a parallelogram. Explain your reasoning. |
> | | **SMP 2** How could point C be moved so that ABCD is a parallelogram? |
>
> **186** CHAPTER 6 Quadrilaterals

> **SMP 2** Three vertices of a parallelogram are (0, 4), (5, 0), and (10, 4). List all possible locations of the fourth vertex.

Lesson Overview

Each **lesson** gives you ample opportunity to explore concepts and deepen understanding.

A strong emphasis on modeling helps you connect mathematics to the real world. Throughout the program you will develop, test, and refine models to more accurately represent a real-world situation.

> ### Representations of Three-Dimensional Figures
> *Use with Lesson 11-1*
>
> **Objectives**
> • Identify the shape of the two-dimensional cross section of a three-dimensional object.
> • Identify the shape of a three-dimensional object generated by the rotation of a two-dimensional object.
>
> A **cross-section** is the intersection of a three-dimensional figure and a plane. The shape of the cross-section is determined by the type of solid and the angle of the plane.
>
> **EXAMPLE 1** Investigate Cross Sections
>
> **a. REASON ABSTRACTLY** Julio has a piece of modeling clay that is molded into a perfect sphere. Julio cuts the sphere so that he has two shapes with flat surfaces remaining. Identify the cross sections formed by horizontal cuts, vertical cuts, and diagonal cuts.
>
> **b. REASON ABSTRACTLY** Are any other cross sections possible?
>
> **c. REASON ABSTRACTLY** Julio also had a right cylinder made of clay and set it on one of its circular bases. Name the shapes for two possible cross sections and describe how you would slice the cylinder into two pieces to obtain each shape.
>
> **d. CRITIQUE REASONING** Julio believes that he can slice the cylinder in a way that would result in the shape to the right. Do you agree? Explain.
>
> **350** CHAPTER 11 Extending Volume

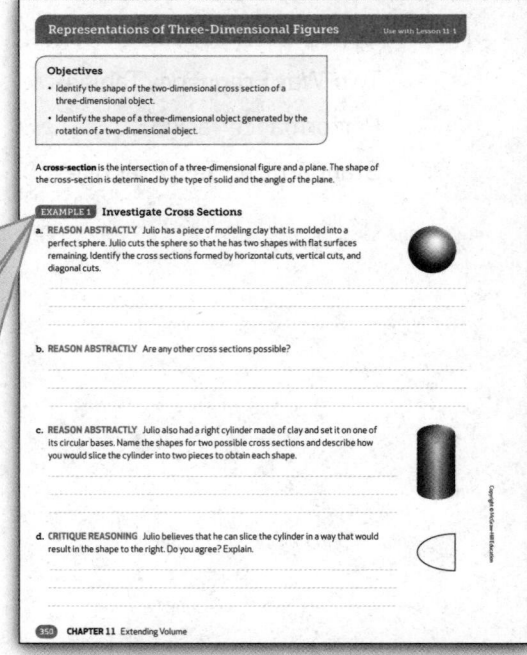

> **a. USE A MODEL** His first stack is made of 8 coins. Each coin is 1 unit high and 20 units across. What is the area of the face of each coin in terms of π? **SMP 4**

Assessment Practice

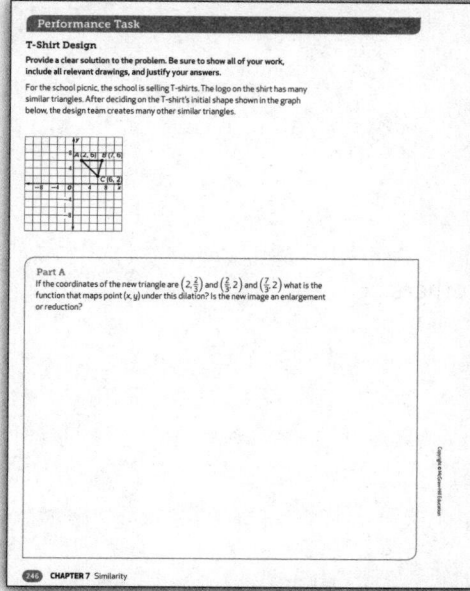

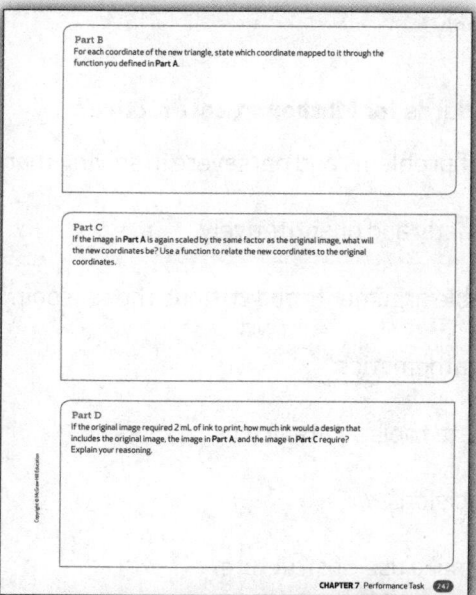

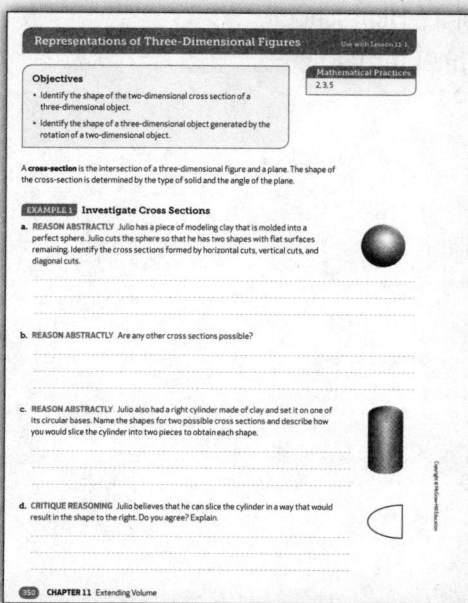

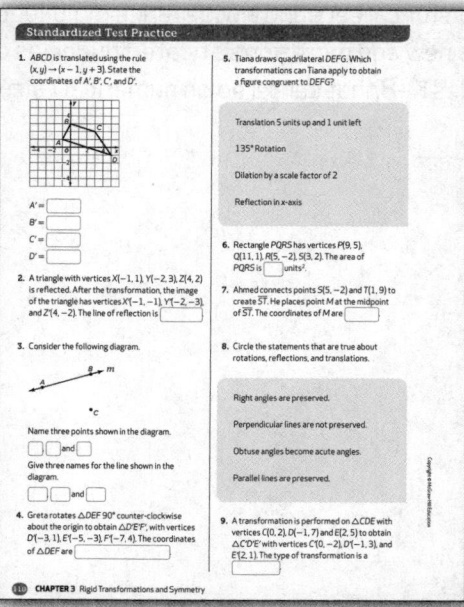

Guide to Developing the Standards for Mathematical Practice

This guide provides a standard-by-standard analysis of the approach taken to each Mathematical Practice, including its meaning and the types of questions students can use to enhance mathematical development.

The goal of the practice standards is to instill in ALL students the abilities to be mathematically literate and to create a positive disposition for the importance of using math effectively.

What are the Standards for Mathematical Practice?

1. Make sense of problems and persevere in solving them.

2. Reason abstractly and quantitatively.

3. Construct viable arguments and critique the reasoning of others.

4. Model with mathematics.

5. Use appropriate tools strategically.

6. Attend to precision.

7. Look for and make use of structure.

8. Look for and express regularity in repeated reasoning.

Why are the Standards for Mathematical Practice important?

The Standards for Mathematical Practice set expectations for using mathematical language and representations to reason, solve problems, and model in preparation for careers and a wide range of college majors. High school mathematics builds new and more sophisticated fluencies on top of the earlier fluencies from grades K–8 that centered on numerical calculation.

1 Make sense of problems and persevere in solving them.

Mathematically proficient students start by explaining to themselves the meaning of a problem and looking for entry points to its solution. They analyze givens, constraints, relationships, and goals. They make conjectures about the form and meaning of the solution and plan a solution pathway rather than simply jumping into a solution attempt. They consider analogous problems, and try special cases and simpler forms of the original problem in order to gain insight into its solution. They monitor and evaluate their progress and change course if necessary. Older students might, depending on the context of the problem, transform algebraic expressions or change the viewing window on their graphing calculator to get the information they need. Mathematically proficient students can explain correspondences between equations, verbal descriptions, tables, and graphs or draw diagrams of important features and relationships, graph data, and search for regularity or trends. Younger students might rely on using concrete objects or pictures to help conceptualize and solve a problem. Mathematically proficient students check their answers to problems using a different method, and they continually ask themselves, "Does this make sense?" They can understand the approaches of others to solving complex problems and identify correspondences between different approaches.

What does it mean?	What questions do I ask?
Solving a mathematical problem takes time. Use a logical process to make sense of problems, understand that there may be more than one way to solve a problem, and alter the process if needed.	• What am I being asked to do or find? What do I know? • How does the given information relate to each other? Does a graph or diagram help? • Is this problem similar to any others I have solved? • What is my plan for solving the problem? • What should I do if I get "stuck"? • Does the answer make sense? • Is there another way to solve the problem? • Now that I've solved the problem, what did I do well? How would I approach a similar problem next time?

2 Reason abstractly and quantitatively.

Mathematically proficient students make sense of quantities and their relationships in problem situations. They bring two complementary abilities to bear on problems involving quantitative relationships: the ability to decontextualize—to abstract a given situation and represent it symbolically and manipulate the representing symbols as if they have a life of their own, without necessarily attending to their referents—and the ability to contextualize, to pause as needed during the manipulation process in order to probe into the referents for the symbols involved. Quantitative reasoning entails habits of creating a coherent representation of the problem at hand; considering the units involved; attending to the meaning of quantities, not just how to compute them; and knowing and flexibly using different properties of operations and objects.

What does it mean?	What questions do I ask?
You can start with a concrete or real-world context and then represent it with abstract numbers or symbols (decontextualize), find a solution, then refer back to the context to check that the solution makes sense (contextualize).	• What do the numbers represent? What are the variables, and how are they related to each other and to the numbers? • How can the relationships be represented mathematically? Is there more than one way? • How did I choose my method? • Does my answer make sense in this problem? • Does my answer fit the facts given in the problem? If not, why not?

3 Construct viable arguments and critique the reasoning of others.

Mathematically proficient students understand and use stated assumptions, definitions, and previously established results in constructing arguments. They make conjectures and build a logical progression of statements to explore the truth of their conjectures. They are able to analyze situations by breaking them into cases, and can recognize and use counterexamples. They justify their conclusions, communicate them to others, and respond to the arguments of others. They reason inductively about data, making plausible arguments that take into account the context from which the data arose. Mathematically proficient students are also able to compare the effectiveness of two plausible arguments, distinguish correct logic or reasoning from that which is flawed, and—if there is a flaw in an argument—explain what it is. Elementary students can construct arguments using concrete referents such as objects, drawings, diagrams, and actions. Such arguments can make sense and be correct, even though they are not generalized or made formal until later grades. Later, students learn to determine domains to which an argument applies. Students at all grades can listen or read the arguments of others, decide whether they make sense, and ask useful questions to clarify or improve the arguments.

What does it mean?	What questions do I ask?
Sound mathematical arguments require a logical progression of statements and reasons. You can clearly communicate their thoughts and defend them.	• How did I get that answer? • Is that always true? • Why does that work? What mathematical evidence supports my answer? • Can I use objects in the classroom to show that my answer is correct? • Can I give a "nonexample" or a counterexample? • What conclusion can I draw? What conjecture can I make? • Is there anything wrong with that argument?

4 Model with mathematics.

Mathematically proficient students can apply the mathematics they know to solve problems arising in everyday life, society, and the workplace. In early grades, this might be as simple as writing an addition equation to describe a situation. In middle grades, a student might apply proportional reasoning to plan a school event or analyze a problem in the community. By high school, a student might use geometry to solve a design problem or use a function to describe how one quantity of interest depends on another. Mathematically proficient students who can apply what they know are comfortable making assumptions and approximations to simplify a complicated situation, realizing that these may need revision later. They are able to identify important quantities in a practical situation and map their relationships using such tools as diagrams, two-way tables, graphs, flowcharts and formulas. They can analyze those relationships mathematically to draw conclusions. They routinely interpret their mathematical results in the context of the situation and reflect on whether the results make sense, possibly improving the model if it has not served its purpose.

What does it mean?	What questions do I ask?
Modeling links classroom mathematics and statistics to everyday life, work, and decision-making. High school students at this level are expected to apply key takeaways from earlier grades to high-school level problems.	• How might I represent the situation mathematically? • How does my equation or diagram model the situation? • What assumptions can I make? Should I make them? • What is the best way to organize the information? What other information is needed? • Can I make a good estimate of the answer? • Does my answer make sense?

5 Use appropriate tools strategically.

Mathematically proficient students consider the available tools when solving a mathematical problem. These tools might include pencil and paper, concrete models, a ruler, a protractor, a calculator, a spreadsheet, a computer algebra system, a statistical package, or dynamic geometry software. Proficient students are sufficiently familiar with tools appropriate for their grade or course to make sound decisions about when each of these tools might be helpful, recognizing both the insight to be gained and their limitations. For example, mathematically proficient high school students analyze graphs of functions and solutions generated using a graphing calculator. They detect possible errors by strategically using estimation and other mathematical knowledge. When making mathematical models, they know that technology can enable them to visualize the results of varying assumptions, explore consequences, and compare predictions with data. Mathematically proficient students at various grade levels are able to identify relevant external mathematical resources, such as digital content located on a website, and use them to pose or solve problems. They are able to use technological tools to explore and deepen their understanding of concepts.

What does it mean?	What questions do I ask?
Certain tools, including estimation and virtual tools, are more appropriate than others. You should understand the benefits and limitations of each tool.	• What tools would help to visualize the situation? • What are the limitations of using this tool? • Is an exact answer needed? • How can I use estimation as a tool? • Can I find additional information on the Internet? • Can I solve this problem using another tool?

6 Attend to precision.

Mathematically proficient students try to communicate precisely to others. They try to use clear definitions in discussion with others and in their own reasoning. They state the meaning of the symbols they choose, including using the equal sign consistently and appropriately. They are careful about specifying units of measure, and labeling axes to clarify the correspondence with quantities in a problem. They calculate accurately and efficiently, express numerical answers with a degree of precision appropriate for the problem context. In the elementary grades, students give carefully formulated explanations to each other. By the time they reach high school they have learned to examine claims and make explicit use of definitions.

What does it mean?	What questions do I ask?
Precision in mathematics is more than accurate calculations. It is also the ability to communicate with the language of mathematics. In high school mathematics, precise language makes for effective communication and serves as a tool for understanding and solving problems.	• How can the everyday meaning of a math term help me remember the math meaning? • Can I give some examples and nonexamples of that term? • Is this term similar to something I already know? • What does the math symbol mean? How do I know? • How do the terms in the problem help to solve it? • What does the variable represent, and in what units? • Does the question require a precise answer or is an estimate sufficient? If the answer needs to be precise, how precise? • Have I checked my answer for the correct labels?

7 Look for and make use of structure.

Mathematically proficient students look closely to discern a pattern or structure. Young students, for example, might notice that three and seven more is the same amount as seven and three more, or they may sort a collection of shapes according to how many sides the shapes have. Later, students will see 7×8 equals the well remembered $7 \times 5 + 7 \times 3$, in preparation for learning about the distributive property. In the expression $x^2 + 9x + 14$, older students can see the 14 as 2×7 and the 9 as $2 + 7$.

They recognize the significance of an existing line in a geometric figure and can use the strategy of drawing an auxiliary line for solving problems. They also can step back for an overview and shift perspective. They can see complicated things, such as some algebraic expressions, as single objects or as being composed of several objects. For example, they can see $5 - 3(x - y)^2$ as 5 minus a positive number times a square and use that to realize that its value cannot be more than 5 for any real numbers x and y.

What does it mean?	What questions do I ask?
Mathematics is based on a well-defined structure. Mathematically proficient students look for that structure to find easier ways to solve problems.	• Can I think of an easier way to find the solution? • How can using what I already know help solve this problem? • How are numerical expressions and algebraic expressions the same? How are they different? • Can the terms of this expression be grouped in a way that would allow it to be simplified or give us more information? • How can what I know about integers help with polynomials? • What shapes do I see in the figure? Could a line or segment be added to the figure that would give us more information?

8 Look for and express regularity in repeated reasoning.

Mathematically proficient students notice if calculations are repeated and look for general methods and shortcuts. Upper elementary students might notice when dividing 25 by 11 that they are repeating the same calculations over and over again, and conclude they have a repeating decimal. By paying attention to the calculation of slope as they repeatedly check whether points are on the line through $(1, 2)$ with slope 3, middle school students might abstract the equation $\frac{y - 2}{x - 1} = 3$. Noticing the regularity in the way terms cancel when expanding $(x - 1)(x + 1)$, $(x - 1)$ $(x^2 + x + 1)$, and $(x - 1)(x^3 + x^2 + x + 1)$ might lead them to the general formula for the sum of a geometric series. As they work to solve a problem, mathematically proficient students maintain oversight of the process, while attending to the details. They continually evaluate the reasonableness of their intermediate results.

What does it mean?	What questions do I ask?
Mathematics has been described as the study of patterns. Recognizing a pattern can lead to results more quickly and efficiently.	• Is there a pattern? • Is this pattern like one I've seen before? How is it different? • What does this problem remind me of? • Is this problem similar to something already known? • What would happen if I…? • How would I prove that? • How would this work with other numbers? Does it work all the time? How do I know? • Would technology help model this situation? How?

1 Tools of Geometry

CHAPTER FOCUS Learn about some of the objectives that you will explore in this chapter. Answer the preview questions. As you complete each lesson, return to these pages to check your work.

What You Will Learn	Preview Question
Undefined Terms	
• Identify undefined terms. • Know the precise definitions of geometric terms.	**SMP 4** What undefined geometric term is modeled by the intersection of a wall and the floor? **A line is formed by the intersection.**
Length	
• Find the length of a segment. • Construct a congruent segment. • Find the distance between two points on a coordinate plane. • Find the coordinates of a point on a directed line segment.	**SMP 6** Point B is between point A and point C. If $AC = 10$ and $AB = 6$, how can you determine BC? **Because B is between A and C, $AB + BC = AC$,** **$6 + BC = 10$. So, $BC = 10 - 6 = 4$.**
Directed Line Segments and Vectors	
• Apply the Law of Sines and Law of Cosines to solve vector problems. • Find the point on a directed line segment that partitions the segment in a given ratio.	**SMP 4** A hiker walks in the direction 50° east of north for 2 hours at a speed of 3.5 miles per hour. How far east does the hiker travel? Round your answer to the nearest hundredth. **5.36 miles** **SMP 6** Find the coordinates of the point that partitions the directed line segment from $A(1, 3)$ to $B(5, 11)$ in the ratio 1 to 3. Then write the component form of the vector from the initial point to the partition point. **(2, 5); $\langle 1, 2 \rangle$**

What You Will Learn	Preview Question
Angles	
• Measure and classify angles. • Identify and use special properties of angles. • Construct a copy of an angle.	**SMP 1** $\angle QSR$ and $\angle RST$ are complementary. $\angle SU$ is the bisector of $\angle RST$ and $m\angle QSU = 72$. What is $m\angle RST$?
Polygons	
• Find the perimeters of polygons on the coordinate plane. • Find the areas of triangles and rectangles on the coordinate plane.	**SMP 6** A square in the coordinate plane has a diagonal with endpoints at the origin and $(0, 5)$. What is the area of the square? Explain your reasoning. _____ _____ _____ _____

Objectives

- Identify undefined terms.
- Know the precise definitions of geometric terms.

In geometry, most of the terms you will use will be defined precisely. However, there are a few **undefined terms** that will only be explained using examples and descriptions.

EXAMPLE 1 Identify Undefined Terms

EXPLORE Cayden saw the figure at the right in his geometry textbook. He was not sure what the term *parallel lines* meant, so he looked up the term in the glossary.

m

n

Lines *m* and *n* are parallel lines.

a. **USE TOOLS** Look up the definition of *parallel lines* online or in your glossary and write it here.

_____

b. **COMMUNICATE PRECISELY** What geometry terms do you need to understand in order to understand the definition of *parallel lines*?

_____

c. **USE TOOLS** Now look up the definitions of the terms you identified in **part b** and write them here.

_____

_____

_____

d. **COMMUNICATE PRECISELY** Based on the definitions you wrote above, which terms do you think are the most basic ones in geometry? Why?

_____

_____

e. **COMMUNICATE PRECISELY** Write your own descriptions of the basic terms you identified in **part d**.

_____

_____

f. **COMMUNICATE PRECISELY** How many points are on a line? How many points are in a plane?

_____

_____

The Key Concept box summarizes the undefined terms you may have identified in the exploration.

KEY CONCEPT Undefined Terms

Description	Figure	Name
A point is a location. It has neither shape nor size.	• P	point P
A line is made up of points and has no thickness or width.	B, A, ℓ	line ℓ, line AB, line BA, $\overleftrightarrow{AB}$, $\overleftrightarrow{BA}$
A plane is a flat surface made up of points that extends infinitely in all directions.	J, L, K, $\mathcal{R}$	plane $\mathcal{R}$, plane JKL, plane JLK, etc.

A **definition** is an explanation of a term that uses undefined terms and/or previously defined terms.

EXAMPLE 2 Write Precise Definitions

Follow these steps to write your own definitions of some geometric terms.

a. **INTERPRET PROBLEMS** The table shows an example and a nonexample of collinear points. Describe the collinear points.

Example	Nonexample
A, B, C	S, R, T
A, B, and C are collinear points.	R, S, and T are not collinear points.

b. **COMMUNICATE PRECISELY** Write your own definition of *collinear points*.

c. **COMMUNICATE PRECISELY** Which undefined terms did you use in your definition?

d. **MAKE A CONJECTURE** What do you think the term *coplanar* means? Write your own definition of *coplanar* and then check your definition by looking up the term online or in a glossary.

1. The table shows an example and a nonexample of concurrent lines.

Example	Nonexample
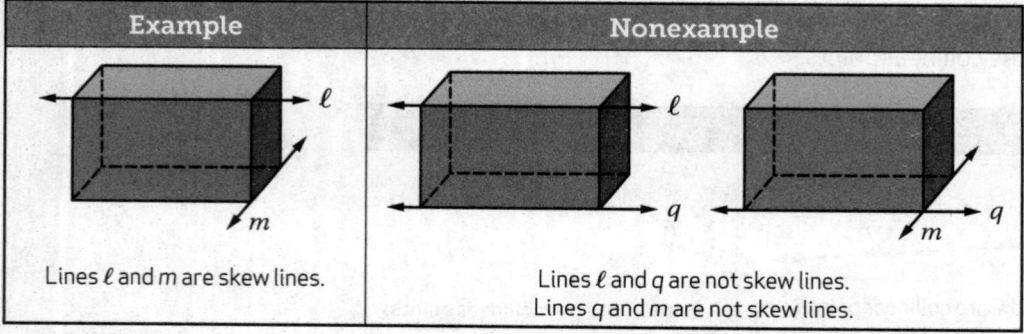 Lines ℓ, m, and n are concurrent lines.	Lines p, q, and r are not concurrent lines.

   a. **INTERPRET PROBLEMS** Describe the concurrent lines.

   b. **COMMUNICATE PRECISELY** Write your own definition of *concurrent lines*.

   c. **COMMUNICATE PRECISELY** Which undefined terms did you use in your definition?

2. The table shows pairs of lines created by the edges of a box that are an example and two nonexamples of skew lines.

Example	Nonexample
Lines ℓ and m are skew lines.	Lines ℓ and q are not skew lines. Lines q and m are not skew lines.

   a. **INTERPRET PROBLEMS** How is the relationship between lines ℓ and m different from the relationship between lines ℓ and q?

   b. **INTERPRET PROBLEMS** How is the relationship between lines ℓ and m different from the relationship between lines q and m?

   c. **COMMUNICATE PRECISELY** Use your answers to **parts a** and **b** to write your own definition of skew lines.

   d. **COMMUNICATE PRECISELY** Which undefined terms did you use in your definition?

e. COMMUNICATE PRECISELY Which defined terms did you use in your definition?

_____

3. CRITIQUE REASONING Tessa was asked to write a definition for the term _square_. She wrote, "A square is a geometric figure with four sides."

a. Do you think this is a good definition? Explain why or why not.

_____

_____

b. Write a better definition for "square."

_____

_____

4. COMMUNICATE PRECISELY Eliza is a programmer for a company that makes video games. She is working on a new science-fiction game that includes some unusual creatures. She prepares the following table to help other programmers work with the creatures.

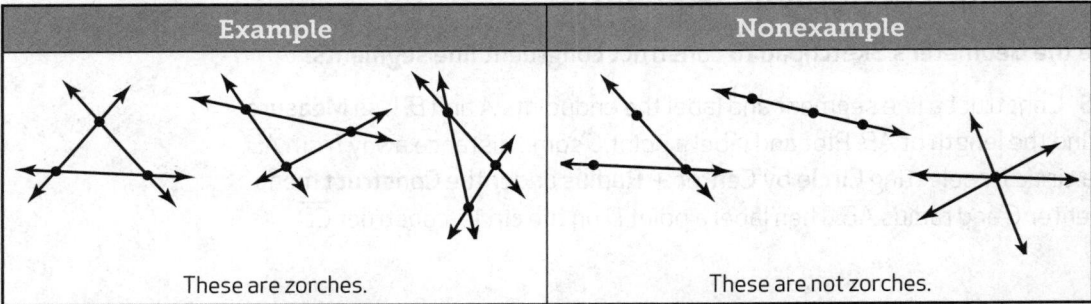

Example	Nonexample
These are zorches.	These are not zorches.

a. Write a definition of a _zorch_.

_____

_____

b. One of Eliza's colleagues drew the creature at the right for the game and claimed that it is a zorch. Use your definition to explain why it is or is not a zorch.

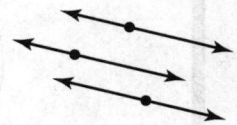

_____

_____

5. INTERPRET PROBLEMS If three lines are coplanar, how many points of intersection can there be? Draw figures in the box below to support your answer.

_____

```




```

Objectives

- Find the length of a segment.
- Construct a congruent segment.
- Find the distance between two points on a coordinate plane.
- Find the coordinates of a point on a directed line segment.

A part of a line consisting of two endpoints and all the points between them is called a **line segment**. The segment with endpoints P and Q is called $\overline{PQ}$ or $\overline{QP}$, and it has a measurable **length** designated by PQ. The length always includes a unit of measure. **Congruent** segments have the same length. Many tools can be used to construct a segment congruent to a given segment.

EXAMPLE 1   Create a Congruent Segment

EXPLORE   Use the Geometer's Sketchpad to construct congruent line segments.

a. **USE TOOLS**   Construct a line segment and label the endpoints A and B. Use **Measure Length** to find the length of $\overline{AB}$. Plot and label a point C some distance away from $\overline{AB}$. Construct a circle by selecting **Circle by Center + Radius** under the **Construct** menu and using center C and radius $\overline{AB}$. Then label a point D on the circle, construct $\overline{CD}$, and find CD.

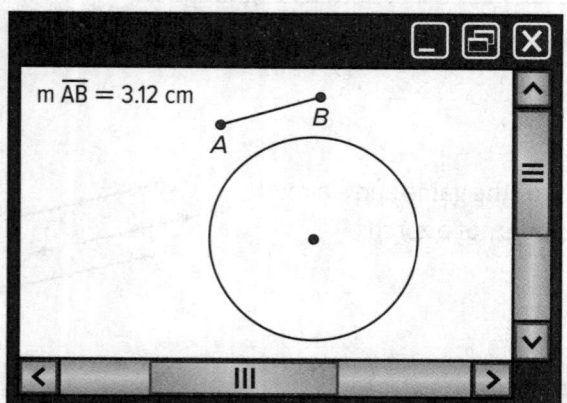
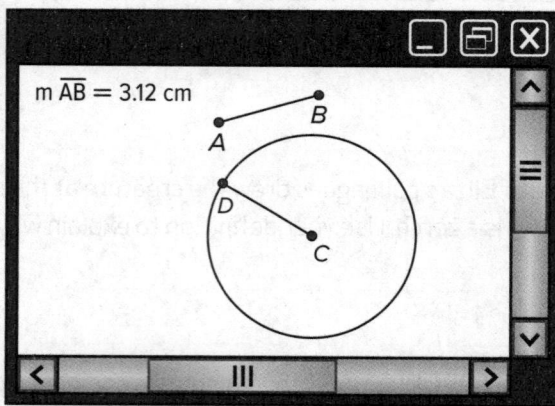

b. **REASON ABSTRACTLY**   What is the relationship of the two segments? If another point E is selected somewhere else on the circle, will $\overline{CE}$ have the same relationship with $\overline{AB}$?

A point lies on a segment if the point is between the endpoints of the segment.

Point C is **between** points A and B if and only if A, B, and C are collinear and $AC + CB = AB$. This definition allows us to write and solve equations to find the length of a segment.

EXAMPLE 2 **Write and Solve Equations to Find Measurements**

a. **REASON QUANTITATIVELY** Point D is between points C and E. Find CE.

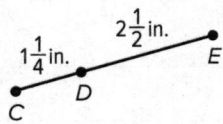

b. **REASON ABSTRACTLY** If $JK = 2x - 3$ and $KL = x - 1$, find the value of x and the lengths of $\overline{JK}$ and $\overline{KL}$.

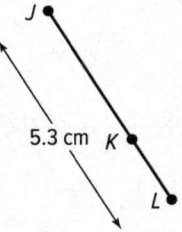

5.3 cm

c. **MAKE A CONJECTURE** Is there a point B on $\overline{AC}$, for which $2(AB) = AC$? Explain.

When line segments are used to show a movement, they are often shown as a directed line segment on a coordinate plane. Recall that the length of the segment on a coordinate plane is found using the distance formula. That is, if M has coordinates (x_1, y_1) and N has coordinates (x_2, y_2), then $MN = \sqrt{(x_2 - x_1)^2 + (y_2 - y_1)^2}$. While a line segment has two endpoints, a **directed** line segment has a starting endpoint and a terminal endpoint. To find the coordinates of a point that partitions a directed line segment into a given ratio, add the fraction of the horizontal and vertical movement to the coordinates of the starting point.

EXAMPLE 3 **Find Points on a Segment**

a. **CALCULATE ACCURATELY** Use the distance formula to find the exact length of $\overline{AB}$.

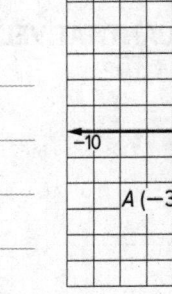

b. COMMUNICATE PRECISELY Find the coordinates of point C on directed line segment AB that partitions it into two segments in a ratio of 2 to 1. Explain your solution.

_____

_____

_____

c. CONSTRUCT ARGUMENTS Use the distance formula to verify that the ratio of $AC:CB$ is equal to 2:1.

_____

_____

_____

d. COMMUNICATE PRECISELY Find the coordinates of point D on $\overline{AB}$ that divides it into a 1 to 1 ratio. Show your work.

_____

_____

_____

e. EVALUATE REASONABLENESS Find the midpoint of $\overline{AC}$ using the formula $M = \left(\frac{x_1 + x_2}{2}, \frac{y_1 + y_2}{2} \right)$ where (x_1, y_1) and (x_2, y_2) are the endpoints of the segment. How does this compare to the coordinates for point D in $\frac{x_1 + x_2}{2}$ **part d**? What conclusion can you make?

_____

_____

_____

PRACTICE

1. a. REASON QUANTITATIVELY For line segment $\overline{AC}$, write and solve an equation to find AB.

_____

_____

_____

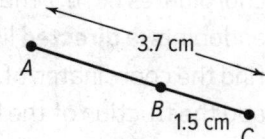

b. REASON QUANTITATIVELY What length would EF need to be for $\overline{DE}$ to be congruent to $\overline{AB}$?

_____

_____

_____

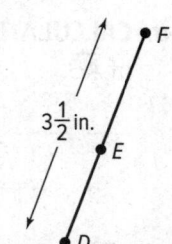

2. USE STRUCTURE Consider a rectangle $QRST$ with $QR = ST = 4$ cm and $RS = QT = 2$ cm. If point U is on $\overline{QR}$ such at $QU = UR$ and point V is on $\overline{RS}$ such that $RV = VS$, then is $\overline{QU}$ congruent to $\overline{RV}$? Explain your reasoning.

_____

_____

_____

3. CALCULATE ACCURATELY What is the exact length of $\overline{RQ}$, shown at right?

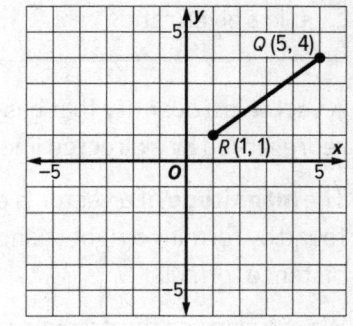

_____

_____

_____

_____

4. a. REASON QUANTITATIVELY If you were to add a point T to RQ from **Exercise 3** such that the ratio of $\overline{RT}$ to $\overline{TQ}$ is 3 to 2, what would be the coordinates of T?

_____

_____

_____

b. COMMUNICATE PRECISELY Find the midpoint M of $\overline{RQ}$. Without using the distance formula, calculate MT. Explain your reasoning.

_____

_____

_____

_____

c. Use the distance formula to confirm your result from **part b.**

_____

_____

_____

_____

_____

_____

_____

Objectives

- Apply the Law of Sines and Law of Cosines to solve vector problems.

- Find the point on a directed line segment that partitions the segment in a given ratio.

A **vector** is a quantity that has both a *direction* and a *magnitude*. A vector can be represented by a directed line segment that has an initial point and a terminal point.

The **magnitude** of a vector is equal to its distance. Therefore, the magnitude may be found by forming a right triangle with the vector as the hypotenuse and using the distance formula.

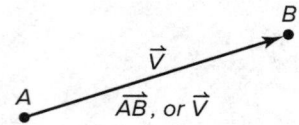

We can also use this triangle to find the *direction* of the vector. The **direction** of a vector can be described in terms of the angle formed with the horizontal similar to identifying angles in a unit circle. Therefore, we can use the tangent relationship to determine the direction.

| EXAMPLE 1 | **Determine Size and Direction of Vectors** |

EXPLORE Find the magnitude and direction of vector $\vec{v}$.

a. **USE STRUCTURE** Sketch a right triangle with EF as the hypotenuse. Label the side opposite E, e. Label the side opposite F, f.

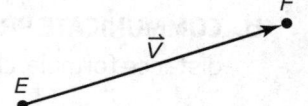

b. **PLAN A SOLUTION** How can you use the triangle you sketched to find the magnitude of $\vec{v}$?

_____

_____

_____

c. **PLAN A SOLUTION** Explain how to use the triangle you sketched to find the direction of $\vec{v}$.

_____

_____

_____

A vector is in standard position if its initial point is at the origin on the coordinate plane. A vector in standard position may be easily described by the coordinates of a single point, written in the format (x, y), referred to as **component form**. Any vector can be written in this form by determining its vertical and horizontal components. This form is convenient for addition and subtraction of vectors. In order to add or subtract two vectors written in this form, simply add or subtract each coordinate separately. If a vector is described by the directed line segment from (a_1, b_1) to (a_2, b_2), then its component form is $(a_2 - a_1, b_2 - b_1)$.

A **resultant vector** is the sum of two or more vectors. You can use the Law of Sines and the Law of Cosines to determine the direction and magnitude of resultant vectors.

Determine a Resultant Vector

Alicia is piloting a small plane and flying at a constant rate of 300 miles per hour in the direction 50° west of north. The wind is blowing due south at 60 miles per hour. Alicia wants to know the plane's resultant speed and direction.

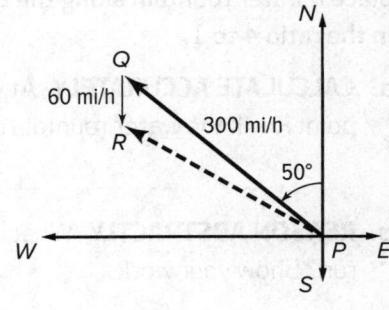

a. **USE STRUCTURE** The figure represents the situation. $\overrightarrow{PR}$ is the resultant vector. What is $m\angle PQR$ in the figure? How do you know?

_____

_____

_____

b. **PLAN A SOLUTION** What property, formula, or theorem can you use to find the magnitude of $\overrightarrow{PR}$? Explain.

_____

_____

c. **REASON ABSTRACTLY** Find the magnitude of $\overrightarrow{PR}$ to the nearest tenth and explain the key steps in your solution.

_____

_____

_____

d. **REASON ABSTRACTLY** Show how you can find $m\angle QPR$ to the nearest degree.

_____

_____

_____

e. **USE A MODEL** Apply and interpret your results. What is the resultant speed and direction of Alicia's plane? Explain. Find the component form of the resultant vector in terms of the direction west and north.

_____

_____

_____

f. **INTERPRET PROBLEMS** Draw a figure with a dashed vector representing the speed and direction Alicia should fly to compensate for the wind where she has a resultant speed and direction of 300 mi/h in the direction 50° west of north.

_____

_____

_____

EXAMPLE 3 Partition a Directed Line Segment

The manager of a botanical garden is planning a new nature trail that can be represented by the directed line segment from $J(-7, -6)$ to $K(8, 4)$. She wants to place a water fountain along the trail at a location W that partitions the segment in the ratio 4 to 1.

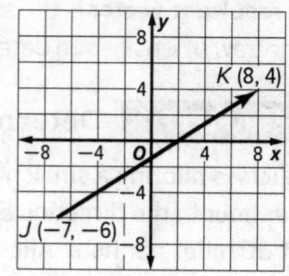

a. **CALCULATE ACCURATELY** At what percent of the distance from point J to point K will the water fountain be located? Explain.

_____

b. **REASON ABSTRACTLY** What is the rise for the directed line segment? What is the run? Show your work.

_____

c. **REASON ABSTRACTLY** Explain why the slope of $\overline{JW}$ must be the same as the slope of $\overline{JK}$.

_____

_____

d. **EVALUATE REASONABLENESS** Using the percent p that you calculated in **part b** and the rise and run from **part c**, add p% of the run to the x-coordinate of point J and add p% of the rise to the y-coordinate of point J. Give the coordinates of the resulting point W. Explain how you know point W is the required point.

_____

_____

_____

PRACTICE

CALCULATE ACCURATELY Find the coordinates of the point that partitions each vector in the given ratio. Then write the component form of the vector.

1. the directed line segment from $A(-3, 1)$ to $B(6, 4)$; 2 to 1

2. the directed line segment from $C(-2, 3)$ to $D(6, -1)$; 1 to 3

_____

3. the directed line segment from $P(-3, -1)$ to $Q(4, 6)$; 4 to 3

4. the directed line segment from $R(-4, 6)$ to $S(0, -6)$; 3 to 1

_____

5. **USE STRUCTURE** Find the magnitude and direction to the nearest tenth of a degree of the vector given in component form.

   a. $\langle 2, 5 \rangle$

   b. $\langle -4, 1 \rangle$

   _____

   c. $\langle -1, -3 \rangle$

   d. $\langle 3, -6 \rangle$

   _____

6. **CALCULATE ACCURATELY** Find the magnitude and direction to the nearest tenth of a degree of the vector given as a directed line segment.

 a. from $P(-2, -1)$ to $Q(1, 6)$

 b. from $P(3, 5)$ to $Q(-1, 2)$

7. Antoine is paddling a canoe at a constant rate of 3 miles per hour in the direction 25° east of north. There is a current moving due east at 1 mile per hour. Antoine wants to know the canoe's resultant speed and direction.

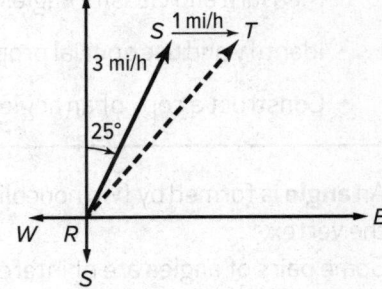

 a. **USE STRUCTURE** Explain how to find $m\angle RST$ in the figure.

 b. **CALCULATE ACCURATELY** Find the magnitude of the resultant vector $\overrightarrow{RT}$ to the nearest tenth. What relationship or theorem did you use to find the magnitude?

 c. **COMMUNICATE PRECISELY** Explain how to find the canoe's resultant direction to the nearest degree.

8. **USE A MODEL** Two forces are applied to an object A at the same time. As shown in the figure, a 10-pound force is applied at an angle of 45° to the horizontal and a 5-pound force is applied at an angle of 120° to the horizontal. The resultant force is represented by $\overrightarrow{AD}$ in the figure. Find the magnitude of the resultant force to the nearest tenth and the direction of the resultant force to the nearest degree.

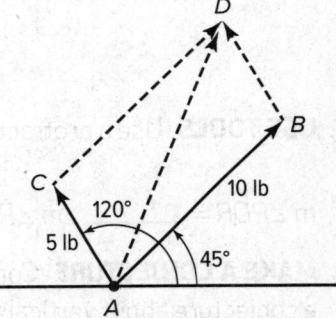

9. **INTERPRET PROBLEMS** Omar and Julie are in a plane that is to fly southwest at a speed of 600 miles per hour. The pilot tells the passengers that there is a crosswind blowing due east at 40 miles per hour. Calculate the speed and heading the plane will need to fly at in order to compensate for the crosswind. Sketch a figure in the space provided to help you explain your answer.

Objectives

- Measure and classify angles.
- Identify and use special properties of angles.
- Construct a copy of an angle.

Mathematical Practices
1, 2, 3, 5, 6, 8

An **angle** is formed by two noncollinear rays that have a common endpoint, called the vertex.

Some pairs of angles are of interest because of the relationship between their angle measures. These include: vertical angles, complementary angles, supplementary angles, and linear pairs.

EXAMPLE 1 **Exploring Vertical Angles**

Fold a piece of patty paper twice so that the two creases form vertical angles. Label the angles ∠PQR, ∠RQS, ∠SQT, and ∠PQT.

a. REASON ABSTRACTLY Compare your figure with those of other students. Which point is labeled the same in every figure? Explain.

_____

_____

b. USE TOOLS Use a protractor to measure the four angles.

m ∠PQR = _____   m ∠RQS = _____   m ∠SQT = _____   m ∠PQT = _____

c. MAKE A CONJECTURE Compare your results with those of other students. Then make a conjecture about vertical angles.

_____

_____

EXAMPLE 2 **Exploring Linear Pairs**

Use the figure you constructed in Example 1 to answer these questions.

a. INTERPRET PROBLEMS Identify any pairs of opposite rays in the figure.

_____

b. INTERPRET PROBLEMS Identify a pair of adjacent angles with noncommon sides that are opposite rays.

_____

c. MAKE A CONJECTURE Compare your results with those of other students. Then make a conjecture about linear pairs.

_____

d. DESCRIBE A METHOD Compare your results with those of other students. Then explain how you can use your conjecture to determine the measures of $\angle CEF$, $\angle DEF$, and $\angle DEG$ if $\overleftrightarrow{CD}$ and $\overrightarrow{FG}$ intersect at E and $m \angle CEG = 65$.

$m \angle CEF =$ _____ $m \angle DEF =$ _____ $m \angle DEG =$ _____

_____

_____

_____

_____

EXAMPLE 3 **Copying an Angle**

Follow the steps to create $\angle QRS$, a copy of $\angle J$.

a. **USE TOOLS** Place the tip of your compass at point J and draw a large arc that intersects both sides of $\angle J$. Label the points of intersection K and L.

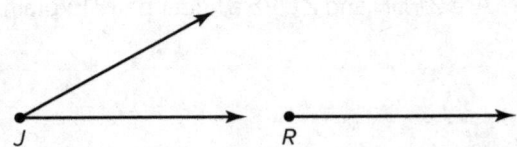

b. **USE TOOLS** Using the same compass setting, place the tip of your compass on point R and draw a large arc that starts above the ray and intersects the ray. Label the point of intersection S.

c. **USE TOOLS** Place the tip of your compass at point L and adjust your compass so that the pencil is at point K. Using this setting, place the tip of your compass at point S and draw an arc to intersect the larger arc drawn in **part b**. Label the point of intersection Q.

d. **USE TOOLS** Use a straightedge to connect point R and point Q to form $\angle QRS$.

e. **MAKE A CONJECTURE** Compare your figures with those of other students. Then make a conjecture about the segment(s) congruent to $\overline{JK}$. List the segment(s).

_____

_____

f. **USE TOOLS** Use a protractor to find the measures of $\angle J$ and $\angle QRS$. Then classify the angles.

_____

_____

g. **MAKE A CONJECTURE** Compare your results with those of other students. Then make a conjecture about the measures of $\angle J$ and $\angle QRS$.

_____

_____

1. In the figure to the right, $m\angle PVQ = 10x - 25$, $m\angle QVR = 5x - 5$, and $\angle PVR$ is a right angle. Mary constructed a copy of $\angle PVQ$ and labeled it $\angle FGH$.

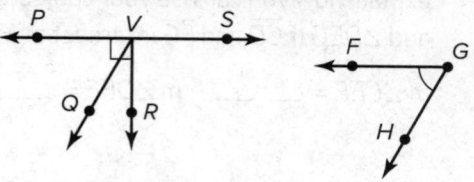

   a. **CALCULATE ACCURATELY** Find the measure of each angle, and classify the angle by its measure.

   $\angle PVQ =$ _____ $\angle QVR =$ _____ $\angle SVR =$ _____

   b. **COMMUNICATE PRECISELY** Are $\angle FGH$ and $\angle QVR$ adjacent angles? Explain.

   _____

   _____

   c. **COMMUNICATE PRECISELY** Are $\angle FGH$ and $\angle QVR$ complementary angles? Explain.

   _____

   _____

   _____

   d. **COMMUNICATE PRECISELY** Are $\angle FGH$ and $\angle QVS$ a linear pair? Explain.

   _____

   _____

   _____

   e. **CRITIQUE REASONING** Mary must also copy $\angle QVS$. Sal says she will create a copy of $\angle QVS$ if she extends $\overrightarrow{GH}$ past G. Mona says Mary can create a copy of $\angle QVS$ by extending $\overrightarrow{GF}$ past G. Who is correct? Justify your answer.

   _____

   _____

   _____

2. $\overrightarrow{AB}$ and $\overrightarrow{AC}$ are opposite rays. G is in the interior of $\angle DAC$. F is in the interior of $\angle DAG$. $\angle BAF$ is a right angle. $m\angle DAC = 134$. $m\angle GAC = 56$. Sketch the figure in the space to the right, and answer the questions.

   a. **REASON QUANTITATIVELY** Find the measure of each angle, and classify the angle by its measure.

   $\angle BAD =$ _____ $\angle CAF =$ _____ $\angle DAG =$ _____

   b. **COMMUNICATE PRECISELY** Does the figure contain any linear pairs? If so, state an example. If not, explain why not.

   _____

   _____

   c. **COMMUNICATE PRECISELY** Does the figure contain vertical angles? If so, state an example. If not, explain why not.

   _____

   _____

3. **CRITIQUE REASONING** Zach's work copying ∠ABC is shown to the right. He drew an arc, centered at B, which intersected both sides of ∠ABC. He drew an arc of the same radius centered at Q and labeled the point of intersection with the ray, R. Next, he adjusted the compass so its tips were where his first arc intersected ∠ABC, and used the compass, to draw a second arc centered at Q. What error did Zach make?

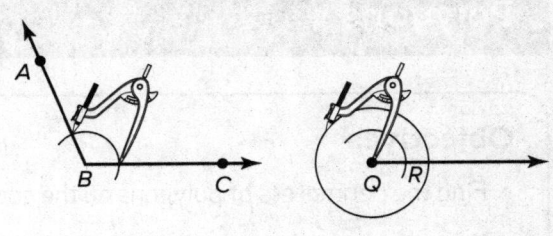

_____

_____

4. Answer the questions about copying an angle.

a. **USE TOOLS** Begin with the given ray to create a copy of ∠A. Describe each step of the process.

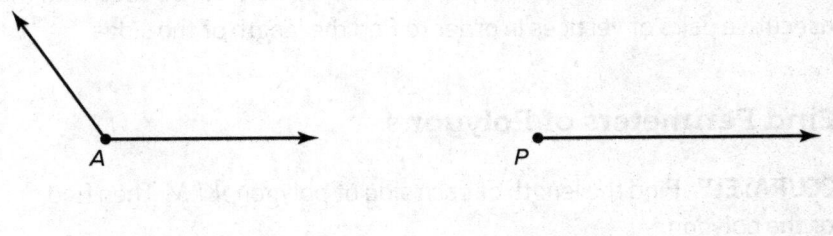

_____

_____

_____

_____

_____

_____

_____

b. **REASON ABSTRACTLY** How are ∠A and ∠BAC related? Explain.

_____

_____

c. **COMMUNICATE PRECISELY** Describe how the copy of an obtuse angle can always be found by using the copy of an acute angle.

_____

_____

_____

5. **INTERPRET PROBLEMS** Identify any vertical angles, complementary angles, supplementary angles, or linear pairs in the figure.

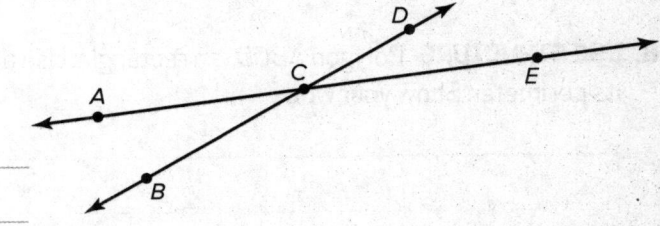

_____

_____

_____

Objectives

Mathematical Practices
1, 3, 6, 7

- Find the perimeters of polygons on the coordinate plane.
- Find the areas of triangles and rectangles on the coordinate plane.

A **polygon** is a closed figure formed by a finite number of coplanar segments called sides such that the sides that have a common endpoint are not collinear and each side intersects exactly two other sides, but only at their endpoints. The **perimeter** of a polygon is the sum of the lengths of the sides of the polygon. The perimeter of a rectangle is $P = \ell + w + \ell + w = 2\ell + 2w$. Since all side lengths are the same for a regular polygon, the perimeter is the product of the side length, s, and the number of sides, n, or $P = ns$. When polygons are drawn on a coordinate plane, the distance formula can be used with the coordinates of consecutive pairs of vertices in order to find the length of the sides.

EXAMPLE 1 Find Perimeters of Polygons

a. **CALCULATE ACCURATELY** Find the length of each side of polygon *JKLM*. Then find the perimeter of the polygon.

_____

_____

_____

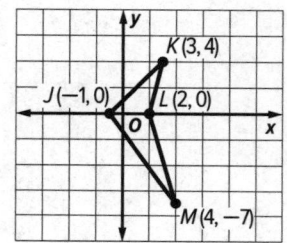

b. **CALCULATE ACCURATELY** Find the length of each side of polygon *QRSTUV*. Then find its perimeter.

_____

_____

_____

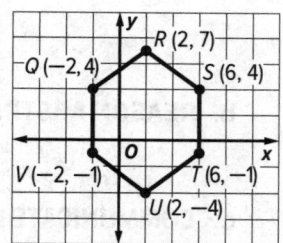

c. **INTERPRET PROBLEMS** What type of figure is polygon *QRSTUV*? How else could you have found its perimeter?

_____

_____

d. **USE STRUCTURE** Polygon *ABCD* is a rectangle. Use a formula to find its perimeter. Show your work.

_____

_____

_____

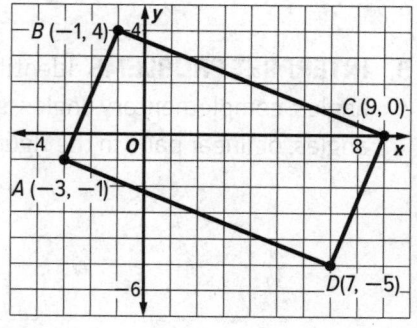

EXAMPLE 2 **Using Coordinates to Check Perimeter**

Kaylee found the perimeter of $\triangle EFG$ as shown below.

$EF = \sqrt{(-2-4)^2 + (-1-7)^2} = \sqrt{(-6)^2 + (-8)^2} = \sqrt{36 + 64} = \sqrt{100} = 10$

$FG = \sqrt{(-1-4)^2 + (8-7)^2} = \sqrt{(-5)^2 + (1)^2} = \sqrt{25+1} = \sqrt{26} \approx 5.1$

$EG = \sqrt{(-2-8)^2 + (-1-(-1))^2} = \sqrt{(-10)^2 + (0)^2} = \sqrt{100} = 10$

$P = 10 + 5.1 + 10 = 25.1$

CRITIQUE REASONING She knew by looking at the diagram that her solution was incorrect. How could she tell? Find and correct her mistake.

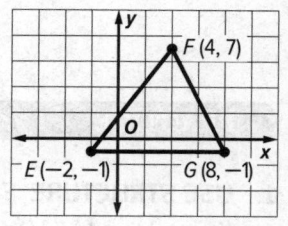

_____

_____

_____

The **area** of a figure is the number of square units needed to cover a surface. The area of a triangle is calculated by multiplying $\frac{1}{2}$ by the product of the base and height of the triangle, $A = \frac{1}{2}bh$. The base and height are perpendicular to each other. Any side of the triangle can be chosen as a base. Then the height is represented by a segment from the vertex opposite the base, perpendicular to the base. The area of a rectangle is the product of its length and width, $A = \ell w$. Since a square is a rectangle with equal length and width, the area of a square is the square of its side length, $A = s^2$.

EXAMPLE 3 **Areas of Triangles and Rectangles**

a. **USE STRUCTURE** Find the length and width of rectangle $WXYZ$. Then find its area.

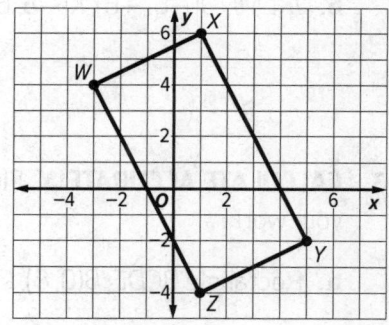

_____

_____

_____

b. **INTERPRET PROBLEMS** Which base and height can be found for $\triangle PQR$? Why are you able to find these?

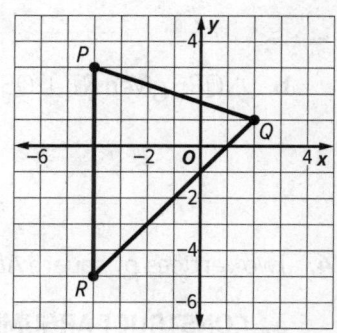

_____

_____

_____

_____

c. Find the length of the base and height from **part b** and use them to calculate the area of $\triangle PQR$.

_____

_____

d. **CALCULATE ACCURATELY** Find the lengths of the three sides of right triangle *TRA*. Which two sides would you use for the base and the height? Explain your choice.

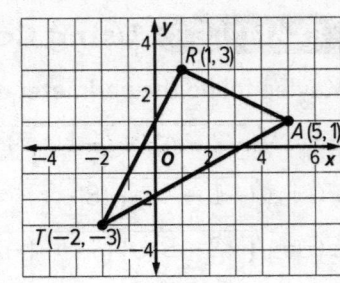

_____

_____

e. What is the area of △*TRA*?

_____

1. **USE STRUCTURE** Find the perimeter of equilateral triangle *KLM* given the vertices *K*(−2, 1) and *M*(10, 6).

_____

2. **CALCULATE ACCURATELY** Find the lengths of the sides of each polygon to the nearest tenth. Then find its perimeter.

   a. △*EFG*; *E*(−18, −15), *F*(−12, 10), *G*(15, 2)

   _____

   b. *JKLMN*; *J*(−6, −6), *K*(−8, 6), *L*(2, 12), *M*(13, 8), *N*(12, −4)

   _____

   _____

3. **CALCULATE ACCURATELY** Find the area of each given polygon. Show your work.

   a. Rectangle *BCDE*; *B*(0, 8), *C*(20, −4), *D*(11, −19), *E*(−9, −7)

   _____

   _____

   b. △*QRS* given $\overline{RT} \perp \overline{QS}$.

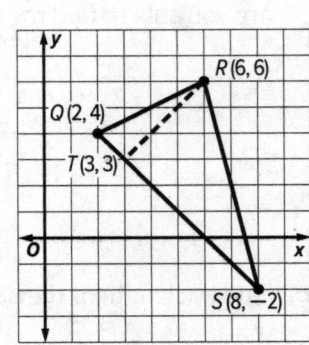

   _____

   _____

4. Two vertices of square *ABCD* are *C*(5, 8) and *D*(2, 4).

   a. **CONSTRUCT ARGUMENTS** Do you need to find the coordinates for the other two vertices in order to find the perimeter and area of the square? Why or why not?

   _____

   _____

b. USE STRUCTURE Find the perimeter and area of square *ABCD*.
Show your work.

_____

5. **USE STRUCTURE** The figure shows Derek's house and his backyard on a
coordinate grid. Derek is planning to fence in the play area in his backyard.
Part of the play area is enclosed by the house and does not need to be
fenced. Each unit on the coordinate grid represents 5 feet. The cost for the
fencing material and installation is $10 per foot. How much will it cost Derek
to install the fence?

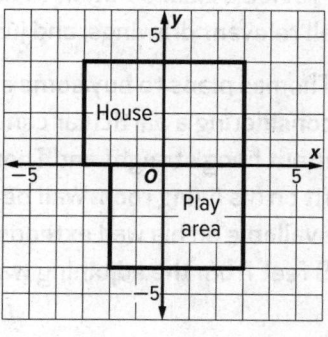

_____

_____

_____

_____

_____

_____

6. **USE STRUCTURE** Rectangle *ABCD* has vertices $A(-2, -5)$, $B(0, 4)$, $C(4, 3)$,
and $D(2, -6)$. Find the perimeter and area of rectangle *ABCD*. Round to the
nearest tenth.

_____

_____

_____

_____

_____

7. **USE STRUCTURE** The coordinate gird shows an equilateral triangle that fits
inside a square.

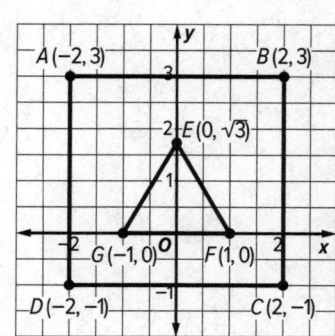

a. Find the area of the square. Show your work.

_____

b. Find the area of the triangle. Show your work.

_____

c. Find the area of the square that is not covered by the triangle. Write an
exact value, then round to the nearest tenth. Explain.

_____

_____

_____

Picture Perfect

Provide a clear solution to the problem. Be sure to show all of your work, include all relevant drawings, and justify your answers.

Thomas plans to buy some artwork at a local gallery. He is considering a particular canvas displayed high on a wall. The canvas is not hung straight and Thomas must determine if the canvas will fit on his living room wall before he makes his purchase. The space available on his wall extends from the ceiling 6 feet and horizontally 8 feet from the adjoining wall.

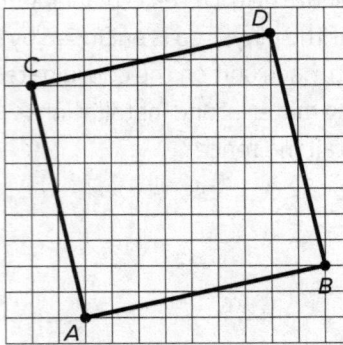

Part A

Thomas plans to use the wall tiles behind the canvas to estimate its dimensions. If each square tile is 6 inches wide, what are the dimensions of the canvas? Justify your solution.

Part B

Before hanging the canvas, Thomas wants to mat and frame the artwork. If he wants a 2-inch mat border around the perimeter of the canvas, what is the area of the mat border he must order? What is the total length of the inside perimeter of the frame that he must order to fit the canvas plus mat border? If the frame is 4 inches wide, what is the new area of the framed artwork?

Part C

The gallery recommended that Thomas install hardware on the back of the framed canvas that will support the artwork at both ends and then at points $\frac{1}{3}$ and $\frac{2}{3}$ along the length of the frame. Where should Thomas install the hardware? Justify your answer.

Part D

Thomas wants to center the canvas horizontally on his wall. He draws a line that runs the entire length of the available space. At what approximate distances from the adjoining wall should Thomas place holes to accommodate the four supports that he added in Part C? Justify your answer.

2 **Logical Arguments and Line Relationships**

CHAPTER FOCUS Learn about some of the objectives that you will explore in this chapter. Answer the preview questions. As you complete each lesson, return to these pages to check your work.

What You Will Learn	Preview Question
Definitions, Postulates, Theorems, and Conjectures	
• Use definitions and undefined terms to make conjectures. • Use postulates to verify conjectures. • Distinguish among definitions, postulates, theorems, and conjectures.	SMP 6 How does a postulate differ from a theorem? **A postulate is a statement that is accepted as true without proof.** **Theorems are considered to be true once they are proven.**
Two-Column and Paragraph Proofs	
• Write two-column proofs about lines and line segments. • Write paragraph proofs about lines and line segments.	SMP 1 Label each statement as *always*, *sometimes*, or *never* true. If two points lie in the same plane, the line containing them lies in that plane. **always** The intersection of a line and a plane is a line. **sometimes** If two planes are perpendicular, their intersection is a line. **always**
Proving Theorems About Line Segments	
• Construct the bisector of a segment. • Prove theorems about line segments using the Segment Addition Postulate.	SMP 3 Write a paragraph proof. **Given:** $XY = 2WX$ and $WX = YZ$ **Prove:** $2XY = WZ$ W X Y Z It is given that **$XY = 2WX$ and $WX = YZ$. By the Seg. Add. Post.,** **$WX + XY + YZ = WZ$. By substitution, $WX + XY + WX = WZ$ and** **$2WX + XY = WZ$. By substitution, $XY + XY = WZ$ and so $2XY = WZ$.**
Proving Theorems About Angles	
• Construct an angle bisector. • Prove theorems about angles.	SMP 3 Complete the following proof. **Given:** $\ell \parallel m$; $\angle 1 \cong \angle 2$ **Prove:** $\ell \perp r$

Statements	Reasons
1. $\ell \parallel m$	1. **Given**
2. $\angle 1 \cong \angle 2$	2. **Given**
3. $\angle 2 \cong \angle 3$	3. **Alt. Ext. ∠s Thm.**
4. $\angle 1 \cong \angle 3$	4. **Trans Prop. of ≅**
5. $\ell \perp r$	5. **If 2 intersecting lines form a linear pair, then the lines are ⊥.**

What You Will Learn	Preview Question

Angles and Parallel Lines

- Identify parallel and skew lines and parallel planes.
- Identify types of angle pairs with a transversal.
- Prove and apply theorems about parallel lines with a transversal.

SPM 3 In the figure, $m\angle 9 = 60$ and $m\angle 2 = 35$. Find $m\angle 3$. Explain how you know.

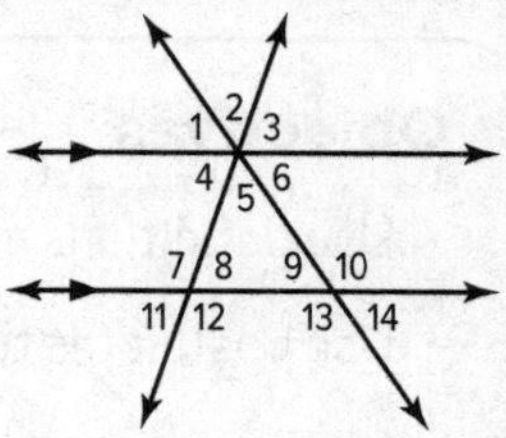

_____

_____

_____

Slopes of Lines

- Use the slope criterion for parallel lines to solve problems.
- Use the slope criterion for perpendicular lines to solve problems.

SMP 7 A line containing the points $(2, 0)$ and $(-1, 6)$ is perpendicular to a line containing the points $(5, 2)$ and $(4, y)$. What is the value of y?

_____

Equations of Lines

- Find the equation of a line parallel to a given line that passes through a given point.
- Find the equation of a line perpendicular to a given line that passes through a given point.

SMP 7 What is the equation of a line parallel to the line $4x + 3y = -12$ through the point $(2, -1)$? What is the slope of a line perpendicular to both lines?

_____

Find the equation of a line that has a y-intercept of 9 and is perpendicular to the line graphed.

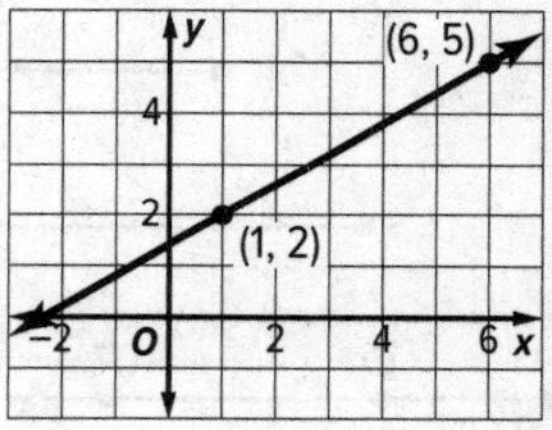

_____

Proving Lines Parallel

- Prove theorems about angle pairs that guarantee parallel lines.
- Construct a line parallel to a given line through a point not on the line.

SMP 3 Write a paragraph proof.
Given: $\angle 1$ and $\angle 2$ are supplementary.
Prove: $\ell \parallel m$

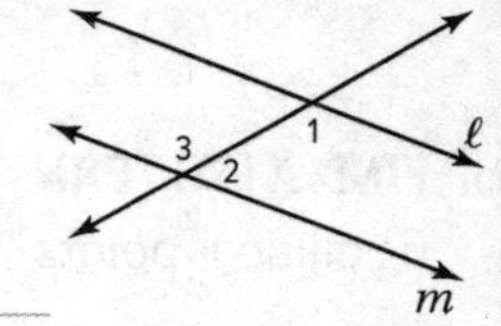

_____

_____

_____

_____

Perpendicular Lines

- Construct perpendicular lines, including the perpendicular bisector of a line segment.
- Prove the Perpendicular Bisector Theorem and its converse.

SMP 1 A line perpendicular to a pair of parallel lines intersects the lines at $(3, 2)$ and $(x, 0)$. If the distance between the lines is $\sqrt{5}$, what could be the equation of the perpendicular line?

_____

Objectives

- Use definitions and undefined terms to make conjectures.
- Use postulates to verify conjectures.
- Distinguish among definitions, postulates, theorems, and conjectures.

Inductive reasoning is the process of looking at a number of specific examples and identifying a pattern in order to arrive at a conclusion. A concluding statement that you reach via inductive reasoning is called a **conjecture**.

EXAMPLE 1 Make a Conjecture

EXPLORE The figure shows the number of line segments that are determined by different numbers of collinear points.

2 points; 1 line segment: $\overline{AB}$

3 points; 3 line segments: $\overline{JL}, \overline{JK}, \overline{KL}$

4 points; 6 line segments: $\overline{PS}, \overline{PR}, \overline{QS}, \overline{PQ}, \overline{QR}, \overline{RS}$

a. **INTERPRET PROBLEMS** What defined terms are used to describe the figure? What undefined terms are used?

_____

b. **FIND A PATTERN** How many line segments are determined by 5 collinear points? 6 collinear points? Explain how you found your answer.

_____

_____

c. **MAKE A CONJECTURE** Make a conjecture regarding the number of line segments determined by n collinear points. Use the fact that $1 + 2 + 3 + \ldots + k = k\dfrac{(k+1)}{2}$. Explain your reasoning.

_____

_____

_____

d. **CONSTRUCT ARGUMENTS** How would you convince someone that the conjecture you made in **part c** is correct?

_____

_____

2. **USE A MODEL** Katie owns a company that makes patios and garden paths out of square tiles. The figure shows the pattern that is used to make paths of different lengths.

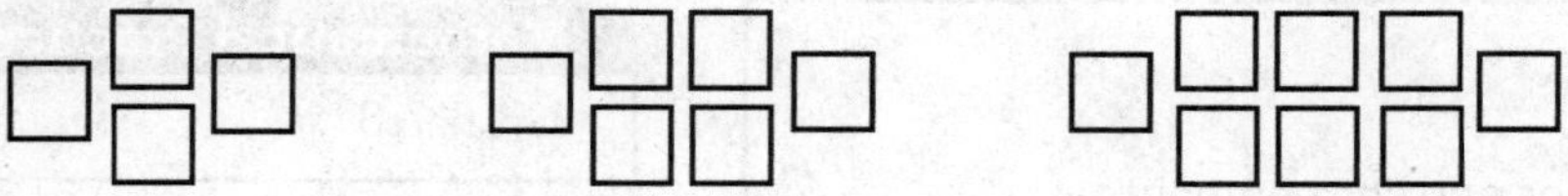

Length 1   Length 2   Length 3

a. **FIND A PATTERN** Katie would like to find the number of tiles needed to make a path of any length. Look for a pattern and make a conjecture about the number of tiles needed to make a path of length *n*.

_____

b. **USE STRUCTURE** Katie said one of the paths her company made today required 44 tiles. What was the length of the path? Explain.

_____

c. **CRITIQUE REASONING** Katie said one of the paths her company made last week required exactly 103 tiles. Is this possible? Justify your response.

_____

COMMUNICATE PRECISELY In Exercises 3 and 4, use undefined terms, definitions, and/or postulates to explain why each conjecture is true.

3. Wei drew the figure shown here. Then she stated the following conjecture: A plane contains at least two lines.

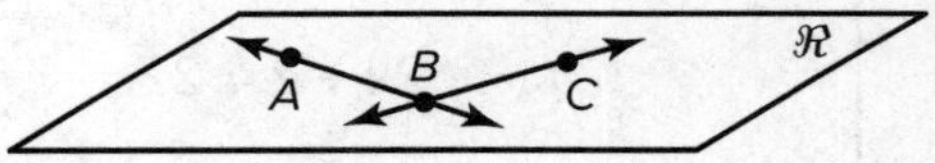

_____
_____
_____

4. Aiden drew the figure shown here. Then he stated the following conjecture: Every line contains at least one line segment.

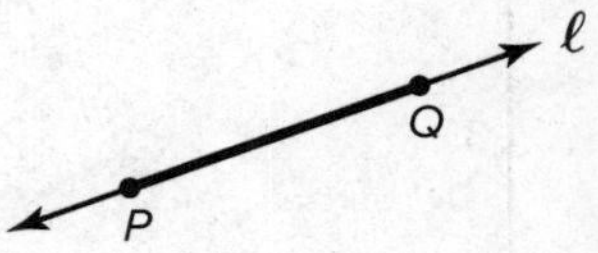

_____
_____
_____

5. **FIND A PATTERN** A line segment of length 1 is repeatedly shortened by removing one-third its remaining length as shown.

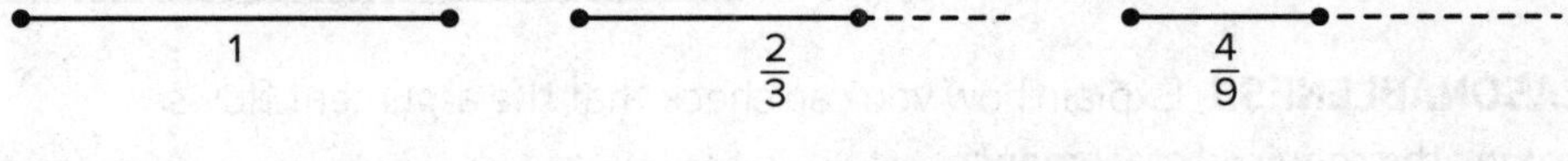

Find and use a pattern to make a conjecture about the length of the line segment after being shortened *n* times.

_____
_____

Two-Column and Paragraph Proofs

Objectives

- Write two-column proofs about lines and line segments.
- Write paragraph proofs about lines and line segments.

A **deductive argument** is a logical chain of statements that links the information you are given to the statement that you are trying to prove.

EXAMPLE 1   Organize a Deductive Argument

EXPLORE Lisette was asked to write a deductive argument for the situation shown here.

Given: $\overline{AB} \cong \overline{CD}$; $AB = 5x + 2$; $CD = 3x + 22$

Prove: $x = 10$

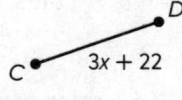

a. CONSTRUCT ARGUMENTS Lisette used the statements shown below on the left to write her deductive argument. Write the statements in the boxes so the argument flows logically.

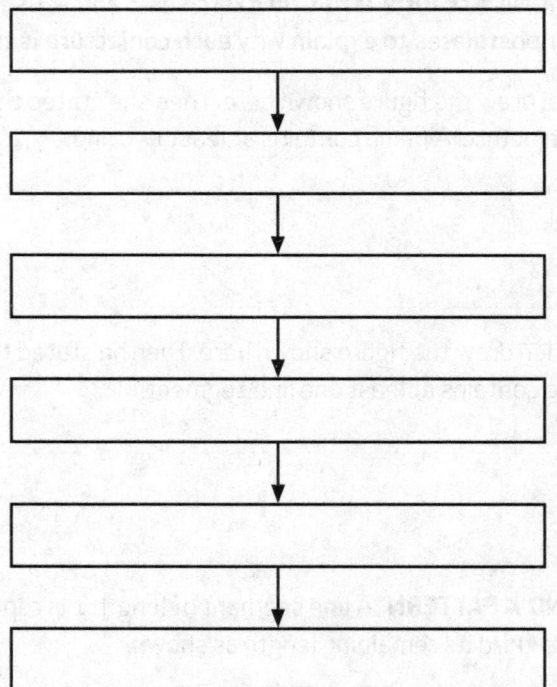

Statements
$2x + 2 = 22$
$x = 10$
$\overline{AB} \cong \overline{CD}$
$AB = CD$
$2x = 10$
$5x + 2 = 3x + 22$

b. EVALUATE REASONABLENESS Explain how you can check that the argument flows logically and proves the required statement.

_____

_____

_____

b. CONSTRUCT ARGUMENTS Write a two-column proof for your conjecture.

Statements	Reasons
1.	1.
2.	2.
3.	3.
4.	4.
5.	5.
6.	6.
7.	7.

2. The figure shows a straight portion of the course for a city marathon. The water station W is located at the midpoint of $\overline{AB}$.

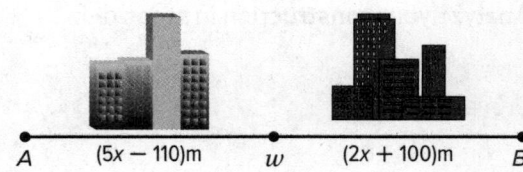

   a. **INTERPRET PROBLEMS** What is the length of the course from point A to point W?

   _____

   b. **CONSTRUCT ARGUMENTS** Write a paragraph proof for your answer to **part a**.

   _____

   _____

   _____

   _____

   c. **COMMUNICATE PRECISELY** Explain how you used a definition in your paragraph proof.

   _____

3. **CONSTRUCT ARGUMENTS** Write a paragraph proof that if $\overline{MN} \cong \overline{PQ}$, $MN = 5x - 10$, and $PQ = 4x + 10$, then $MN = 90$.

   _____

   _____

   _____

   _____

4. **COMMUNICATE PRECISELY** Write a paragraph proof to prove that if $PQ = 4(x - 3) + 1$, $QR = x + 10$, and $x = 7$, then $\overline{PQ} \cong \overline{QR}$.

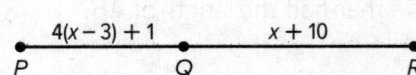

   _____

   _____

   _____

Objectives

- Construct the bisector of a segment.
- Prove theorems about line segments using the Segment Addition Postulate.

EXAMPLE 1 **Bisect a Segment**

EXPLORE Follow Steps a–c to use a compass and straightedge to bisect $\overline{AB}$. Analyze your construction in steps d–f.

a. Open the compass to a little more than half the length of $\overline{AB}$. Place the tip of the compass on point A and make an arc as shown below.

b. Without changing the compass setting, place the tip of the compass on point B. Make an arc that intersects the first arc at points P and Q, as shown.

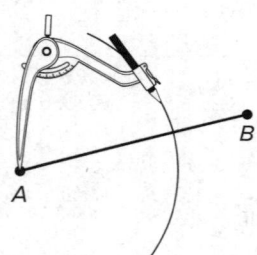

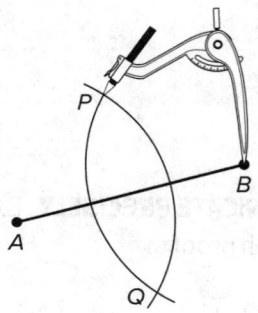

c. Use the straightedge to draw $\overleftrightarrow{PQ}$. Label the intersection of $\overline{AB}$ and $\overleftrightarrow{PQ}$ as point M.

d. **USE TOOLS** Is M the midpoint of $\overline{AB}$? Explain.

_____

_____

e. **CONSTRUCT ARGUMENTS** In **step a**, why do you need to open the compass to more than half the length of $\overline{AB}$?

_____

_____

f. **CONSTRUCT ARGUMENTS** How are the lengths AM, MB, and AB related? Write one or more equations to express the relationships.

_____

In the exploration, you may have noticed a relationship among the lengths of the original segment and the lengths of the two shorter segments you created. The Segment Addition Postulate states that this relationship is true.

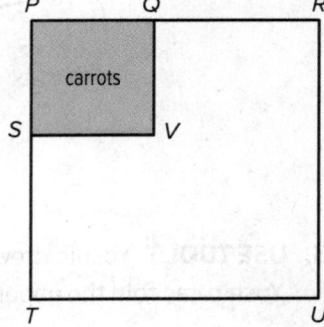

KEY CONCEPT **Segment Addition Postulate**

Complete the following statement.
If A, B, and C are collinear, then point B is between

A and C if and only if _____ + _____ = _____.

EXAMPLE 2 **Use the Segment Addition Postulate**

Dima has a square vegetable plot in her garden. She wants to set aside a corner of the plot for planting carrots, as shown. Based on her measurements, she knows that $\overline{PR} \cong \overline{PT}$ and $\overline{QR} \cong \overline{ST}$. She wants to know if she can conclude that $\overline{PQ} \cong \overline{PS}$.

a. **CONSTRUCT ARGUMENTS** Complete the two-column proof below.

Given: $\overline{PR} \cong \overline{PT}$ and $\overline{QR} \cong \overline{ST}$

Prove: $\overline{PQ} \cong \overline{PS}$

Statements	Reasons
1. $\overline{PR} \cong \overline{PT}$ and $\overline{QR} \cong \overline{ST}$	1.
2.	2. Congruent segments have equal lengths.
3. $PR = PQ + QR$ and $PT = PS + ST$	3.
4. $PQ + QR = PS + ST$	4.
5. $PQ + QR = PS + QR$	5.
6. $PQ = PS$	6.
7.	7. Segments with equal lengths are congruent.

b. **COMMUNICATE PRECISELY** In the second step of the proof, why is it necessary to change the given statements about congruent segments into statements about lengths of segments?

_____

_____

c. **COMMUNICATE PRECISELY** Extend the line segment $\overline{SV}$ to the line segment $\overline{RU}$ and let W be the point where they intersect. Write a paragraph proof to prove that if $\overline{PR} \cong \overline{SW}$ and $\overline{QR} \cong \overline{VW}$, then $\overline{PQ} \cong \overline{SV}$.

_____

_____

_____

_____

_____

USE TOOLS Use a compass and straightedge to bisect $\overline{AB}$. Label the midpoint of the segment point M.

1.

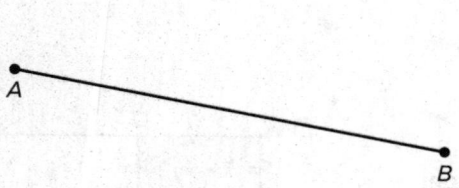

2.

3. **USE TOOLS** Yoshio drew a line segment, $\overline{JK}$, on a sheet of tracing paper. Explain how Yoshio can fold the paper in order to bisect $\overline{JK}$.

_____

_____

4. **CONSTRUCT ARGUMENTS** Molly wants to use a compass and straightedge to bisect a segment. She finds that she cannot change the setting. Will she be able to do the construction anyway? Explain.

_____

5. a. **CONSTRUCT ARGUMENTS** Complete the two-column proof below.

   Given: $\overline{PQ} \cong \overline{RS}$

   Prove: $\overline{PR} \cong \overline{QS}$

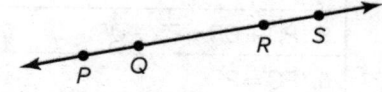

Statements	Reasons
1. $\overline{PQ} \cong \overline{RS}$	1.
2.	2. Congruent segments have equal lengths.
3. $PQ + QR = PR$ and $QR + RS = QS$	3.
4. $RS + QR = PR$	4.
5. $QR + RS = PR$	5.
6. $PR = QS$	6.
7.	7. Segments with equal lengths are congruent.

   b. **INTERPRET PROBLEMS** Can it also be proven that $\overline{PQ} \cong \overline{RS}$ if $\overline{PR} \cong \overline{QS}$? Explain.

_____

_____

6. A city planner is designing a new park. The park has two straight paths, $\overline{AB}$ and $\overline{CD}$, which are the same length. A monument, M, is located at the midpoint of both paths.

a. **INTERPRET PROBLEMS** The city planner thinks that the length of $\overline{AM}$ will be the same as the length of $\overline{CM}$. Explain why this makes sense.

_____

_____

_____

b. **CONSTRUCT ARGUMENTS** Complete the two-column proof.

Given: $\overline{AB} \cong \overline{CD}$; M is the midpoint of $\overline{AB}$ and $\overline{CD}$.

Prove: $\overline{AM} \cong \overline{CM}$

Statements	Reasons
1.	1. Given
2. $AB = CD$	2.
3. $\overline{AM} \cong \overline{MB}$; $\overline{CM} \cong \overline{MD}$	3.
4.	4. Congruent segments have equal lengths.
5. $AM + MB = AB$; $CM + MD = CD$	5.
6. $AM + MB = CM + MD$	6.
7. $AM + AM = CM + CM$	7.
8. $2AM = 2CM$	8.
9.	9. Division Property of Equality
10.	10. Segments with equal lengths are congruent.

7. **CRITIQUE REASONING** Justin knows that point R is the midpoint of $\overline{QS}$, and he knows that this means $QR = RS$. He says that $PR = PQ + QR$ by the Segment Addition Postulate. So $PR = PQ + RS$ by substitution. Do you agree with Justin's reasoning? Explain.

_____

_____

_____

8. **CONSTRUCT ARGUMENTS** Write a paragraph proof to prove that if P, Q, R, and S are collinear and $\overline{PQ} \cong \overline{RS}$ and Q is the midpoint of $\overline{PR}$, then R is the midpoint of $\overline{QS}$.

_____

_____

_____

_____

Objectives

- Construct an angle bisector.
- Prove theorems about angles.

Mathematical Practices
2, 3, 5, 6

You have learned about the Segment Addition Postulate. A similar relationship exists between the measures of angles. The Angle Addition Postulate states that D is in the interior of $\angle ABC$ if and only if $m\angle ABD + m\angle DBC = m\angle ABC$.

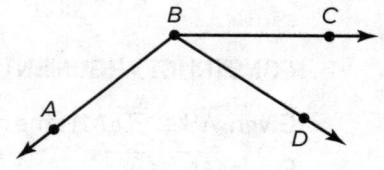

EXAMPLE 1 **Constructing an Angle Bisector**

a. **USE TOOLS** Follow these steps to create $\overrightarrow{JM}$, a bisector of $\angle J$.

- In the space to the right, draw obtuse angle $\angle J$.

- Place your compass at point J and draw an arc that intersects both sides of $\angle J$. Label the points of intersection K and L.

- With the compass at point K, draw an arc in the interior of $\angle J$ that has a radius equal to more than half the distance from K to L.

- Keeping the same compass setting, place the compass at L and draw an arc that intersects the arc centered at K. Label the point of intersection M.

- Draw $\overrightarrow{JM}$.

b. **MAKE A CONJECTURE** Compare your results with those of other students. Then make a conjecture about the measures of $\angle LJM$ and $\angle KJM$.

_____

_____

c. **USE TOOLS** Follow these steps to create $\overrightarrow{WZ}$, a bisector of $\angle W$.

- Copy $\angle J$ from **part a** onto a piece of patty paper. Label its vertex W. Place point X on one side of the angle and point Y on the other ray.

- Fold the paper so that the two rays of the angle lie on top of each other. The crease should pass through W.

- Place point Z on the crease, in the interior of $\angle W$.

- Draw $\overrightarrow{WZ}$.

- Tape the patty paper in the space to the right.

d. **MAKE A CONJECTURE** Compare your results with those of other students. Then make a conjecture about the measures of $\angle XWZ$ and $\angle YWZ$.

e. **USE TOOLS** Use a protractor to measure the angles.

$m\angle KJM = $ _____ $m\angle LJM = $ _____ $m\angle XWZ = $ _____ $m\angle YWZ = $ _____

f. **CONSTRUCT ARGUMENTS** Write a paragraph proof.

Given: $\overrightarrow{JM}$ bisects $\angle KJL$, $\overrightarrow{WZ}$ bisects $\angle XWY$, and $\angle KJL \cong \angle XWY$

Prove: $\angle KJM \cong \angle XWZ$

_____

_____

_____

_____

_____

_____

EXAMPLE 2 **Using Angle Bisectors to Prove Angle Relationships**

Let $\angle QRS$ be the angle formed by the wings of an airplane, and $\overrightarrow{RT}$ be a ray on the body of the plane.

a. **CONSTRUCT ARGUMENTS** Write a two-column proof.

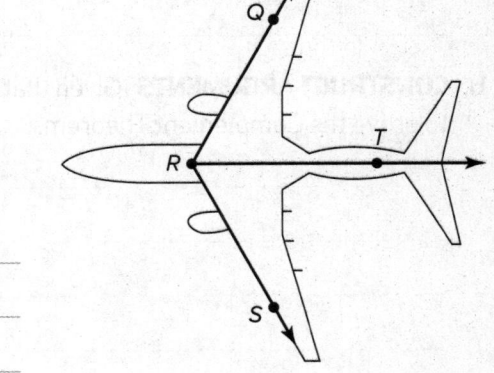

Given: $\overrightarrow{RT}$ bisects $\angle QRS$

Prove: $m\angle TRS = \frac{1}{2}(\angle QRS)$

_____

_____

_____

_____

_____

_____

b. **COMMUNICATE PRECISELY** Suppose $m\angle QRS = 120$. Write a paragraph proof to show that if $m\angle QRT = 60$, then $\overrightarrow{RT}$ bisects $\angle QRS$.

_____

_____

Supplement Theorem If two angles form a linear pair, then they are supplementary angles.

Complement Theorem If the noncommon sides of two adjacent angles form a right angle, then the angles are complementary angles.

Vertical Angles Theorem If two angles are vertical angles, then they are congruent

Congruent Supplements Theorem Angles supplementary to the same angle or to congruent angles are congruent.

Congruent Complements Theorem Angles complementary to the same angle or to congruent angles are congruent.

EXAMPLE 3 Identifying and Proving Angle Relationships

In the figure to the right, ∠AQC and ∠BQD are right angles.

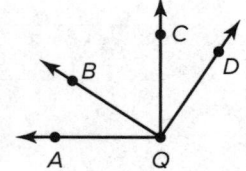

a. **COMMUNICATE PRECISELY** Explain what the Complement Theorem implies about the adjacent angles that make up ∠AQC and ∠BQD. Then, explain what the Congruent Complements Theorem implies about ∠AQB and ∠DQC.

_____

_____

_____

b. **CONSTRUCT ARGUMENTS** Given that ∠AQC is a right angle, write a paragraph proof to prove the Complement Theorem.

_____

_____

_____

c. **CONSTRUCT ARGUMENTS** Given that ∠AQB and ∠BQC are complementary angles and ∠BQC and ∠DQC are complementary angles, write a two-column proof of the Congruent Complements Theorem.

_____

_____

_____

_____

_____

_____

_____

d. REASON QUANTITATIVELY Find $m\angle BQC$ and $m\angle CQD$ when $m\angle AQB = 32$ without using the Congruent Complements Theorem. Justify your answer. How does this compare to finding the measures using the Congruent Complements Theorem?

PRACTICE

1. In the figure to the right, $\angle MRQ$ and $\angle MRN$ are a linear pair and $\angle MRN$ and $\angle NRP$ are a linear pair.

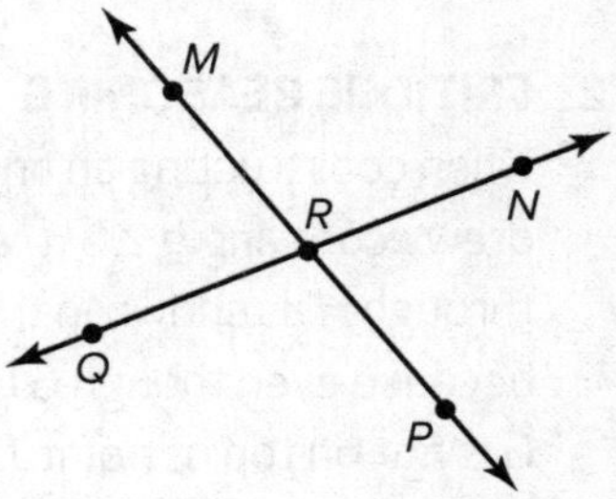

a. CONSTRUCT ARGUMENTS Write a paragraph proof of the Supplement Theorem.

b. CONSTRUCT ARGUMENTS Explain what the Vertical Angles Theorem says about $\angle MRQ$ and $\angle NRP$. Then write a paragraph proof of the Vertical Angles Theorem using this figure.

c. REASON ABSTRACTLY Use the Vertical Angles Theorem to find $m\angle MRQ$ and $m\angle NRP$ if $m\angle MRQ = 6x + 24$ and $m\angle NRP = 18x - 72$. Justify each step.

d. CONSTRUCT ARGUMENTS A line is drawn through the point R as shown. Write a two-column proof to prove that if $\overrightarrow{RT}$ bisects $\angle MRQ$, then $\overrightarrow{RS}$ bisects $\angle NRP$.

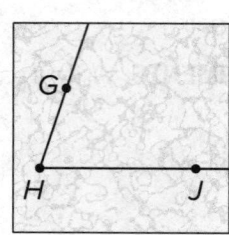

_____

_____

_____

_____

_____

_____

_____

_____

2. CRITIQUE REASONING David believes he has made an error when constructing an angle bisector by paper folding. He drew acute angle $\angle GHJ$ and folded the paper with a crease through H as shown in the figure to the right. He is concerned because even though $\overrightarrow{GJ}$ and $\overrightarrow{HJ}$ are on top of each other, point G is not on top of point J. Is the crease from David's fold a bisector of $\angle GHJ$?

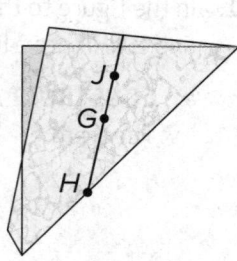

_____

_____

_____

3. Answer the questions about bisecting an angle.

a. CRITIQUE REASONING Amber has made an error when constructing an angle bisector with a compass and straightedge. Her work is shown to the right. Explain the source of Amber's error and suggest a way she can complete the construction.

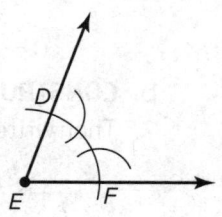

_____

_____

_____

_____

b. USE TOOLS Use a compass and straightedge to bisect $\angle B$ shown in the figure.

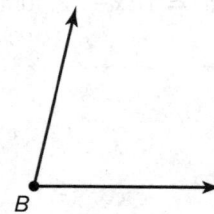

4. CRITIQUE REASONING Geri's proof that all right angles are congruent is shown below.

Given: ∠1, ∠2, ∠3, and ∠4 are right angles.

Prove: ∠1 ≅ ∠2 ≅ ∠3 ≅ ∠4

Statements	Reasons
1. ∠1, ∠2, ∠3, and ∠4 are right angles.	1. Given
2. m∠1 = m∠2 = m∠3 = m∠4	2. Def. of rt. ∠
3. ∠1 ≅ ∠2 ≅ ∠3 ≅ ∠4	3. Def. of ≅

Explain the error Geri made in her proof. Then write a correct two-column proof in the space provided.

_____

_____

_____

_____

_____

_____

5. CONSTRUCT ARGUMENTS Write a proof.

Given: ∠FGK and ∠HGK are a linear pair. $\overrightarrow{GK}$ bisects ∠FGH.

Prove: ∠FGJ and ∠JGK are complementary.

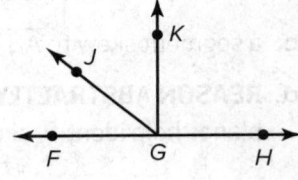

_____

_____

_____

_____

_____

_____

_____

_____

_____

_____

6. CONSTRUCT ARGUMENTS Write a paragraph proof to prove that if ∠A and ∠B are a linear pair and ∠A ≅ ∠B, then ∠A and ∠B are right angles.

_____

_____

_____

Objectives

- Identify parallel and skew lines and parallel planes.
- Identify types of angle pairs with a transversal.
- Prove and apply theorems about parallel lines with a transversal.

Parallel lines are coplanar lines that do not intersect. **Skew lines** are lines that do not intersect and are not coplanar. **Parallel planes** are planes that do not intersect.

EXAMPLE 1   Identify Parallel and Skew Relationships

Identify an example of each relationship in the diagram of the home shown.

a. a segment parallel to $\overline{GH}$ _____

b. a plane parallel to $ABCD$ _____

c. a segment skew to $\overline{AB}$ _____

d. **REASON ABSTRACTLY** How does identifying different planes help identify skew lines?

_____

_____

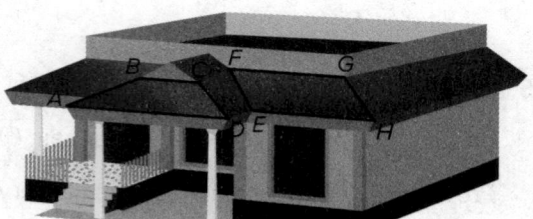

e. **CONSTRUCT ARGUMENTS** Can the Transitive Property be applied to parallel lines? In other words, if a line $\overleftrightarrow{QS}$ is parallel to line $\overleftrightarrow{RT}$ and $\overleftrightarrow{RT}$ is parallel to line $\overleftrightarrow{UW}$, must $\overleftrightarrow{QS}$ be parallel to $\overleftrightarrow{UW}$? Explain your reasoning.

_____

A **transversal** is a line that intersects two or more coplanar lines at two different points. A transversal of two lines creates four **interior angles** between the two lines and four **exterior angles** not between them.

KEY CONCEPT   Name the angle pairs for each type of angle.

Consecutive interior angles are interior angles lying on the same side of transversal ℓ.		
Alternate interior angles are interior angles lying on opposite sides of transversal ℓ.		
Alternate exterior angles are exterior angles lying on opposite sides of transversal ℓ.		
Corresponding angles lie on the same side of transversal ℓ and on the same side of lines a and b.		

Explore Angles Created by Parallel Lines and a Transversal

EXPLORE Using dynamic geometry software, compare angles created by a pair of parallel lines and a transversal.

a. **USE TOOLS** Plot two points and label them A and B. Construct the line through A and B. Then, plot and label a third point C, not on $\overleftrightarrow{AB}$, as shown.

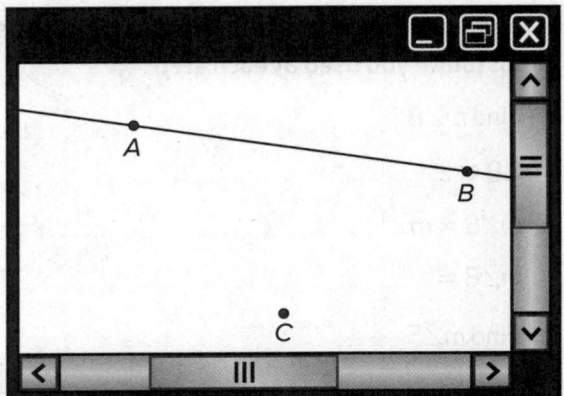

b. Construct the line through C parallel to $\overleftrightarrow{AB}$.

c. Plot and label point D on the original line, and construct transversal $\overleftrightarrow{CD}$.

d. Plot and label points E and F on $\overleftrightarrow{CD}$, outside the two parallel lines. Also plot points G and H on the line parallel to $\overleftrightarrow{AB}$, on either side of $\overleftrightarrow{CD}$, as shown.

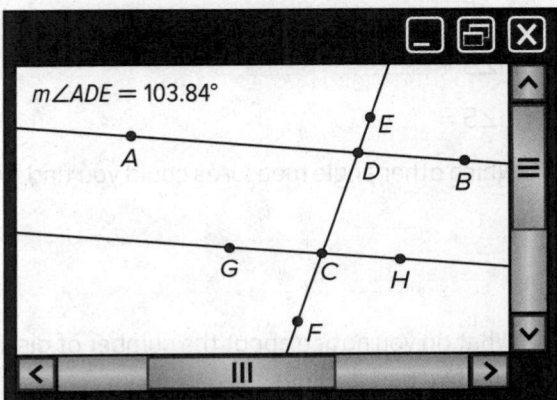

e. Measure all eight angles created, and record the results in the first row of the table. Drag point E or F to change the angle measures; record the new results in the second row.

Angle	∠ADE	∠BDE	∠BDC	∠ADC	∠GCD	∠HCD	∠FCH	∠FCG
1st measure								
2nd measure								

f. **COMMUNICATE PRECISELY** What do you observe about each type of angle pair?

Corresponding ∠s are _____ ; consecutive interior ∠s are _____ ;

alternate interior ∠s are _____ ; alternate exterior ∠s are _____ .

g. **MAKE A CONJECTURE** Drag point E or F until the measure of any of the angles is 90. Make a conjecture about a transversal that is perpendicular to one of two parallel lines.

_____

_____

h. **REASON QUANTITATIVELY** If ∠ADE is a right angle, what can be determined regarding each of the other seven angles? Explain.

_____

_____

_____

The **Corresponding Angles Postulate** states that If two parallel lines are cut by a transversal, then each pair of corresponding angles is congruent.

EXAMPLE 3   **Use the Corresponding Angles Postulate**

In the figure, $m\angle 3 = 107$. Find each angle measure. State which theorem, property, or postulate you used at each step.

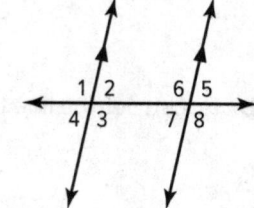

a. Find $m\angle 8$.

$\angle 8 \cong \angle$ _____   _____ Post.

$m\angle 8 = m\angle$ _____   _____

$m\angle 8 =$ _____   _____ Prop. of _____

b. Find $m\angle 5$.

$\angle 5$ and $\angle$ _____ are a _____ pair

$m\angle 5 + m\angle$ _____ = _____   _____

$m\angle 5 +$ _____ = _____   _____ Prop. of _____

$m\angle 5 =$ _____   _____ Prop. of _____

c. Which other angle measures could you find with the given information? Explain.

_____

_____

d. What do you notice about the number of distinct angle measures when a transversal cuts two parallel lines?

_____

Alternate Interior Angles Theorem If two parallel lines are cut by a transversal, then each pair of alternate interior angles is congruent.
Consecutive Interior Angles Theorem If two parallel lines are cut by a transversal, then each pair of consecutive interior angles is supplementary.
Alternate Exterior Angles Theorem If two parallel lines are cut by a transversal, then each pair of alternate exterior angles is congruent.

EXAMPLE 4   **Prove the Alternate Interior Angles Theorem**

Given: ℓ is a transversal of q and r, and $q \parallel r$.

a. Complete the proof that $\angle 3 \cong \angle 5$.

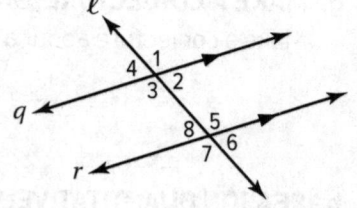

Statement	Reason
1. $q \parallel r$	Given
2. $\angle 1 \cong \angle 5$	
3. $\angle 1$ and $\angle 3$ are vert. $\angle$s	Given
4.	Vert. $\angle$s Thm.
5. $\angle 3 \cong \angle 5$	

Copyright © McGraw-Hill Education

b. DESCRIBE A METHOD How could you use the Corresponding Angles Postulate with the Vertical Angles Theorem or the Supplement Theorem to prove other theorems about parallel lines?

_____

_____

_____

| EXAMPLE 5 | **Use Theorems About Parallel Lines** |

Union Street and Filbert Street are parallel streets in San Francisco that intersect Columbus Avenue at Washington Square Park.

a. CONSTRUCT ARGUMENTS If $m\angle 1 = 49$, find $m\angle 3$. State a reason for each step.

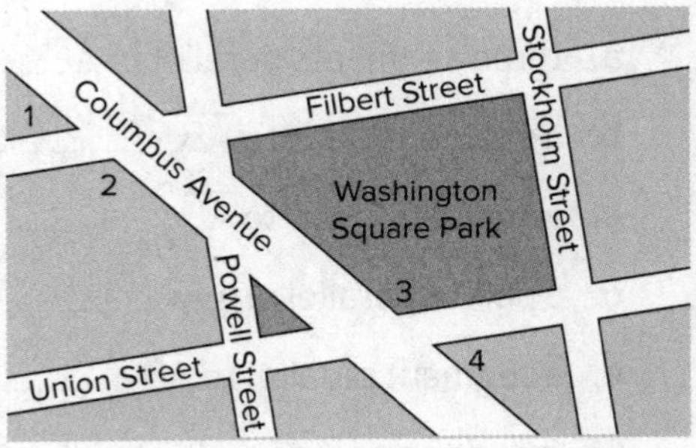

_____

_____

_____

_____

b. PLAN A SOLUTION Powell and Stockholm Streets are perpendicular to Filbert Street. How could you determine all the remaining angles of the triangular and pentagonal pieces of Washington Square Park?

_____

_____

_____

c. CONSTRUCT ARGUMENTS If $\angle 1$ and $\angle 4$ are complementary, prove that $m\angle 2 = 135$.

_____

_____

_____

_____

| EXAMPLE 6 | **State and Prove the Perpendicular Transversal Theorem** |

Given that $q \parallel r$ and $\ell \perp q$, what is true about ℓ and r?

a. MAKE A CONJECTURE Given that $q \parallel r$ and $\ell \perp q$, state a conclusion that you believe to be true about ℓ and r.

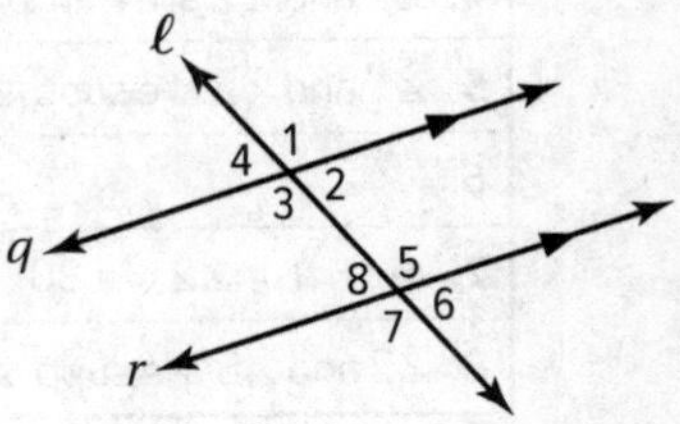

_____

_____

b. Complete this paragraph proof of the Perpendicular Transversal Theorem.

$\ell \perp r$ (given), so $\angle 2$ is a(n) _____ (def. of $\perp$ lines) and, therefore, _____

(def. of right $\angle$s). _____ (given), so by the Cons. Int. $\angle$s Thm. and the def. of supp.

$\angle$s, _____ . By the Substitution and Subtraction Props. of Equality,

$m\angle 5 =$ _____ , so $\angle 5$ is a(n) _____ (def. of _____ $\angle$s) and,

therefore, _____ (def. of $\perp$ lines).

PRACTICE

1. Identify an example of each relationship.

    a. three segments parallel to $\overline{AE}$ _____

    b. a segment skew to $\overline{AB}$ _____

    c. a segment skew to $\overline{AE}$ _____

    d. a pair of parallel planes _____

    e. a segment parallel to $\overline{AD}$ _____

    f. three segments parallel to $\overline{HG}$ _____

    g. five segments skew to $\overline{BC}$ _____

    h. **USE STRUCTURE** How could you characterize the relationship between planes
    ABCD and *DCGH*? Explain.

    _____

    _____

2. **USE STRUCTURE** Plane *A* contains lines *p* and *q*. Line *r* intersects
plane *A* at point *X*. Given that $p \perp q$ and that *r* is skew to *q* but not *p*,
draw and label a figure that fits this description.

3. a. **CONSTRUCT ARGUMENTS** Given: ℓ is a transversal of *q* and *r* and
    $q\|r$. **Prove:** $\angle 2$ and $\angle 5$ are supplementary.

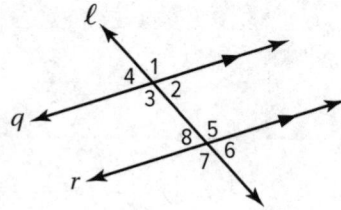

Statement	Reason
1. $q\|r$	Given
2.	Corr. $\angle$s Post.
3. $m\angle 1 = m\angle 5$	Def. of cong. $\angle$s
4. $\angle 1$ and $\angle 2$ are a lin. pair	Given
5. $\angle 1$ and $\angle 2$ are supp. $\angle$s	
6.	Def. of supp. $\angle$s
7. $m\angle 5 + m\angle 2 = 180$	Subst. Prop. of =
8. $\angle 2$ and $\angle 5$ are supp. $\angle$s	

b. DESCRIBE A METHOD Could you also prove that ∠1 and ∠6 are supplementary? Explain your answer.

_____

_____

4. **REASON QUANTITATIVELY** In the figure, $m\angle 4 = 118$. Find each angle measure. Justify each step.

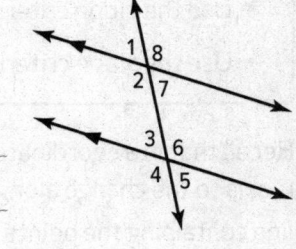

   a. Find $m\angle 8$.

   _____

   _____

   b. Find $m\angle 7$.

   _____

   _____

   _____

   _____

5. **REASON QUANTITATIVELY** Find the values of x and y in the trapezoid. Justify your answer.

   _____

   _____

   _____

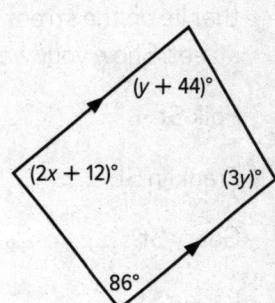

6. **CONSTRUCT ARGUMENTS** Write a paragraph proof of the Alternate Exterior Angles Theorem. Given: $q \| r$; Prove: $\angle 1 \cong \angle 7$.

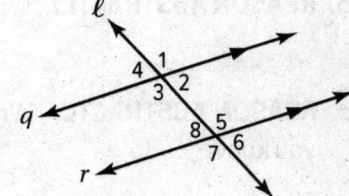

   _____

   _____

   _____

7. **CONSTRUCT ARGUMENTS** In the figure, the lines **m** and **n** are parallel and the lines **p** and **q** are parallel. Write a paragraph proof to prove that if $m\angle 1 - m\angle 4 = 25$, then $m\angle 9 - m\angle 12 = 25$.

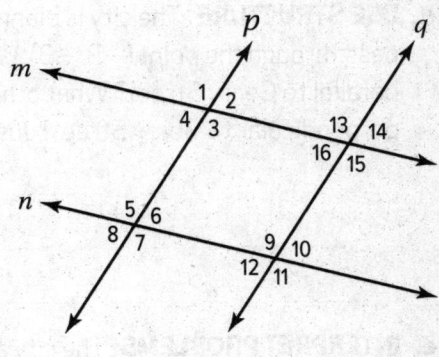

   _____

   _____

   _____

   _____

   _____

   _____

   _____

Objectives

- Use the slope criterion for parallel lines to solve problems.
- Use the slope criterion for perpendicular lines to solve problems.

Mathematical Practices
1, 2, 4, 6, 7, 8

Recall that in a coordinate plane, the **slope** of a line is the ratio of the change along the y-axis to the change along the x-axis between any two points on the line. The slope m of a line containing the points (x_1, y_1) and (x_2, y_2) is $m = \frac{y_2 - y_1}{x_2 - x_1}$, where $x_1 \neq x_2$.

EXAMPLE 1   Investigate Slope Criteria

EXPLORE The coordinate plane shows a map of the streets in downtown Rockland.

a. **CALCULATE ACCURATELY** For each street, identify a pair of points that lie on the street. Then use the coordinates to find the slope of the street. Show your work.

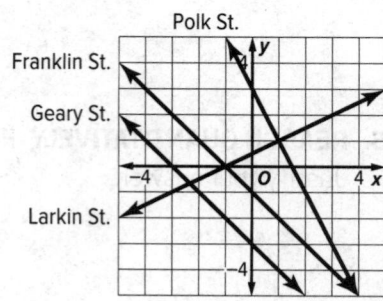

Polk St. _____

Franklin St. _____

Geary St. _____

Larkin St. _____

b. **REASON ABSTRACTLY** Which of the streets, if any, are parallel? Explain how you know.

_____

c. **REASON ABSTRACTLY** Which of the streets, if any, are perpendicular? Explain how you know.

_____

d. **USE STRUCTURE** The city is planning to add a new street to the neighborhood. It will pass through the point $(-2, -3)$. What other point will the street pass through if it is parallel to Geary Street? What other point will the street pass through if it is perpendicular to Geary Street? Justify your responses.

_____

_____

_____

e. **INTERPRET PROBLEMS** The city adds a new street to the neighborhood that passes through the points $(-4, -1)$ and $(-2, -3)$. Is this new street parallel to any streets currently in the neighborhood? Explain.

_____

_____

The Key Concept box summarizes the criteria for parallel and perpendicular lines that you used in the exploration.

KEY CONCEPT **Parallel and Perpendicular Lines**

Complete each criterion. Then use the lines in the figure to give an example of each criterion.

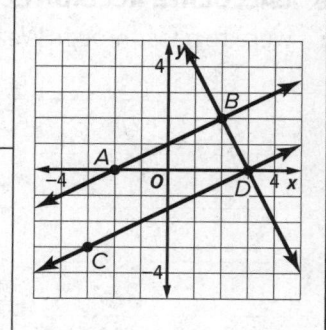

Parallel Lines
Two nonvertical lines are parallel if and only if _____. All vertical lines are parallel.
Example: _____
Perpendicular Lines
Two nonvertical lines are perpendicular if and only if _____
_____. Vertical and horizontal lines are perpendicular.
Example: _____

EXAMPLE 2 **Apply the Slope Criteria**

The figure shows a map of two straight highways. Regional planners want to build a new connector road from Highway 332 to Route 1. They want the road to be perpendicular to Highway 332 and to pass through Orland. Follow these steps to determine where the road should intersect Route 1.

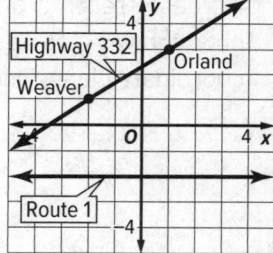

a. **INTERPRET PROBLEMS** What should the slope of the connector road be? Explain how you know.

_____

_____

_____

b. **USE STRUCTURE** You need to find the point along Route 1 where the connector road intersects it. What can you conclude about the coordinates of any point along Route 1? Why?

_____

_____

c. **CALCULATE ACCURATELY** Use the general coordinates for a point on Route 1, the coordinates of Orland, and what you know about the slope of the connector road to write and solve an equation that allows you to find the coordinates of the required point. Explain your reasoning.

_____

_____

_____

_____

d. EVALUATE REASONABLENESS Explain how you can use the graph to check that your answer is reasonable.

_____

_____

_____

e. CALCULATE ACCURATELY Determine where the road would intersect Route 1 if it were to pass through Weaver instead of Orland. Explain your reasoning.

_____

_____

_____

_____

_____

f. INTERPRET PROBLEMS Suppose the traffic on Highway 332 has significantly increased recently and a new highway it to be built to compensate for the change. The new highway is to be parallel to Highway 332 and pass through the point (4, 0) on the coordinate plane. Find the point at which the new highway intersects Route 1.

_____

_____

_____

_____

PRACTICE

CALCULATE ACCURATELY Determine whether any of the lines in each figure are parallel or perpendicular. Justify your answers.

1.

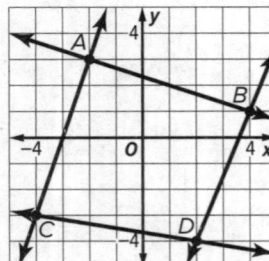

2.

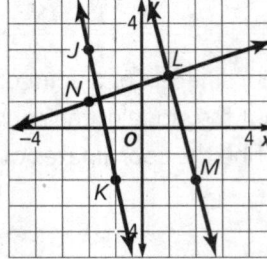

3.

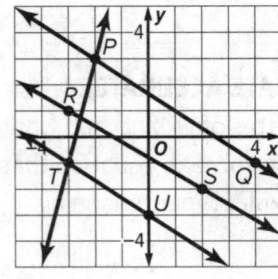

_____ _____ _____

_____ _____ _____

_____ _____ _____

4. REASON ABSTRACTLY $\overleftrightarrow{AB}$ is parallel to $\overleftrightarrow{CD}$. The coordinates of A, B, and C are $A(-3, 1)$, $B(6, 4)$, and $C(1, -1)$. What is a possible set of coordinates for point D? Describe the reasoning you used to find the coordinates.

_____

_____

_____

5. A video game designer is using a coordinate plane to plan the path of a helicopter. She has already determined that the helicopter will move along straight segments from P to Q to R.

a. USE A MODEL The designer wants the next part of the path, $\overline{RS}$, to be perpendicular to $\overline{QR}$, and she wants point S to lie on the y-axis. What should the coordinates of point S be? Justify your answer.

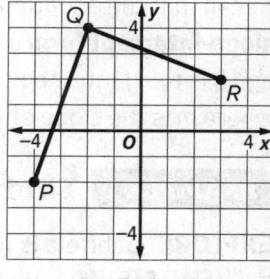

_____

_____

_____

b. REASON ABSTRACTLY Would your answer be different if the designer wanted the next part of the helicopter's path to be parallel to $\overline{PQ}$? Explain.

_____

_____

_____

6. DESCRIBE A METHOD Line ℓ passes through $(0, 2)$ and is parallel to $\overleftrightarrow{AB}$, where A and B have coordinates $A(2, 0)$ and $B(4, 4)$.

a. Complete the table of values showing coordinates of points on line ℓ.

x	0	1	2	3	4
y	2				

b. Look for a pattern in the table. In general, how can you find the y-coordinate of a point on line ℓ if you know the x-coordinate?

_____

c. Does the point $(18, 40)$ lie on line ℓ? Explain.

_____

7. REASON QUANTITATIVELY Line p passes through $(1, 3)$ and $(4, 7)$ and line q passes through $(0, -2)$ and (a, b).

a. Find the slope of lines p and q.

_____

b. Find possible values of a and b if $p \parallel q$.

_____

Objectives

- Find the equation of a line parallel to a given line that passes through a given point.

- Find the equation of a line perpendicular to a given line that passes through a given point.

Recall that the equation of a nonvertical line can be written in different forms. The **slope-intercept form** of a linear equation is $y = mx + b$, where m is the slope of the line and b is the y-intercept. The **point-slope form** of a linear equation is $y - y_1 = m(x - x_1)$, where m is the slope of the line and (x_1, y_1) is a point on the line.

EXAMPLE 1   Find the Equation of a Line Parallel to a Given Line

EXPLORE Gabriela is using a coordinate plane to design a large model train layout. In the part of the layout she is working on, Track 1 and Track 2 are straight tracks that are parallel to each other. Track 1 passes through $A(0, -1)$ and $B(3, 3)$. Track 2 passes through $C(1, 4)$. Gabriela wants to find an equation for Track 2.

a. **INTERPRET PROBLEMS** Graph and label points A, B, and C and Track 1 on the coordinate plane.

b. **CALCULATE ACCURATELY** What is the slope of Track 1? What does this tell you about Track 2? Why?

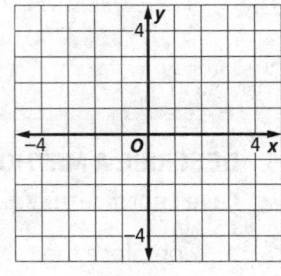

_____

_____

c. **USE STRUCTURE** Explain how you can use the slope-intercept form of a line to find the equation of Track 2. Find the point-slope form of the equation for Track 2.

_____

_____

_____

_____

d. **USE A MODEL** In the layout, the area above the line $y = -4$ will be covered with gravel and the area below the line $y = -4$ will be covered with artificial grass. At what point on the coordinate plane will Track 2 cross from gravel to artificial grass? Explain how you used your mathematical model to find the coordinates of the point.

_____

_____

_____

EXAMPLE 2 **Find the Equation of a Line Perpendicular to a Given Line**

Find the equation of the line through point R that is perpendicular to the line through points P and Q.

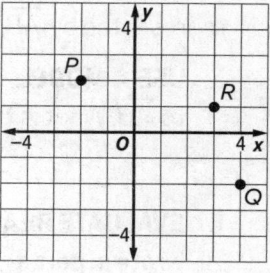

a. INTERPRET PROBLEMS What are the main steps you will use in solving this problem?

_____

_____

b. CALCULATE ACCURATELY Show how to use the steps you outlined above, and the point-slope form of a line, to find the equation of the required line.

_____

_____

_____

c. EVALUATE REASONABLENESS Rewrite the equation from **part b** in slope-intercept form and use this form of the line to check that your answer is reasonable.

_____

_____

_____

d. REASON QUANTITATIVELY A line goes that through the point R and $\left(-2, -\frac{32}{5}\right)$ appears to be perpendicular to the line through P and Q. Explain how to check if it is in fact perpendicular by using the equation you found in **part b**.

_____

_____

_____

_____

EXAMPLE 3 **Find the Equation of a Line**

Find the equation of the line through $(6, 8)$ that is perpendicular to the line whose equation is $3x + 4y = 1$.

a. USE STRUCTURE What is the slope of the line whose equation is $3x + 4y = 1$? How do you know?

_____

_____

b. CALCULATE ACCURATELY Explain how to use your answer to **part a** to find the equation of the required line.

_____

_____

_____

1. Gavin is working on an animated film about skiing. The figure shows a ski slope, represented by $\overleftrightarrow{AB}$, and one of the chairs on the chair lift, represented by point C.

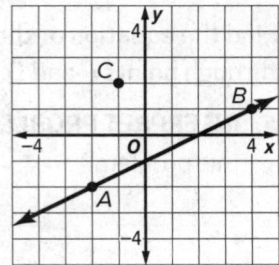

   a. **USE A MODEL** The chair needs to move along a straight line that is parallel to $\overleftrightarrow{AB}$. What is the equation of this line?

   _____

   b. **EVALUATE REASONABLENESS** How can you be sure that the equation you wrote in **part a** is correct?

   _____

   _____

   c. **USE A MODEL** The top of the chair lift occurs at $y = 20$. Explain how Gavin can find the coordinates of the chair when it reaches the top of the chair lift.

   _____

   _____

   _____

2. **CALCULATE ACCURATELY** Write an equation in the indicated form for each line described. Explain your reasoning.

   a. Find the equation of the line through the point $J(2, -1)$ that is perpendicular to the line through $K(-2, 4)$ and $L(3, -1)$. Write the equation in slope-intercept form.

   _____

   b. Find the equation of the line through the point $M(6, 2)$ that is parallel to the line through $P(1, 0)$ and $Q(4, 9)$. Write the equation in point-slope form.

   _____

   c. Find the equation of the line through the point $R(0, 3)$ that never intersects the line through $S(0, 4)$ and $T(4, 1)$. Write the equation in point-slope form.

   _____

3. The equation of line ℓ is $3y - 2x = 6$.

   a. **USE STRUCTURE** Line m is perpendicular to line ℓ and passes through the point $P(6, -2)$. Find the equation of line m.

   _____

   b. **CONSTRUCT ARGUMENTS** Line n is parallel to line m. Is it possible to write the equation of line n in the form $2x + 3y = k$ for some constant k? Give an argument to justify your answer.

   _____

   _____

   _____

4. **CRITIQUE REASONING** A student was asked to find the equation of the line perpendicular to $\overleftrightarrow{AB}$ that passes through point P, given that A, B, and P have coordinates $A(0, 3)$, $B(2, 2)$, and $P(1, 4)$. The student's work is shown at the right. Do you agree with the student's solution? If not, identify any errors and find the correct equation.

_____

_____

_____

> Slope of $\overleftrightarrow{AB} = \frac{2-3}{2-0} = -\frac{1}{2}$.
>
> So, the slope of the required line is 2. The equation of this line is $y = 2x + b$. The line passes through $P(1, 4)$.
>
> To find b: $1 = 2(4) + b$
> $1 = 8 + b$
> $-7 = b$
>
> So, the equation is $y = 2x - 7$.

CONSTRUCT ARGUMENTS Determine whether each statement is *always, sometimes,* or *never* true. Explain.

5. If p and q are real numbers, then the line $2x + 4y = p$ is parallel to the line $2x + y = q$.

_____

6. A line through the origin is perpendicular to the line whose equation is $y = 3x + 2$.

_____

7. If $m \neq n$, then the line through the points $(m, 4)$ and $(n, 4)$ is perpendicular to the y-axis.

_____

8. The director of a marching band uses a coordinate plane to design the band's formations. During one formation, a drummer marches from point A to point B, then turns 90° to her right and marches until she reaches the x-axis.

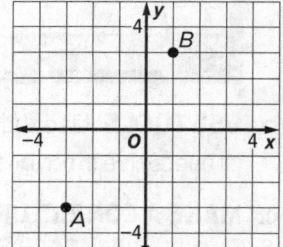

    a. **USE A MODEL** When the drummer marches from point B to the x-axis, what is the equation of the line she marches along?

    _____

    b. **REASON ABSTRACTLY** The director wants to know if the drummer will cross the x-axis at a point where the x-coordinate is greater than or less than 5. Explain how the director can answer this question.

    _____

    _____

9. **REASON QUANTITATIVELY** Let a and b be nonzero real numbers. The line p has the equation $y = ax + b$.

    a. Find the equation of the line through $(5, 1)$ that is parallel to p. Write the equation in point-slope form. Explain your reasoning.

    _____

    b. Find the equation of the line through $(2, 3)$ that is perpendicular to p. Write the equation in slope-intercept form. Explain your reasoning.

    _____

Copyright © McGraw-Hill Education

Objectives

- Prove theorems about angle pairs that guarantee parallel lines.
- Construct a line parallel to a given line through a point not on the line.

EXAMPLE 1 Investigate Conditions for Parallel Lines

EXPLORE Use the Geometer's Sketchpad to investigate lines cut by a transversal and the facts about angle pairs that guarantee that the lines are parallel.

a. USE TOOLS Use the Geometer's Sketchpad to draw two lines, *j* and *k*, that are intersected by a transversal, *t*. Label the lines and angles that are formed, as shown on the left below.

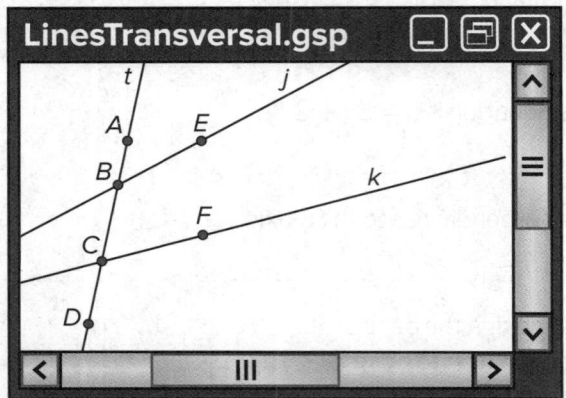

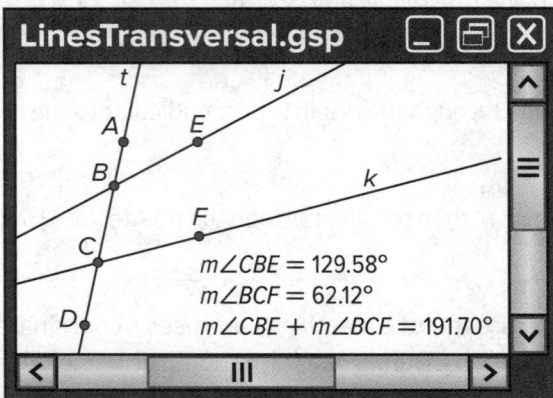

b. USE TOOLS Use the Geometer's Sketchpad to find the sum of the measures of the consecutive interior angles, as shown on the right above.

c. MAKE A CONJECTURE Adjust the lines as needed so that the consecutive interior angles are supplementary. What appears to be true about lines *j* and *k*? Compare your results with other students and make a conjecture about the relationship you observed.

_____

_____

d. MAKE A CONJECTURE Now use the Geometer's Sketchpad to adjust the lines so each of the following conditions is true. Note the relationship between lines *j* and *k*, compare your results with other students, and make a conjecture about the relationship you observed.

- A pair of corresponding angles is congruent.
- A pair of alternate exterior angles is congruent.
- A pair of alternate interior angles is congruent.

_____

_____

_____

e. MAKE A CONJECTURE Do you expect the three conditions listed in **part d** to be equivalent? In other words, do you expect all of them to be true if one of them is known to be true? Explain your reasoning.

_____

_____

_____

The Key Concept box summarizes conditions you can use to prove lines are parallel.

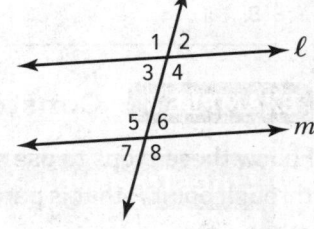

KEY CONCEPT **Proving Lines Parallel**

Use the figure to give an example of each condition that allows you to prove lines are parallel. The first example has been done for you.

Postulate: Consecutive Interior Angles Converse	Example:
If two lines in a plane are cut by a transversal so that a pair of consecutive interior angles is supplementary, then the lines are parallel.	If $m\angle 4 + m\angle 6 = 180$, then $\ell \parallel m$.
Theorem: Corresponding Angles Converse	**Example:**
If two lines in a plane are cut by a transversal so that a pair of corresponding angles is congruent, then the lines are parallel.	
Theorem: Alternate Exterior Angles Converse	**Example:**
If two lines in a plane are cut by a transversal so that a pair of alternate exterior angles is congruent, then the lines are parallel.	
Theorem: Alternate Interior Angles Converse	**Example:**
If two lines in a plane are cut by a transversal so that a pair of alternate interior angles is congruent, then the lines are parallel.	

EXAMPLE 2 **Prove the Corresponding Angles Converse**

Follow these steps to prove the Corresponding Angles Converse.
Given: $\angle 1 \cong \angle 2$
Prove: $\ell \parallel m$

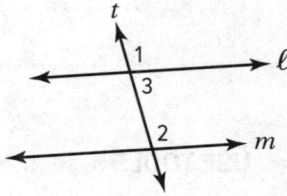

a. PLAN A SOLUTION Which of the conditions in the Key Concept box will you be able to use as a reason in your proof? Why?

_____

b. PLAN A SOLUTION Describe the main steps you will use in the proof.

_____

_____

_____

_____

c. CONSTRUCT ARGUMENTS Complete the two-column proof of the theorem.

Statements	Reasons
1. $\angle 1 \cong \angle 2$	1.
2. $\angle 1$ and $\angle 3$ form a linear pair.	2.
3.	3.
4. $m\angle 1 + m\angle 3 = 180$	4.
5. $m\angle 1 = m\angle 2$	5.
6. $m\angle 2 + m\angle 3 = 180$	6.
7.	7.
8. $\ell \parallel m$	8.

EXAMPLE 3 Construct Parallel Lines

Follow these steps to use a compass and straightedge to construct a line through point P that is parallel to line q. Work directly on the figure at the right.

a. **USE TOOLS** Use a straightedge to draw a line t through point P that intersects line q.

b. **USE TOOLS** Copy the angle formed by line t and line q so that the vertex of the copy is point P and one side of the copy is line t.

c. **USE TOOLS** One side of the angle you made in Step b forms the required line. Use the straightedge to extend this side if necessary and label it line r. Line $q \parallel$ line r.

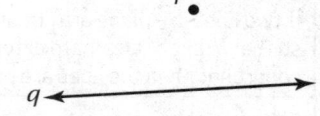

d. **CONSTRUCT ARGUMENTS** Explain why this construction method produces parallel lines.

_____

_____

_____

_____

e. **USE TOOLS** Use the same process to construct a line through point P that is parallel to line q for each of the situations given.

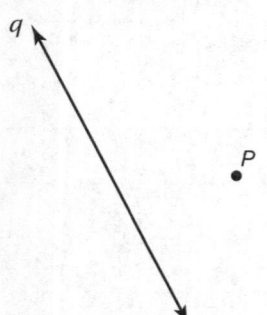

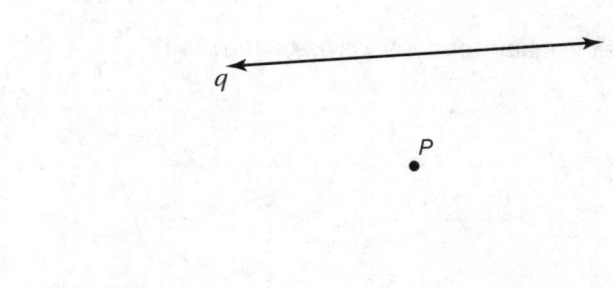

1. Follow these steps to prove the Alternate Exterior Angles Converse.

   Given: $\angle 1 \cong \angle 2$

   Prove: $\ell \parallel m$

   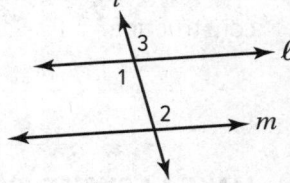

   a. **CONSTRUCT ARGUMENTS** Complete the two-column proof.

Statements	Reasons
1. $\angle 1 \cong \angle 2$	1.
2. $\angle 2 \cong \angle 3$	2.
3.	3.
4.	4.

   b. **COMMUNICATE PRECISELY** Which postulate or theorem from the Key Concept box did you use as a reason in the proof? Why is it permissible to use this postulate or theorem as a reason?

   _____

   _____

2. **CONSTRUCT ARGUMENTS** Write a paragraph proof to prove the Alternate Interior Angles Converse.

   Given: $\angle 1 \cong \angle 2$

   Prove: $\ell \parallel m$

   _____

   _____

   _____

USE TOOLS Use a compass and straightedge to construct the line through point P that is parallel to line q.

3.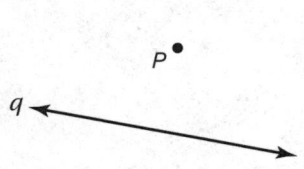

4.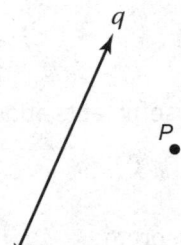

5. **CONSTRUCT ARGUMENTS** Write a paragraph proof to prove that if a transversal cuts two lines and alternate exterior angles are congruent, then alternate interior angles are congruent.

   _____

   _____

   _____

Perpendicular Lines

Objectives

- Construct perpendicular lines, including the perpendicular bisector of a line segment.
- Prove the Perpendicular Bisector Theorem and its converse.

EXAMPLE 1 Construct and Analyze a Perpendicular Bisector

EXPLORE Follow these steps to construct a perpendicular bisector and make a conjecture about points that lie on a perpendicular bisector.

a. **USE TOOLS** The construction you learned in a previous lesson to bisect a line segment also constructs the perpendicular bisector of the segment. Use a compass and straightedge to construct the perpendicular bisector ℓ of $\overline{AB}$.

b. **COMMUNICATE PRECISELY** Use the definition of perpendicular to explain what must be true about line ℓ and $\overline{AB}$.

_____

c. **EVALUATE REASONABLENESS** How can you use a ruler and protractor to check your construction?

_____

_____

d. **MAKE A CONJECTURE** Choose a point P on line ℓ. Use a ruler to measure the distances AP and BP. Then choose a point Q on line ℓ and find AQ and BQ. Finally, choose a point R on line ℓ and find AR and BR. What do you notice? State a conjecture based on your findings.

_____

_____

The Key Concept box summarizes useful facts about perpendicular bisectors.

KEY CONCEPT

Complete the statement of each theorem.

Perpendicular Bisector Theorem

If a point lies on the perpendicular bisector of a line segment, then

---

Converse of the Perpendicular Bisector Theorem

If a point is equidistant from the endpoints of a segment, then

---

CHAPTER 2 Logical Arguments and Line Relationships

You will prove the Perpendicular Bisector Theorem in **Example 2** and prove its converse as an exercise.

EXAMPLE 2 **Prove the Perpendicular Bisector Theorem**

Follow these steps to prove the Perpendicular Bisector Theorem.

Given: Line ℓ is the perpendicular bisector of $\overline{AB}$. Point P lies on line ℓ.
Prove: $AP = BP$

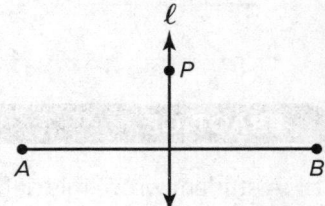

a. **INTERPRET PROBLEMS** Mark the given figure to show that line ℓ is the perpendicular bisector of $\overline{AB}$.

b. **COMMUNICATE PRECISELY** Consider the reflection in line ℓ. What is the image of point P? Why? What is the image of point A? Why? (*Hint:* Use the definition of a reflection.)

_____

_____

c. **CONSTRUCT ARGUMENTS** Based on your answer to **part b**, what can you conclude about AP and BP? Explain your reasoning.

_____

_____

_____

d. **USE TOOLS** Verify that P lies on the perpendicular bisector of $\overline{AB}$. Explain your solution process.

_____

_____

_____

_____

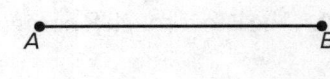

e. **USE TOOLS** Use a ruler to verify the Perpendicular Bisector Theorem.

EXAMPLE 3 **Construct a Perpendicular to a Line**

Follow these steps to construct the perpendicular from point G to line m.

a. **USE TOOLS** Place the point of your compass on point G. Open the compass to a distance greater than the distance from point G to line m and make an arc that intersects line m at points J and K.

b. **USE TOOLS** Now construct the perpendicular bisector of $\overline{JK}$.

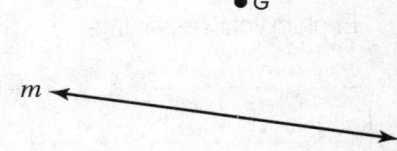

c. **CONSTRUCT ARGUMENTS** How can you use the Converse of the Perpendicular Bisector Theorem to explain why point G will lie on the perpendicular bisector of $\overline{JK}$?

_____

_____

_____

PRACTICE

1. A student wrote this definition: "Two lines are perpendicular if they intersect to form four congruent angles." Do you think this is a good definition of perpendicular lines? Why or why not?

_____

_____

USE TOOLS Construct each of the following.

2. The perpendicular bisector of $\overline{EF}$

3. The perpendicular bisector of $\overline{TV}$

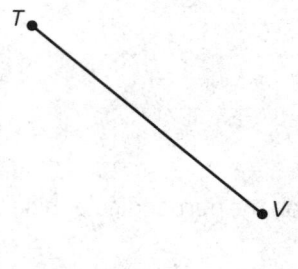

4. The line through R perpendicular to ℓ

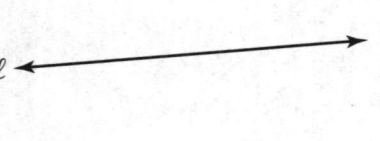

5. The line through P perpendicular to m

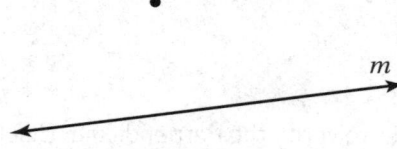

6. **CALCULATE ACCURATELY** $\overline{RQ}$ is the perpendicular bisector of $\overline{FK}$. $QG = GH = HJ = JK$. If $FQ = (6x + 8)$ inches and $GH = (2x - 5)$ inches, find the length of $\overline{HK}$. Explain your reasoning.

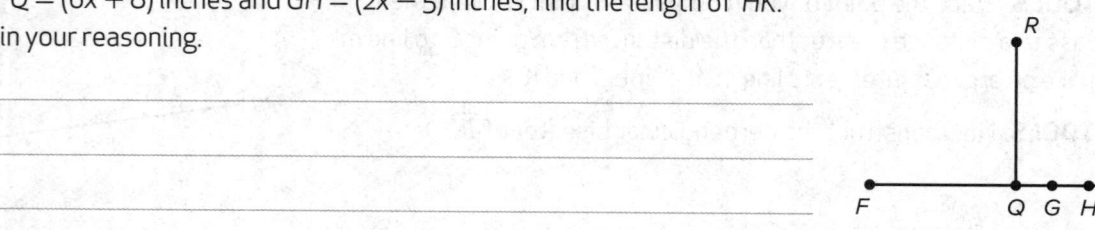

_____

_____

_____

_____

7. **CONSTRUCT ARGUMENTS** Follow these steps to prove the Converse of the Perpendicular Bisector Theorem.

    Given: $AP = BP$

    Prove: Point P lies on the perpendicular bisector ℓ of $\overline{AB}$.

    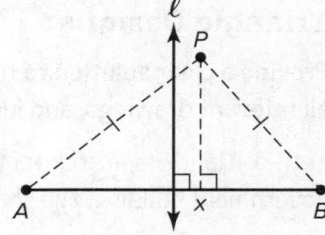

    a. Suppose point P does *not* lie on the perpendicular bisector of $\overline{AB}$. Then when you draw the perpendicular from P to $\overline{AB}$, the point at which the perpendicular intersects $\overline{AB}$ is not the midpoint. Call this point X. You can assume that $AX > BX$. Why?

    _____

    _____

    b. $\triangle APX$ and $\triangle BPX$ are right triangles. Use the Pythagorean Theorem to write a relationship among the side lengths in $\triangle APX$ and a relationship among the side lengths in $\triangle BPX$.

    _____

    c. You wrote two equations in **part b**. What is the resulting equation when you subtract one equation from the other?

    _____

    d. Your answer to **part c** should include the expression $AP^2 - BP^2$. What can you conclude about the value of this expression? Why?

    _____

    e. What does your answer to **part d** tell you about AX and BX? Explain why this leads to a contradiction and why this completes the proof.

    _____

    _____

    _____

    _____

    _____

8. **COMMUNICATE PRECISELY** State the Perpendicular Bisector Theorem as a biconditional.

    _____

    _____

9. **INTERPRET PROBLEMS** Point P lies on the perpendicular bisector of a segment with A as one endpoint. Which of the following points is the other endpoint of the segment? Explain your reasoning.

    _____

    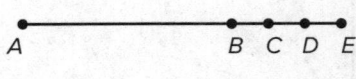

    _____

    _____

Triangle Designs

Provide a clear solution to the problem. Be sure to show all of your work, include all relevant drawings, and justify your answers.

Letitia, a landscape architect, is designing a set of paths for a park in front of a new government building. The shape of the triangular park is as shown.

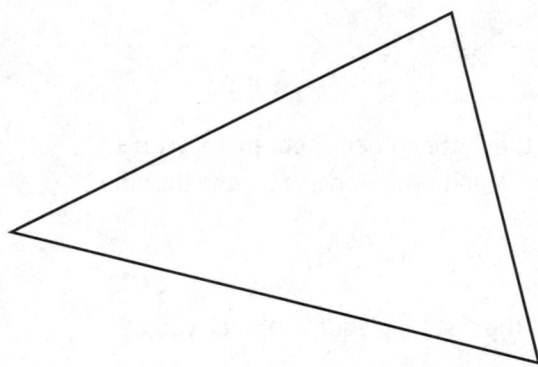

Part A

In her first design, Letitia decides to place three paths in the park, each of which is parallel to a side of the triangle. Write a paragraph proof to prove that $\angle 1 \cong \angle 5$ given that $\angle 1$ and $\angle 2$ are supplementary and $\angle 3$ and $\angle 4$ are supplementary.

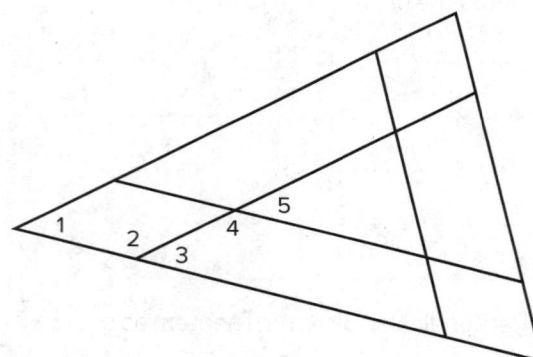

Part B

In her second design, Letitia decides to place three paths in the park as shown in the diagram. Write a paragraph proof to show that if $VW = XZ$ and $YW = YZ$, then $VY = XY$.

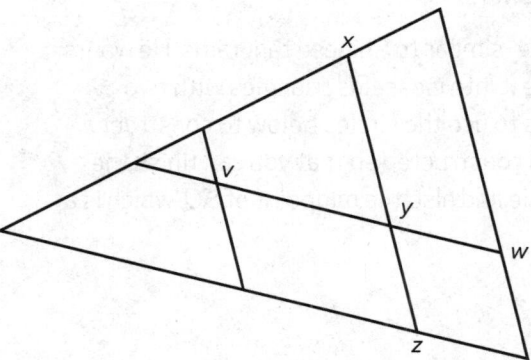

Part C

Letitia wants to draw two different designs of three paths for the park. In each design, a trash receptacle will be placed at the point where the three paths intersect. In one design the paths will be the angle bisectors of the triangle and in the other design the paths will be the perpendicular bisectors of the triangle. Explain the advantage of the placement of the trash receptacle in each design.

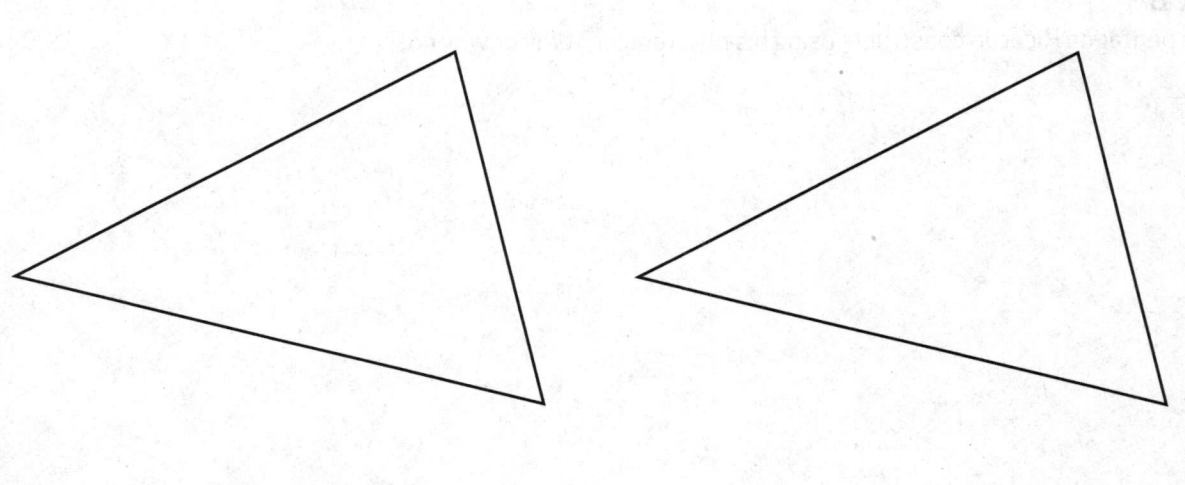

Pen-Tangram Puzzle

Provide a clear solution to the problem. Be sure to show all of your work, include all relevant drawings, and justify your answers.

Ricardo is designing a "pen-tangram puzzle" similar to Chinese tangrams. He wants to construct a regular pentagon and divide it into isosceles triangles with two different sets of angle measures. He plans to use the circles below to construct a regular pentagon. These circles have been constructed so that you can finish the pentagon. R is the center of the larger circle and also the midpoint of $\overline{SQ}$, which is a radius of the smaller circle.

Part A
Construct the perpendicular bisector of $\overline{PS}$; label the points where it intersects the smaller circle A and E. Construct the perpendicular bisector of $\overline{ST}$; label the points where it intersects the smaller circle B and D. Join points A through E to form pentagon $ABCDE$.

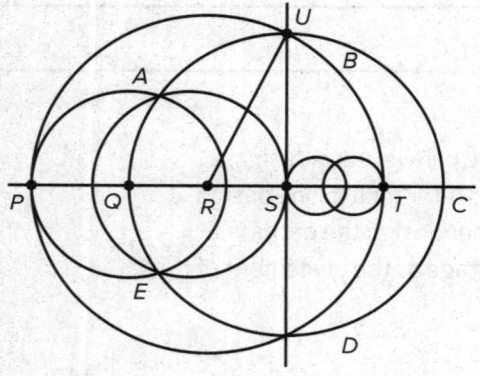 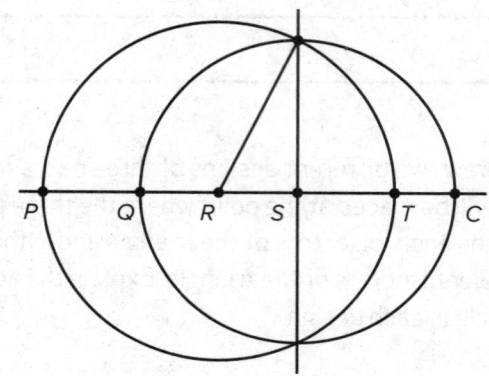

Part B
Is the pentagon Ricardo constructs using his plan regular? Why or why not?

Part C

Draw diagonals $\overline{AC}$, $\overline{BE}$, and $\overline{CE}$. How can you use compass and straightedge constructions to estimate the classification of $\triangle ACE$ and $\triangle BCE$?

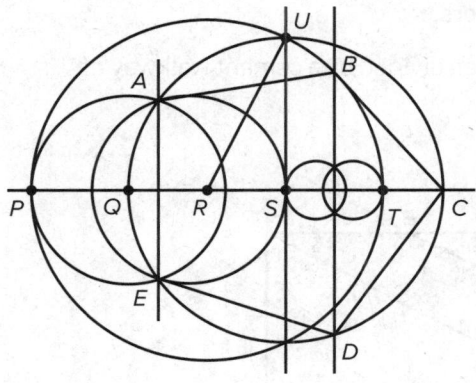

Part D

Label the intersection of $\overline{AC}$ and $\overline{BE}$ as point F. Finally draw in segment $\overline{DF}$. How does $\overline{DF}$ appear to relate to $\angle EDC$? Describe a method to test your conjecture. Explain your reasoning.

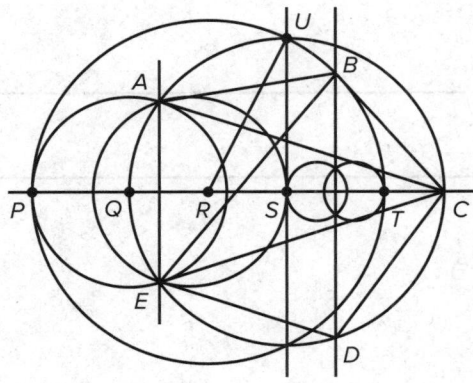

Landscaping a Garden

Provide a clear solution to the problem. Be sure to show all of your work, include all relevant drawings, and justify your answers.

A landscaper is designing a rectangular garden divided by a cement walkway of constant width, indicated by shading.

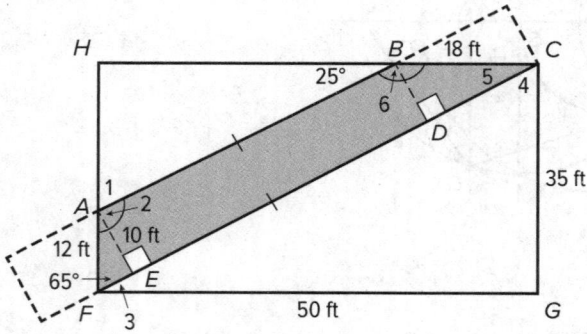

Part A

To stake off the region for the walkway, the landscaper needs to know the measures of angles 1 through 6. What are the measures of these angles? Justify your answers.

Part B

Is *ABDE* a rectangle? Justify your response.

Part C

Find the area of the walkway within the garden. Show your work.

Part D

The landscaper wants to plant wildflowers in the regions on each side of the walkway. Bags of soil cover 1.5 cubic feet. If the soil will be 2 inches thick, how many bags of soil are needed? Show your work.

Down the Drain

Provide a clear solution to the problem. Be sure to show all of your work, include all relevant drawings, and justify your answers.

A 15-gallon tub fills and empties at different rates. One graph represents the relationship between the amount of water in the tub as it fills and the amount of time that the water has been flowing. The other graph represents the relationship between the amount of water in the tub as it empties and the amount of time it has been emptying.

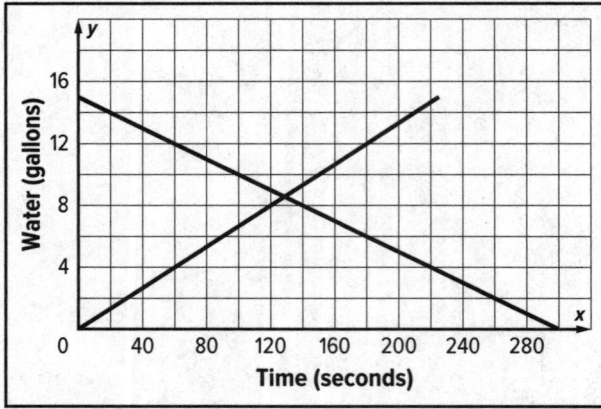

Part A
Find the slope of each line. What is the meaning of these slopes in the context of the problem?

Part B

Write the equation of each line. Are there any restrictions on the domain and range? Explain.

Part C

How long does it take the tub to fill? How long does it take for the tub to drain?

Part D

Sarah begins to fill the tub but forgets to put the plug in the drain. At what rate will the tub fill or empty, and in how many minutes will the tub be full or empty? How much water will go down the drain during that time?

1. An angle is a figure formed by two [ ] with a common [ ].

2. A line segment is a part of a [ ] that is formed by two [ ] and all points between them.

3. Complete the steps in the following proof of the Vertical Angles Theorem.

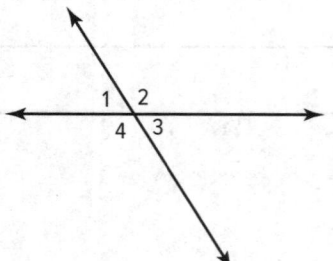

Prove: $\angle 1 \cong \angle 3$

$\angle 1$ and $\angle 2$ form a [ ]. Therefore, by the Linear Pair Theorem, they are [ ].

This means that $m\angle 1 + m\angle 2 =$ [ ].
Similarly, $m\angle 2 + m\angle 3 =$ [ ]. By the [ ] Property of Equality, $m\angle 1 + m\angle 2 = m\angle 2 + m\angle 3$, and by the [ ] Property of Equality, [ ].

4. If $AD = 3BC$, C is the midpoint of $\overline{BD}$, and $AB = 3$, what is the length of $\overline{AD}$?

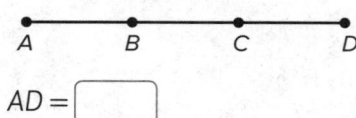

$AD =$ [ ]

5. Which of the following expressions are equivalent to $m\angle BGC$?

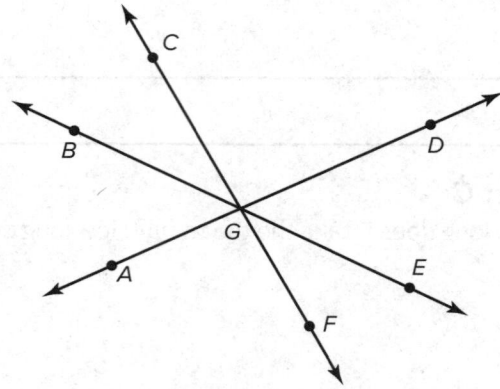

$m\angle FGE$

$180 - m\angle EGD - m\angle CGD$

$m\angle AGF - m\angle AGB$

$m\angle FGD - m\angle EGD$

6. Consider each statement. Determine whether it can be best described as a definition, theorem, or postulate. Select *Definition*, *Theorem*, or *Postulate* in each row.

Statement	Definition	Theorem	Postulate
Two points determine a line.			
Vertical angles are congruent.			
A parallelogram is a rectangle if and only if its diagonals are congruent.			
A figure is a quadrilateral if and only if it is a polygon with four sides.			
A line segment has only one midpoint.			

7. Complete the following proof.

Given: ∠A is supplementary to ∠B.

∠C is supplementary to ∠B.

Prove: ∠A ≅ ∠C

Statement	Reason
∠A is supplementary to ∠B.	
m∠A + m∠B = 180	
	Given
	Definition of Supplementary Angles
m∠A + m∠B = m∠C + m∠B	
	Subtraction Property of Equality
∠A ≅ ∠C	

8. The steps to create ∠XYZ as a copy of ∠A are listed below. In the first column, place the order for each step.

Order	Step
	Without changing the compass width, move the compass point to X and draw a similar arc, crossing $\overrightarrow{XY}$ at S.
	Draw $\overrightarrow{XY}$.
	Draw $\overrightarrow{XZ}$ containing T.
	Set the compass point at B and set its width to BC.
	Mark point X, which will be the vertex of the new angle.
	Without changing the compass width, move the compass point to S and draw an arc crossing the first arc to create point T.
	Set the compass point on A and draw an arc across the angle, creating points B and C.

9. Explain the difference between a *conjecture* and a *theorem*.

_____

_____

10. The Converse of the Angle Bisector Theorem states that if a point in the interior of an angle is equidistant from both sides of the angle, then that point lies on the angle bisector. Explain how this theorem justifies the method used to construct an angle bisector.

_____

_____

_____

_____

_____

1. A line goes through $(2, -5)$ and is parallel to the line $y = \frac{1}{2}x + 3$. In slope-intercept form, the equation of this line is ⬚ .

2. Complete the following proof of the Consecutive Interior Angles Theorem.

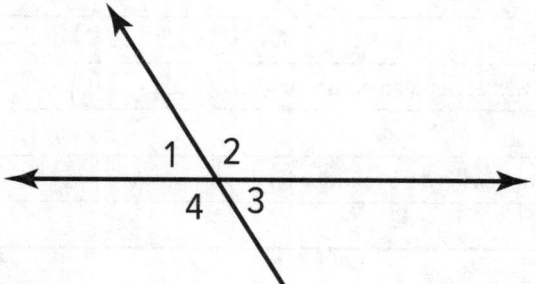

Given: $m \parallel n$

Prove: $\angle 1$ and $\angle 4$ are supplementary.

Proof: Since $m \parallel n$, we know that $\angle 2 \cong \angle 4$ by the ⬚ . By the definition of ⬚ , ⬚ = $m\angle 4$. $\angle 1$ and $\angle 2$ form a linear pair, so they are supplementary by the ⬚ . $m\angle 1 + m\angle 2 = 180$ by the definition of supplementary. By the ⬚ Property of Equality, $m\angle 1 + $ ⬚ $= 180$. Therefore, by the definition of ⬚ angles, $\angle 1$ and $\angle 4$ are supplementary.

3. Allison is building a gate for a fence. She wants to make sure that sides $\overline{AB}$ and $\overline{CD}$ are parallel. She measures $\angle BAC$ and finds that it is $52°$. In order to prove that the sides are parallel, she should measure ⬚ . If its measure is ⬚ , then the sides are parallel.

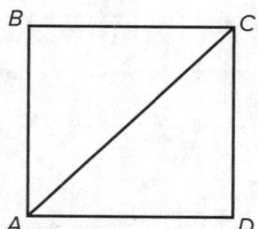

4. Complete the following.

Two lines are perpendicular if and only if they ⬚ at ⬚ angles.

5. Mateo is working on the construction below. Which best describes the construction Mateo is making?

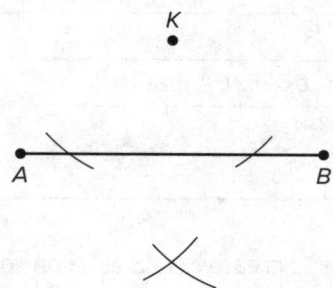

a line through K parallel to $\overline{AB}$

a line through K perpendicular to $\overline{AB}$

perpendicular bisector of $\overline{AB}$

6. Given the diagram below, which of the following statements are true?

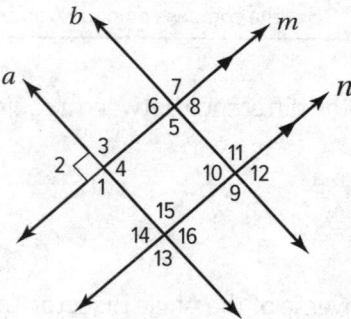

$\angle 1 \cong \angle 4$	$m\angle 9 + m\angle 16 = 180$
$\angle 15 \cong \angle 9$	$\angle 7 \cong \angle 13$
$\angle 7 \cong \angle 11$	$m\angle 7 + m\angle 12 = 180$
$\angle 12 \cong \angle 16$	$\angle 1 \cong \angle 14$

7. The steps to construct a line parallel to $\overline{AB}$ through X not on $\overline{AB}$ are listed below. In the first column, place the order for each step.

Order	Step
	Place the compass point at A. Set the compass width to about half of XA. Draw an arc across both lines. Label the points where the arc crosses the line D and G.
	Draw a straight line through X and Y.
	Without changing the compass width, place the compass point at C and draw an arc. Mark point Y where the arcs intersect.
	Without changing the compass setting, move the compass point to X and make a large arc. Label the point where the arc intersects $\overline{XA}$ as C.
	Draw a line through X that intersects $\overline{AB}$ at A.
	Use the compass to measure the distance between the points where the lower arc intersects $\overline{AB}$ and $\overline{XA}$.

8. The vertices of rectangle $JKLM$ are $J(-3, 0)$, $K(-1, 5)$, $L(9, 1)$, and $M(7, -4)$.

  a. What are the slopes of the sides of the rectangle? Show your work.

  b. How do you know that the opposite sides of the rectangle are parallel?

  c. How do you know that consecutive sides of the rectangle are perpendicular?

9. Consider the following diagram. If $\angle 12$ and $\angle 13$ are supplementary, which lines are parallel? Justify your answer.

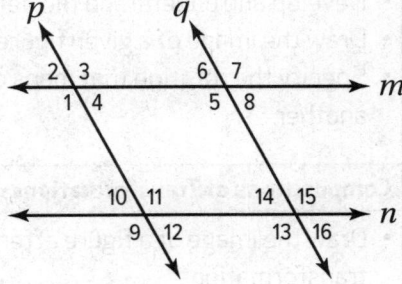

10. The diagonals of a quadrilateral intersect at their midpoints.

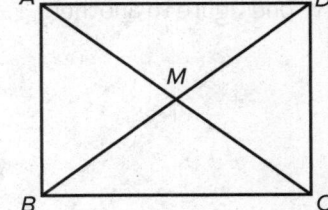

  a. If $ABCD$ is a square, what can you conclude about the four triangles created by the diagonals? How do you know?

  b. Based on your answer from **part a**, find $m\angle AMD$.

  c. What can be concluded about the diagonals if $ABCD$ is a square?

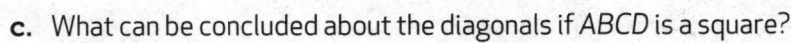

CHAPTER FOCUS Learn about some of the objectives that you will explore in this chapter. Answer the preview questions. As you complete each lesson, return to these pages to check your work.

What You Will Learn	Preview Question
Reflections	
• Develop the definition of a reflection in a line. • Draw reflections using various tools, including geometry software and the coordinate plane. • Identify transformations that are reflections in a line.	**SMP 7** The point $(-2, 7)$ is reflected in the x-axis. How do you find the coordinates of the image? **Multiply the y-coordinate by -1.** **$(-2, 7) \rightarrow (-2, -7)$**
Translations	
• Define, identify, and compare translations on the plane. • Draw translations given the figure and translation vector. • Specify a translation that maps one figure onto another.	**SMP 5** Describe the image of $(3, 5)$ translated along $<a, b>$. **The image is $(3 + a, 5 + b)$.**
Rotations	
• Develop and understand the definition of rotations. • Draw the image of a given figure under a rotation. • Specify the rotation that maps one figure to another.	**SMP 2** A triangle has vertices $(-2, 1), (1, 2),$ and $(3, -1)$. What are the vertices of its image after a 90° counterclockwise rotation about the origin? **$(-1, -2), (-2, 1),$ and $(1, 3)$**
Compositions of Transformations	
• Draw the image of a figure after a composition of transformations. • Specify a sequence of transformations that maps one figure to another.	**SMP 3** A triangle with vertices $(0, 4), (8, 2),$ and $(3, -3)$ is translated along $<2, 4>$ and reflected in the y-axis. A student said the vertices of the image are $(-2, 8), (-10, 6),$ and $(-5, 1)$. Is the student correct? Why or why not? **Yes; each point $(x, y) \rightarrow (x + 2, y + 4)$ and then** **to $(-x, y)$.**

What You Will Learn	Preview Question
Symmetry	
• Describe the rotations and reflections that carry a figure onto itself.	SMP 8 Draw all lines of symmetry for the regular polygon.

Objectives

- Develop the definition of a reflection in a line.
- Draw reflections using various tools, including geometry software and the coordinate plane.
- Identify transformations that are reflections in a line.

A **transformation** is an operation that maps an original geometric figure onto a new figure. A **reflection** is one type of transformation. When reflecting a shape in a line, the shape before the reflection is known as the **preimage**, and the resulting shape after the reflection is known as the **image**. To obtain the image from the preimage, fold the preimage over the line of reflection.

EXAMPLE 1 Model a Tile Border Using Reflections

EXPLORE Mai-Linh wants to create a tile border around her bathroom. She is using a parallelogram for part of her design and will reflect it horizontally in a vertical line of reflection, as shown in the diagram at right.

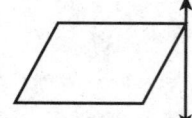

a. **USE A MODEL** Sketch the first four tiles of the pattern. Fold the shape over the vertical line. Then fold the resulting shape over a vertical line through its rightmost vertex. Continue until you have produced four tiles.

b. **USE TOOLS** Explain how could you use two different tools to verify your answer to **part a**.

_____

_____

c. **USE TOOLS** Using the Geometer's Sketchpad, check your sketch for **part a** by constructing a parallelogram and vertical line as shown above and making a reflection. Use the Geometer's Sketchpad to complete the remaining two reflections. Describe the process you used.

_____

_____

_____

_____

d. **FIND A PATTERN** What do you notice about every other image in the pattern? How might you describe reflections in two parallel lines?

_____

_____

_____

EXAMPLE 2 **Define a Reflection**

EXPLORE Using the Geometer's Sketchpad, construct a line of reflection, *m*, and a point not on the line, *A*.

a. Reflect point *A* in the line of reflection and label the image *A'*. Measure the distances between the line and each point. Describe the relationship you find.

_____

b. **COMMUNICATE PRECISELY** Connect point *A* and image *A'* with a line segment. Would you say that the line of reflection bisects $\overline{AA'}$? How do you know? How would you describe the relationship of the line of reflection to $\overline{AA'}$? How do you know?

_____

_____

_____

c. **CONSTRUCT ARGUMENTS** Is there a point *A* for which the observations in **part b** are not true? How would you describe the relationship between the image and preimage for this point?

_____

_____

_____

d. **USE TOOLS** With the Geometer's Sketchpad, construct a triangle and reflect it in the line of reflection. How would you describe the relationship between the image and the preimage? What would you have to do in order to complete this reflection with only a compass and straightedge?

_____

_____

_____

_____

e. Complete the following definition. A reflection in a line is a function that maps a point to its image such that:

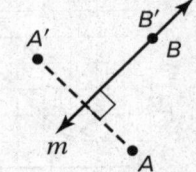

• If the point is not on the line, then the line of reflection is the _____ _____ of the segment joining the point and its image.

• If a point is on the line, then the image and preimage are _____.

In addition to constructing reflections using geometry software or using compass and straightedge, reflections can be represented by graphs on a coordinate plane.

EXAMPLE 3 **Describe Reflections as Functions**

EXPLORE Complete the following exploration to describe reflections as mapping functions.

a. *A′B′C′D′* is a reflection of *ABCD* in the *y*-axis. Fill in the chart at right with the ordered pairs that represent the vertices of the image and preimage in the graph below.

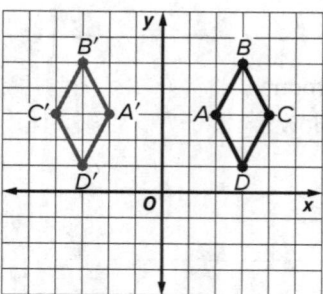

Preimage		Image	
A		A′	
B		B′	
C		C′	
D		D′	

b. REASON ABSTRACTLY Describe the relationship between the coordinates for this reflection in the *y*-axis.

---

c. *E′F′G′H′* is a reflection of *EFGH* in the *x*-axis. Fill in the chart at right with the ordered pairs that represent the vertices of the image and preimage in the graph below.

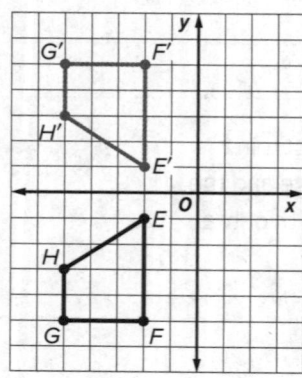

Preimage		Image	
E		E′	
F		F′	
G		G′	
H		H′	

d. REASON ABSTRACTLY Describe the relationship between the coordinates for this reflection in the *x*-axis.

---

e. *J′K′L′M′* is a reflection of *JKLM* in the line *y = x*. Fill in the chart at right with the ordered pairs that represent the vertices of the image and preimage in the graph below.

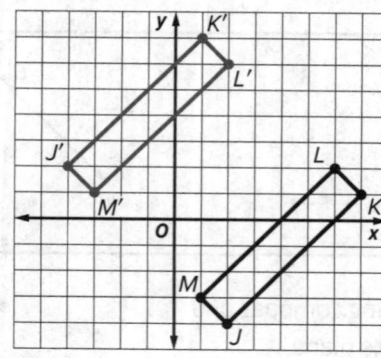

Preimage		Image	
J		J′	
K		K′	
L		L′	
M		M′	

f. REASON ABSTRACTLY Describe the relationship between the coordinates for this reflection in the line $y = x$.

_____

_____

_____

KEY CONCEPT

Reflection in the Coordinate Plane

- When a point is reflected in the y-axis, the coordinates of the image can be given by the mapping function $(x, y) \rightarrow (-x, y)$.
- When a point is reflected in the x-axis, the coordinates of the image can be given by the mapping function $(x, y) \rightarrow (x, -y)$.
- When a point is reflected in the line $y = x$, the coordinates of the image can be given by the mapping function $(x, y) \rightarrow (y, x)$.

EXAMPLE 4 Identify Transformations That Are Reflections

a. CONSTRUCT ARGUMENTS Is $A'B'C'D'$ a reflection of $ABCD$ in the y-axis. Why or why not?

_____

_____

_____

b. CONSTRUCT ARGUMENTS Is $A'B'C'D'$ a reflection of $ABCD$ in the x-axis on the graph to the right? Explain why or why not.

_____

_____

_____

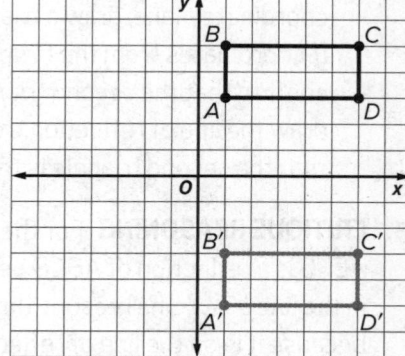

c. COMMUNICATE PRECISELY How could you label the vertices of the image in **part b** so that $A'B'C'D'$ is a reflection of $ABCD$ in the x-axis? What advice can you give to ensure that any reflected polygon is properly labeled?

_____

_____

_____

_____

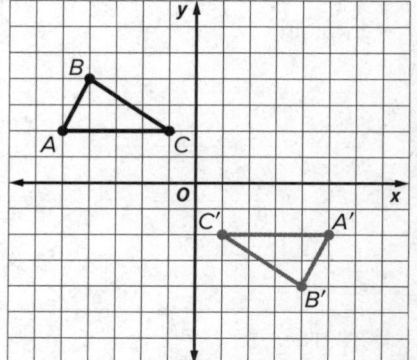

d. PLAN A SOLUTION Can one reflection of ABC result in $A'B'C'$? Describe the reflection(s) needed to make this transformation.

_____

_____

1. **REASON ABSTRACTLY** Make a graph and a chart like the ones in **Example 3** for reflecting a quadrilateral of your choice in the line $y = 2$. Find a mapping function that describes this scenario.

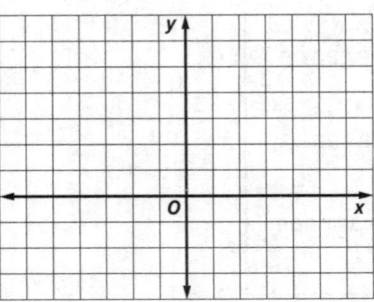

Vertex	Preimage	Image
1		
2		
3		
4		

2. **USE STRUCTURE** You can graph the **inverse of a function** by reversing the values of x and y for each point of the function. The domain becomes the range and the range becomes the domain.

   a. How could you describe the inverse of a function in the context of reflections?

   _____

   _____

   _____

   b. Draw the triangle with vertices (1, 2), (5, 0), and (3, 4). On the same coordinate plane, draw a triangle using vertices that have the x- and y-coordinates from the first triangle reversed. Draw the line $y = x$ and confirm that the second triangle is a reflection of the first. Finally, draw the line of reflection from your conjecture in **part a** to confirm that the second triangle is the reflection of the first.

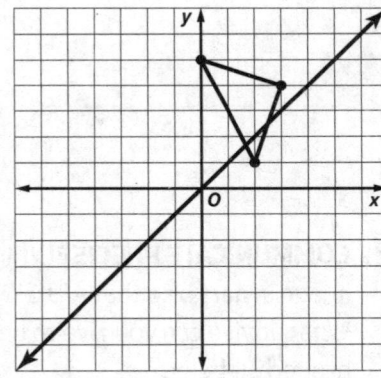

3. **CRITIQUE REASONING** For the graph at the right, Evelyn maintains that *AEFG* is a reflection of *ABCD* because it fits the definition of a reflection in the line $y = x$. She reasons that A is the same point in each figure because it is on the line of reflection and the remaining vertices are equidistant from that line. Do you agree with Evelyn's analysis? Explain.

   _____

   _____

   _____

   _____

   _____

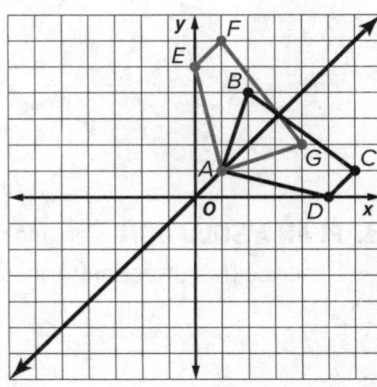

4. **CONSTRUCT ARGUMENTS** Consider the graph at right.

   a. Explain why the graph cannot represent a reflection in the x-axis.

      _____

      _____

   b. Redraw the graph so that it does represent a reflection in the x-axis.

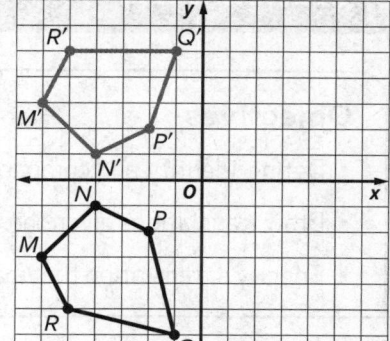

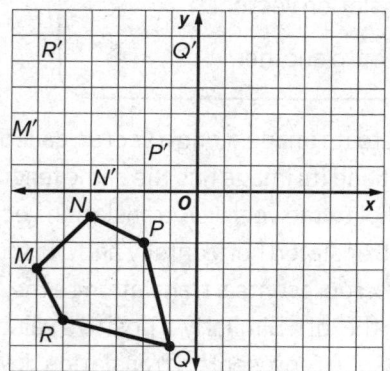

USE A MODEL Graphic designers use transformations to create beautiful designs.

5. Describe the reflections in the design at right.

   _____

   _____

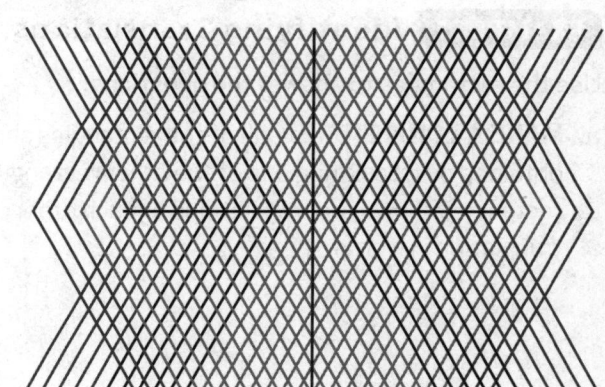

6. a. When graphic designers create logos, they often use reflections of letters of the alphabet. Choose a letter to reflect horizontally, vertically and diagonally and sketch an example of each reflection.

   b. Create five points on the coordinate plane to form the letter M. Find their image under a reflection in the y-axis, under a reflection in the x-axis, and under a reflection in the line $y = x$.

      _____

      _____

      _____

      _____

7. **USE A MODEL** Sketch a tile border pattern created by reflecting two different polygons in two vertical lines of reflection.

Objectives

- Define, identify, and compare translations on the plane.
- Draw translations given the figure and translation vector.
- Specify a translation that maps one figure onto another.

A **translation** is a function that maps each point to its image along a *vector*, called the **translation vector**. Each segment joining a point and its image has the same length as the vector, and each of these segments is also parallel to the vector. A translation vector is written as $\langle x, y \rangle$, where x is the number of units translated horizontally and y is the number of units translated vertically. A positive value of x represents a translation to the right, and a negative value of x represents a translation to the left. Similarly, a positive value of y represents a translation up, and a negative value of y represents a translation down. For example, the vector $\langle 4, -3 \rangle$ represents a translation 4 units to the right and 3 units down.

EXAMPLE 1 Identifying Translations

Use the rectangles to answer the questions.

a. **MAKE A CONJECTURE** One of the rectangles shown can be obtained by translating one of the other rectangles. Which of the rectangles shown are translations of each other? Describe the translation in words.

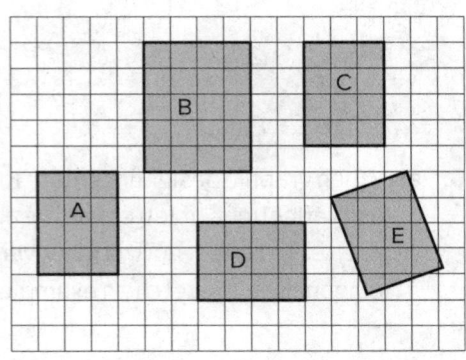

_____

_____

_____

b. **USE TOOLS** Choose two opposite vertices of one of the rectangles you identified in **part a**, and draw segments to the corresponding vertices of the other rectangle you identified. For each segment you drew, form a right triangle with the segment as the hypotenuse. What do you notice about the triangles?

_____

c. **INTERPET PROBLEMS** Write the translation vector that represents the translation from one rectangle to the other. Write the translation vector that represents the translation from the second rectangle to the first.

_____

_____

d. **COMMUNICATE PRECISELY** What do you notice about the translation vectors you wrote in **part c**?

_____

_____

EXAMPLE 2 **Translations on the Coordinate Plane**

Use the quadrilateral *ABCD* on the coordinate plane to answer these questions.

a. **REASON ABSTRACTLY** Predict the effect of translating *ABCD* by the vector $\langle -7, 2 \rangle$. Draw the image *A'B'C'D'*.

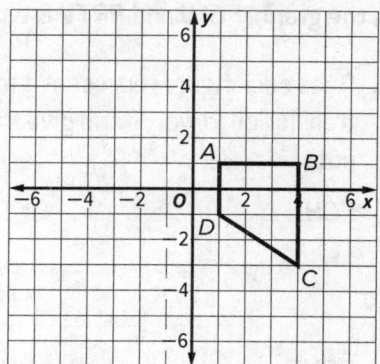

_____

_____

_____

b. **REASON ABSTRACTLY** *ABCD* is translated so that the image of point *C* is point *B*. Draw the image quadrilateral *A'B'C'D'* and write the translation vector. Explain how you determined the vector.

_____

_____

c. **USE STRUCTURE** Complete each mapping function to describe the translations from parts **a** and **b**.

Translation from part a

$A(x, y) \rightarrow A'(x - 7, \boxed{})$

$B(x, y) \rightarrow B'(x - 7, \boxed{})$

$C(x, y) \rightarrow C'(\boxed{}, y + 2)$

$D(x, y) \rightarrow D'(\boxed{}, \boxed{})$

Translation from part b

$A(x, y) \rightarrow A'(\boxed{})$

$B(x, y) \rightarrow B'(\boxed{})$

$C(x, y) \rightarrow C'(\boxed{})$

$D(x, y) \rightarrow D'(\boxed{})$

d. **MAKE A CONJECTURE** Suppose you have a mapping function that describes what happens to the vertex of a figure for a certain translation. Make a conjecture about the function for all the vertices of the figure. Can you determine the translation vector? Explain.

_____

_____

_____

KEY CONCEPT

Complete the table by writing the mapping function for each translation. Assume $a \geq 0$ and $b \geq 0$.

Translation	Mapping Function
Left *a* units and up *b* units	$(x, y) \rightarrow (x - a, y + b)$
Right *a* units and down *b* units	$(x, y) \rightarrow (\boxed{})$
Left *a* units and down *b* units	$(x, y) \rightarrow (\boxed{}, y - b)$
Right *a* units and up *b* units	$(x, y) \rightarrow (\boxed{})$

EXAMPLE 3 **Specify a Transformation**

In the graph, *FGHJ* and *PRTV* are transformations of parallelogram *ABCD*.

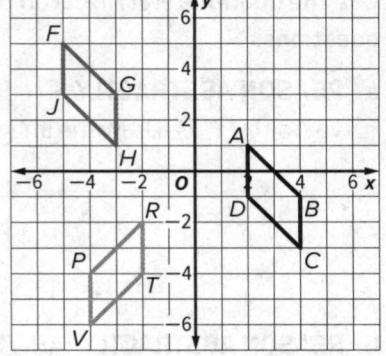

a. Does each figure represent a translation of *ABCD*? If yes, describe the translation in words and give the translation vector. If no, justify why it is not.

FGHJ _____

_____

_____

PRVT _____

_____

b. CRITIQUE REASONING Diego says, "I can use the Distance Formula to prove *PRTV* is not a translation of parallelogram *ABCD*." He then calculates that $AP = \sqrt{61}$ and $DV = \sqrt{61}$. Does this prove his statement? If it does, explain why. If it does not, explain how Diego could modify what he did so that he does prove his statement.

_____

_____

c. USE REASONING Jocelyn draws parallelogram *KLMN* on the coordinate plane. The distances from each vertex of *KLMN* to each corresponding vertex of *ABCD* are the same. Is this sufficient to claim that *KLMN* is a translation of *ABCD*? Justify your answer.

_____

_____

_____

> PRACTICE

1. USE TOOLS Use tracing paper and a centimeter ruler to draw a translation of the figure below. The translation vector is <9, −1>, with units of centimeters. Explain your technique.

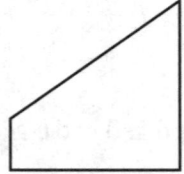

_____

_____

_____

2. **CRITIQUE REASONING** Determine if each statement about translations is *always*, *sometimes*, or *never* true. Justify your answer.

   a. Lengths and angle measures of the image and preimage are preserved.

   _____

   _____

   b. All lines drawn from the vertices of the preimage to the image are parallel.

   _____

   _____

   _____

   c. The vector $<a, b>$ will translate each coordinate of a preimage a units right and b units up.

   _____

   _____

USE TOOLS Draw and label the image of each figure after the given translation.

3. 3 units to the left

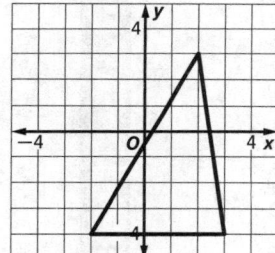

4. Translation vector $<1, -2.5>$

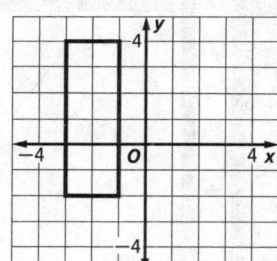

5. Translation vector $<-5, -7>$

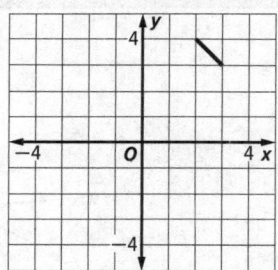

6. **INTERPRET PROBLEMS** A square in the coordinate plane has vertices at $(2, 3)$, $(4, 3)$, $(2, 1)$, and $(4, 1)$. It is translated so that one of the vertices is at the origin. Find the coordinates of each vertex of the image, if the translation vector has the least possible length. Explain your reasoning. Draw the image on the coordinate plane.

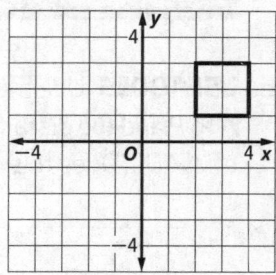

   _____

   _____

   _____

7. **USE A MODEL** The triangle represents the area on a map covered by a fleet of fishing ships, where each square represents a square mile. If this region is translated along the vector $<4, -5>$, then draw the image and list the coordinates of its vertices. What distance has the coverage area been moved?

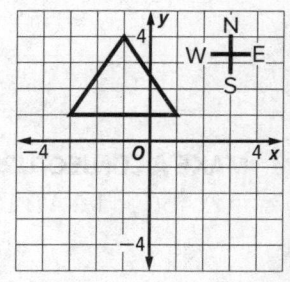

   _____

   _____

Mathematical Practices
1, 3, 5, 6, 8

Objectives

- Develop and understand the definition of rotations.

- Draw the image of a given figure under a rotation.

- Specify the rotation that maps one figure to another.

EXAMPLE 1 **Develop the Definition of Rotations**

EXPLORE Use geometry software to explore rotations. As you do so, think about how you could use angles and distances to define a rotation about a point.

a. **USE TOOLS** Use geometry software to draw a triangle and a point. Label these as △*ABC* and point *P*, as shown below on the left.

b. **USE TOOLS** Draw the image of △*ABC* after a counterclockwise rotation of 100° around point *P*. Label the image △*A'B'C'*, as shown below on the right.

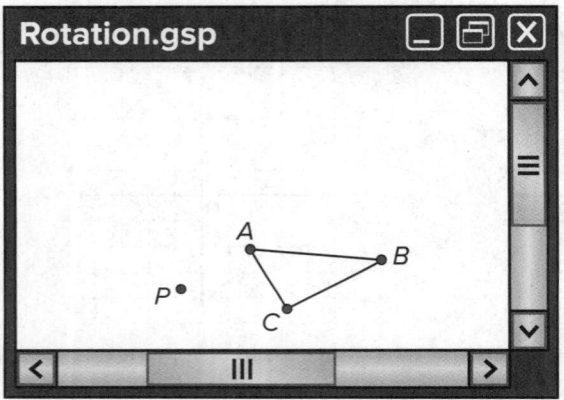
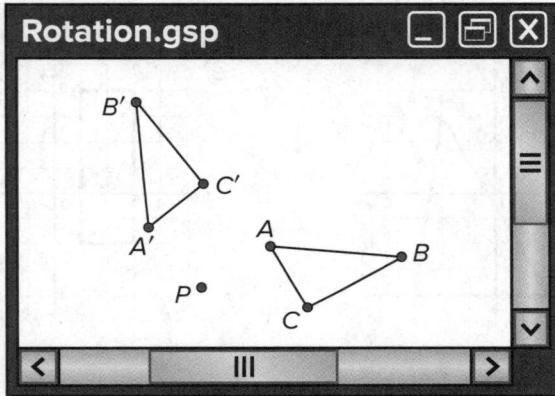

c. **USE TOOLS** Use the measurement tools in the software to measure the distance from *A* to *P* and the distance from *A'* to *P*. What do you notice? Change the shape or location of △*ABC*. Does this relationship remain the same?

_____

_____

d. **USE TOOLS** Use the angle measurement tools in the software to measure ∠*APA'*, ∠*BPB'*, and ∠*CPC'*. Change the shape or location of △*ABC*. What do you notice?

_____

_____

e. **MAKE A CONJECTURE** What can you conclude about the distances or angle measures in △*ABC* and △*A'B'C'*? Use the software to check your conjecture.

_____

A **rotation** about a fixed point, P, called the **center of rotation**, through an angle of $x°$, is a function that maps a point to its image as follows.

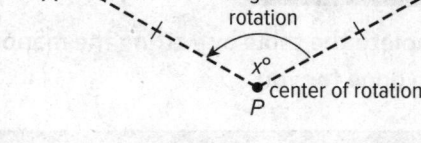

- If the point is the center of rotation, then the image and preimage are the same point.

- If the point is not the center of rotation, then the image and preimage are the same distance from the center of rotation, and the measure of the angle formed by the preimage, the center of rotation, and the image is x.

In the above definition, the **angle of rotation** is $x°$. Unless otherwise stated, you can assume all rotations are counterclockwise.

EXAMPLE 2 Draw a Rotation

Follow the steps below to draw the image of $\triangle JKL$ after a 160° rotation about point Q.

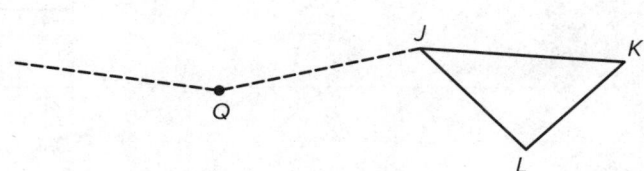

a. **USE TOOLS** Use a straightedge to draw $\overline{JQ}$.

b. **USE TOOLS** Use a protractor to draw a ray that forms a 160° angle with $\overline{JQ}$, as shown.

c. **USE TOOLS** Use a ruler to mark a point J' on the ray so that $J'Q = JQ$.

d. **USE TOOLS** Repeat steps a–c to locate points K' and L'. Then use a straightedge to draw $\triangle J'K'L'$

e. **EVALUATE REASONABLENESS** How can you use a piece of tracing paper to check that your drawing is reasonable?

_____

_____

_____

f. **COMMUNICATE PRECISELY** What would it mean to rotate something $-160°$?

_____

_____

_____

Like other transformations, a rotation is a function that takes points of the plane as inputs and gives other points of the plane as outputs. When you perform rotations on a coordinate plane, you can use mapping functions to specify how a point is mapped to its image.

Complete the table by writing the mapping function for each rotation. The first one has been done for you.

Rotation	Mapping Function
90° about the origin	$(x, y) \rightarrow (-y, x)$
180° about the origin	
270° about the origin	

EXAMPLE 3 Draw a Rotation

Follow these steps to draw the image of quadrilateral *ABCD* after a rotation of 180° about the origin.

a. **INTERPRET PROBLEMS** Predict the effect of the rotation on quadrilateral *ABCD* before drawing the image.

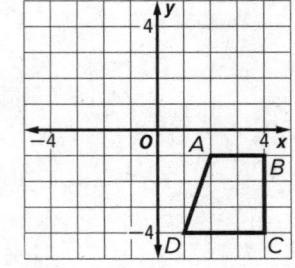

_____

_____

b. **INTERPRET PROBLEMS** Complete the table to find the image of each vertex of quadrilateral *ABCD*.

c. **USE TOOLS** Use the table to help you draw the image of quadrilateral *ABCD* on the coordinate plane above.

Preimage (x, y)	Image (−x, −y)
A(2, −1)	A′(−2, 1)

d. **CRITIQUE REASONING** A student said that another way to map quadrilateral *ABCD* to its final image is by first rotating it 90° about the origin and then rotating the image 90° about the origin. Do you agree? Use the mapping functions to support your answer.

_____

_____

_____

EXAMPLE 4 Specify a Transformation

Manuel is designing a logo for a jewelry store. He used transformations to draw the triangles shown at the right.

a. **COMMUNICATE PRECISELY** Manuel's assistant draws △*RGX*. Explain how he can use a transformation to create △*VCJ*.

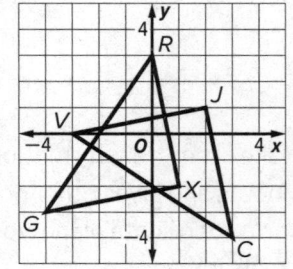

_____

_____

b. EVALUATE REASONABLENESS Explain how you can use a mapping function to check that your answer to **part a** is correct.

_____

_____

_____

_____

c. COMMUNICATE PRECISELY Is there a different rotation that Miguel's assistant can use to create △VCJ? Explain.

_____

d. CONSTRUCT ARGUMENTS Suppose Miguel's assistant starts by drawing △VCJ. What transformation can he use to create △RGX?

_____

PRACTICE

1. **COMMUNICATE PRECISELY** In the figure, △D'E'F' is the image of ∠DEF after a rotation about point Z.

   a. What is the distance from E' to Z? Explain how you know.

   _____

   _____

   b. What is m∠FZF'? Explain how you know.

   _____

   _____

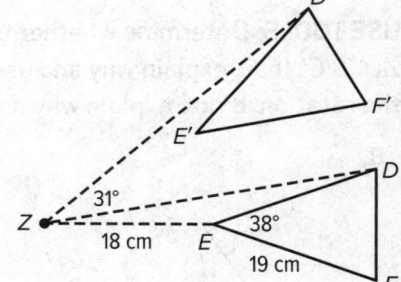

USE TOOLS Draw and label the image of each figure after the given rotation about point P.

2. 75°

3. 140°

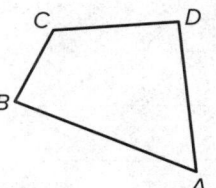

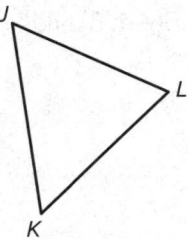

USE TOOLS Draw and label the image of each figure after the given rotation.

4. 270° about the origin

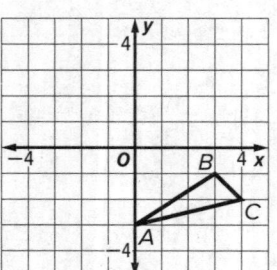

5. 90° about the origin

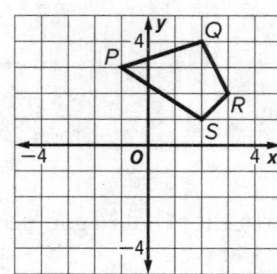

6. 180° about the origin

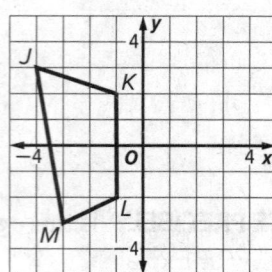

7. **COMMUNICATE PRECISELY** Catherine is using a coordinate plane to design a video game set in outer space. She draws quadrilaterals *DEFG* and *STUV* to represent comets. How can she use a transformation to map *DEFG* to *STUV*? Describe the transformation in words and give a mapping function for the transformation.

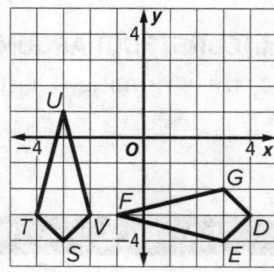

_____

_____

USE TOOLS Determine whether there is a rotation about point *P* that maps △*ABC* to △*A′B′C′*. If so, explain why and use a ruler, protractor, or other tool to help you describe the rotation. If not, explain why not.

8.

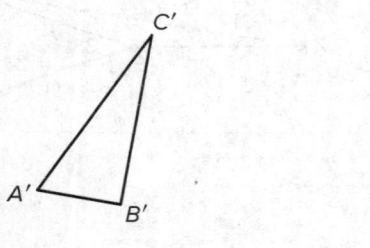

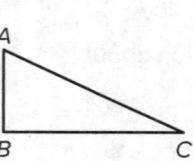

9.

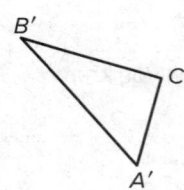

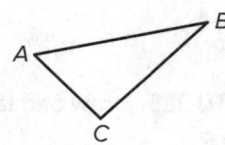

_____

_____

_____

_____

_____

_____

10. **CONSTRUCT ARGUMENTS** Under a rotation around the origin, the point *A*(5, −1) is mapped to the point *A′*(1, 5). What is the image of the point *B*(−4, 6) under this rotation? Explain.

_____

_____

_____

11. An interior designer uses a coordinate plane to position furniture in a room. The designer decides to move a sofa, which is represented by trapezoid *JKLM*, using the rotation $(x, y) \rightarrow (-y, x)$.

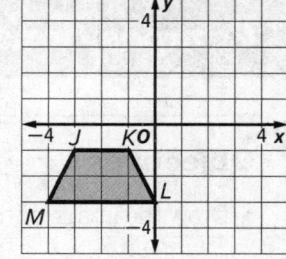

a. **USE TOOLS** Draw the new location of the sofa on the coordinate plane.

b. **COMMUNICATE PRECISELY** Write a mapping function that the designer can apply to the sofa in its new location in order to move it back to its original location. Explain.

_____

_____

12. a. COMMUNICATE PRECISELY What is the result of a rotation followed by another rotation about the same point? Give an example.

_____

_____

_____

b. **COMMUNICATE PRECISELY** Is a reflection followed by another reflection in a parallel line still a reflection? Explain your reasoning.

_____

_____

c. **COMMUNICATE PRECISELY** Thomas claims that a reflection in the *x*-axis followed by a reflection in the *y*-axis is the same thing as a rotation. Is Thomas correct?

_____

_____

_____

_____

13. CRITIQUE REASONING Robert is looking at the figure below, which shows two congruent triangles. He measures the angle that rotates *A* to *A'* around *O* and finds it to be 30 degrees. He measures the angle that rotates *B* to *B'* around *O* and finds it to also be 30 degrees. He then claims that because the two triangles are congruent, a 30 degree rotation has occurred around point *O*. Is Robert correct? Explain.

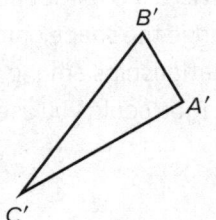
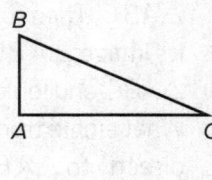

_____

_____

_____

_____

Compositions of Transformations

Objectives

- Draw the image of a figure after a composition of transformations.
- Specify a sequence of transformations that maps one figure to another.

Mathematical Practices
1, 2, 3, 5, 6

A **composition of transformations** is a sequence of transformations in which the first transformation is applied to a given figure and then another transformation is applied to its image.

EXAMPLE 1 Investigate Compositions

EXPLORE Use dynamic geometry software to explore compositions of transformations.

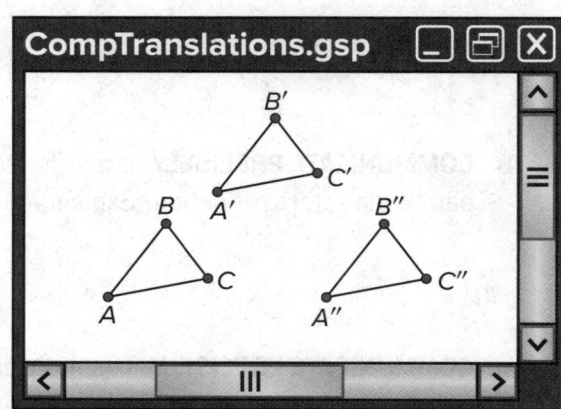

a. **USE TOOLS** Use geometry software to draw a triangle and label it △ABC. Use the software to translate the triangle and label its image △A'B'C'. Then use the software to translate △A'B'C' and label its image △A"B"C". Change the shape or location of △ABC, and look for relationships among the triangles. Which type of transformation could you use to map △ABC directly to △A"B"C"?

b. **MAKE A CONJECTURE** What conjecture can you make about the composition of two translations?

_____

_____

c. **USE TOOLS** Use the software to draw △ABC and a point P. Rotate △ABC 80° about point P and label its image △A'B'C'. Then rotate △A'B'C' 70° about point P and label its image △A"B"C". Change the shape or location of △ABC, and look for relationships among the triangles. What single transformation could you use to map △ABC directly to △A"B"C"?

_____

_____

d. **MAKE A CONJECTURE** What conjecture can you make about the composition of a rotation of m° about point P followed by a rotation of n° about point P?

_____

_____

e. USE TOOLS Use the software to draw △ABC and two parallel lines, j and k. Reflect △ABC in line j and label its image △A'B'C'. Then reflect △A'B'C' in line k and label its image △A"B"C". Which type of transformation could you use to map △ABC directly to △A"B"C"?

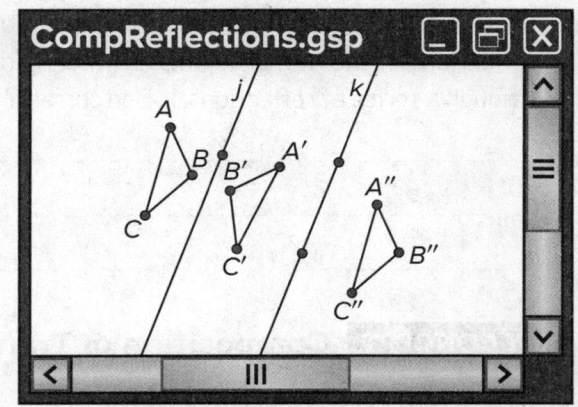

CompReflections.gsp

_____

f. MAKE A CONJECTURE What conjecture can you make about the composition of two reflections in parallel lines?

_____

_____

g. MAKE A CONJECTURE Repeat **part e**, but this time draw lines j and k so that they intersect. What conjecture can you make about the composition of two reflections in intersecting lines?

_____

h. MAKE A CONJECTURE Recall that translations, reflections, and rotations are rigid motions because they do not change the size or shape of a figure. Based on your work above, what conjecture can you make about the composition of two rigid motions?

_____

i. CRITIQUE REASONING Ben claims that translations, reflections, and rotations are all we need to achieve any rigid motion. In other words, every rigid motion can be the composition of translations, reflections, and rotations. Use the following diagram to explain why Ben is correct.

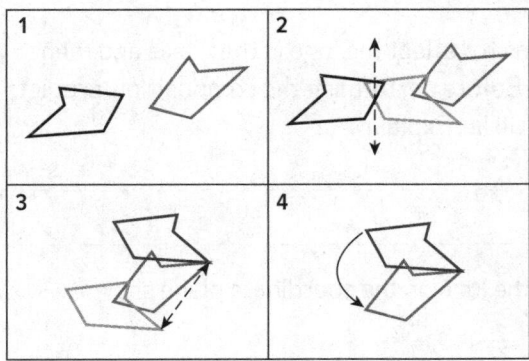

_____

_____

_____

_____

_____

_____

j. CRITIQUE REASONING Ben now claims that the only rigid motion we need is a reflection. In other words, no matter what rigid motion we want, we can always get it through a series of reflections. Is Ben correct? Explain.

_____

_____

_____

KEY CONCEPT Composition of Transformations

Complete each description in the table.

Composition	Description
Two translations	The composition of two translations is a _____.
Two rotations	The composition of a rotation of $m°$ about point P followed by a rotation of $n°$ about point P is a rotation of _____° about point P.
Two reflections in parallel lines	The composition of two reflections in parallel lines is a _____.
Two reflections in intersecting lines	The composition of two reflections in intersecting lines is a _____.
Two rigid motions	The composition of two rigid motions is a _____.

EXAMPLE 2 Draw a Composition

Vanessa is using a coordinate plane to design a logo for a company. The figure shows her first design for the logo.

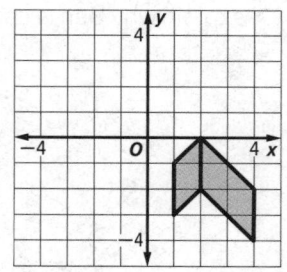

a. INTERPRET PROBLEMS Vanessa plans to reflect the logo in the x-axis and then rotate the image 90° about the origin. Before performing the composition, predict the quadrant in which the final image will lie. Explain.

_____

_____

b. USE TOOLS Draw the final image of the logo on the coordinate plane shown above.

c. COMMUNICATE PRECISELY Could Vanessa have mapped the original logo to its final position using a single transformation? If so, describe the transformation. If not, explain why not.

_____

d. CONSTRUCT ARGUMENTS Show how you can use the mapping function for a reflection in the x-axis and for a rotation of 90° about the origin to justify your answer to **part c**.

_____

_____

e. **CRITIQUE REASONING** Vanessa claims that she can get the same result by performing the composition in the opposite order; that is, by first rotating the original logo 90° about the origin and then reflecting it in the *x*-axis. Do you agree? Explain why or why not.

_____

_____

_____

f. **MAKE A CONJECTURE** In general, does the order in which you perform a sequence of transformations matter? Explain.

_____

_____

EXAMPLE 3 **Specify a Sequence of Transformations**

Given two figures, specify a sequence of transformations that maps one figure to the other.

a. **COMMUNICATE PRECISELY** It is possible to map △*ABC* to △*RST* using a reflection followed by a translation? Describe the reflection and translation.

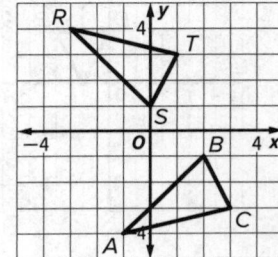

_____

b. **COMMUNICATE PRECISELY** Is there a different sequence of two transformations that maps △*ABC* to △*RST*? Explain.

_____

_____

c. **COMMUNICATE PRECISELY** Why is your answer in **part b** different from your answer in **Example 2 part e**?

_____

_____

_____

d. **COMMUNICATE PRECISELY** Specify a sequence of two or more transformations that maps quadrilateral *EFGH* to quadrilateral *MNPQ*.

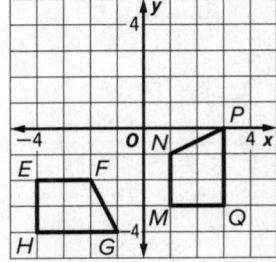

_____

_____

e. **CONSTRUCT ARGUMENTS** Explain how you can use your answer to **part c** to specify a sequence of transformations that maps quadrilateral *MNPQ* to quadrilateral *EFGH*.

_____

_____

f. CONSTRUCT ARGUMENTS Specify a sequence of three translations that maps △JKL to △RST. Is there more than one such sequence? Explain.

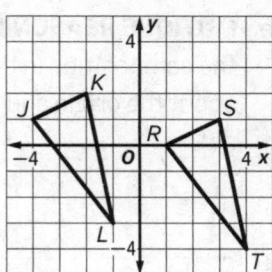

_____

_____

_____

_____

USE TOOLS Draw and label the image of each figure after the given composition of transformations.

1. 270° rotation about the origin followed by translation along ⟨2, −2⟩

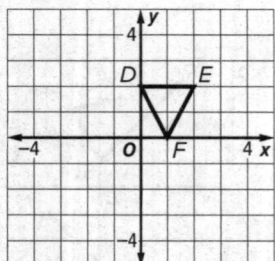

2. reflection in the y-axis followed by 180° rotation about the origin

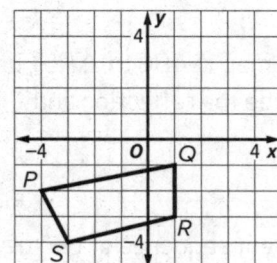

3. translation along ⟨−1, 1⟩ followed by reflection in the line y = x

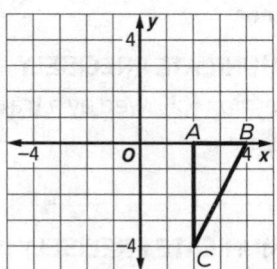

COMMUNICATE PRECISELY Specify a sequence of transformations that maps △JKL to △MNP.

4.

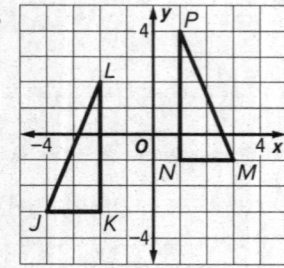

_____

_____

_____

5.

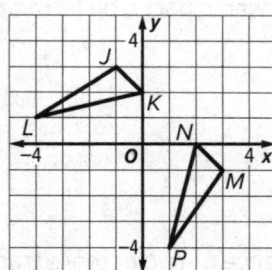

_____

_____

_____

6.

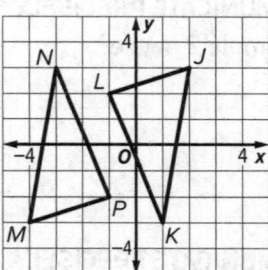

_____

_____

_____

7. REASON ABSTRACTLY Connor transforms a figure by applying the transformation $(x, y) \rightarrow (x, -y)$ followed by the transformation $(x, y) \rightarrow (-x, -y)$. Write a single mapping function that has the same effect as the composition. What transformation does your mapping function represent?

_____

CONSTRUCT ARGUMENTS Determine whether each statement is *always, sometimes,* or *never* true. Explain.

8. A composition of two reflections is a rotation.

   _____

9. A composition of two translations is a rotation.

   _____

10. A reflection in the *x*-axis followed by a reflection in the *y*-axis leaves a point in its original location.

    _____

11. The translation along $\langle a, b \rangle$ followed by the translation along $\langle c, d \rangle$ is the translation along $\langle a + c, b + d \rangle$.

    _____

    _____

12. **COMMUNICATE PRECISELY** Erwin is a programmer for the video game *Rocket Race*. He uses a coordinate plane to program the motion of the rockets on the screen.

    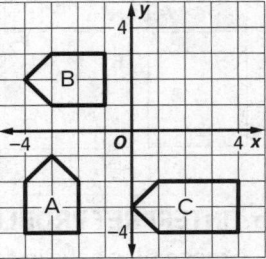

    a. Is it possible for Erwin to use a composition of two rigid motions to map rocket A to rocket B? If so, describe the composition. If not, explain why not.

       _____

       _____

    b. Is it possible for Erwin to use a composition of two rigid motions to map rocket A to rocket C? If so, describe the composition. If not, explain why not.

       _____

       _____

13. **USE TOOLS** Use dynamic geometry software or other tools to explore the composition of two reflections in parallel lines. What can you say about the translation vector for the translation that is equivalent to this composition?

    _____

    _____

14. **REASON ABSTRACTLY** We have seen that the transformation $(x, y) \rightarrow (y, x)$ is a reflection in the line $y = x$. It is also true that the transformation $(x, y) \rightarrow (-y, -x)$ is a reflection in the line $y = -x$. The examples in this lesson tell us that the composition of these two reflections should be a rotation. Give the mapping function that describes the composition, and describe the resulting rotation.

    _____

    _____

Compositions of Transformations **103**

Objectives

- Describe the rotations and reflections that carry a figure onto itself.

Mathematical Practices
1, 3, 6, 7

A figure has **symmetry** if there is a rigid motion (translation, reflection, or rotation) that maps the figure onto itself.

EXAMPLE 1 Identify Symmetry

EXPLORE Nina owns a store that sells hand-painted ceramic tiles. The figure shows eight new square tiles that she received today. Nina is sorting the tiles for her showroom.

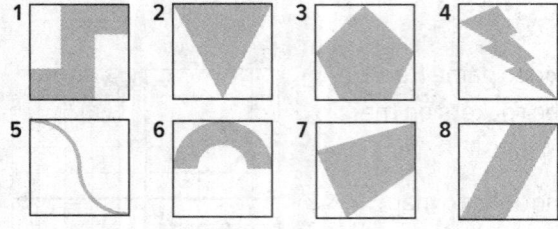

a. **INTERPRET PROBLEMS** Nina places tiles 2, 3, and 6 in the same group. Explain what these tiles have in common.

_____

_____

b. **USE STRUCTURE** What other set of three tiles could Nina group together? Why?

_____

_____

_____

c. **COMMUNICATE PRECISELY** Which of the tiles have symmetry? Justify your answer using the definition of symmetry.

_____

_____

_____

d. **COMMUNICATE PRECISELY** Which of the tiles do not have symmetry? Justify your answer using the definition of symmetry.

_____

_____

_____

A figure has **line symmetry** if it can be mapped onto itself by a reflection in a line, called the **line of symmetry**.

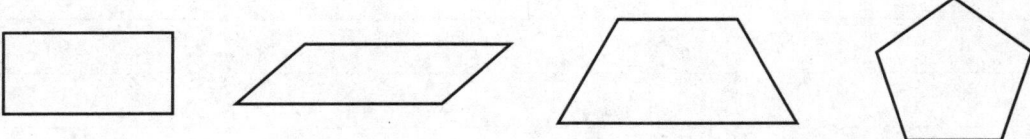

line of symmetry

EXAMPLE 2 Identify Line Symmetry

The figures show a rectangle, a parallelogram, an isosceles trapezoid, and a regular pentagon.

a. **COMMUNICATE PRECISELY** Draw all lines of symmetry on the figures.

b. **COMMUNICATE PRECISELY** Which of the figures have line symmetry? Why?

_____

_____

c. **MAKE A CONJECTURE** How many lines of symmetry do you think a regular 17-gon has? Explain.

_____

d. **MAKE A CONJECTURE** How many lines of symmetry do you think a regular *n*-gon has? Explain.

_____

_____

A figure has **rotational symmetry** if the figure can be mapped onto itself by a rotation between 0° and 360° about the center of the figure, called the **center of symmetry**.

The **order of symmetry** is the number of times a figure maps onto itself as it rotates from 0° to 360°. The **magnitude of symmetry** is the smallest angle through which a figure can be rotated so it maps onto itself.

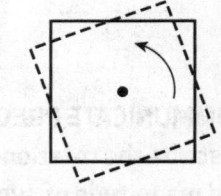

order of symmetry: 4
magnitude of symmetry: 90°

EXAMPLE 3 Identify Rotational Symmetry

Refer to the rectangle, parallelogram, isosceles trapezoid, and regular pentagon from Example 2.

a. **COMMUNICATE PRECISELY** Which of the figures, if any, do not have rotational symmetry? Explain your choice(s).

_____

b. **COMMUNICATE PRECISELY** For each of the figures that have rotational symmetry, give the order of symmetry and magnitude of symmetry.

_____

_____

_____

c. MAKE A CONJECTURE What is the order of symmetry of a regular 20-gon? What is the magnitude of symmetry? Explain.

_____

_____

d. MAKE A CONJECTURE What is the order of symmetry of a regular *n*-gon? What is the magnitude of symmetry? Explain.

_____

_____

_____

PRACTICE

COMMUNICATE PRECISELY State whether each figure has line symmetry and describe the reflections, if any, that map the figure onto itself. Draw any lines of reflection on the figure.

1. Square

2. Regular hexagon

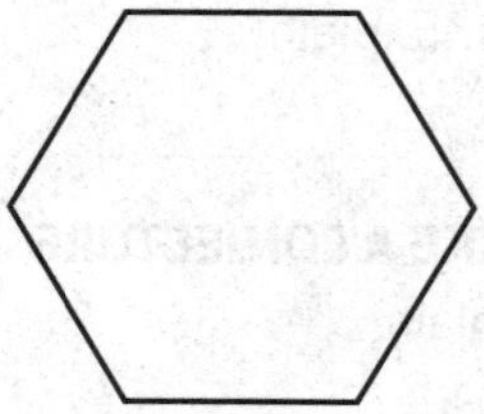

_____

COMMUNICATE PRECISELY State whether each figure has rotational symmetry. If so, describe the rotations that map the figure onto itself by giving the order of symmetry and magnitude of symmetry.

3. Equilateral triangle

_____

4. Scalene triangle

_____

5. Regular hexagon

_____

6. COMMUNICATE PRECISELY The figure shows the floor plan for a new gallery in an art museum. Describe every reflection or rotation that maps the gallery onto itself.

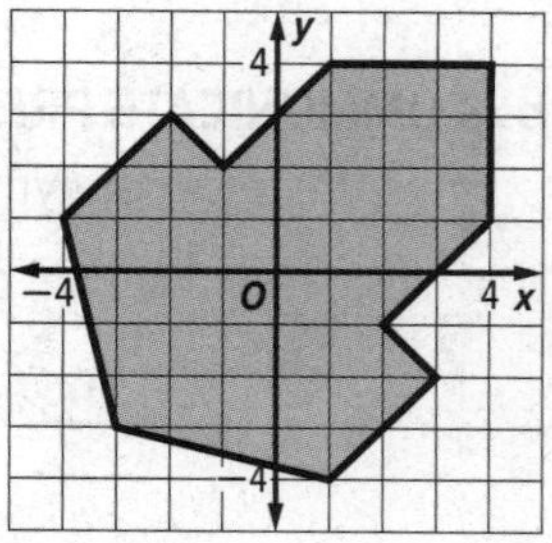

_____

7. a. **CONSTRUCT ARGUMENTS** How many lines of symmetry does a circle have? Explain your reasoning.

_____

_____

_____

b. **CONSTRUCT ARGUMENTS** What is the order of rotation for a circle? Explain your reasoning.

_____

_____

_____

USE STRUCTURE Sketch a figure in the space provided with the described symmetry.

8. No line symmetry; rotational symmetry with order of symmetry 2

9. Exactly one line of symmetry; no rotational symmetry

10. Exactly 3 lines of symmetry; rotational symmetry with order of symmetry 3

11. No line symmetry; rotational symmetry with magnitude of symmetry 120°

12. **CRITIQUE REASONING** The figure shows a black-and-white version of the Union Jack, which is the flag of the United Kingdom. A student said that the flag has line symmetry. Do you agree or disagree? Justify your response.

_____

_____

_____

13. **CONSTRUCT ARGUMENTS** A regular polygon has magnitude of symmetry 15°. How many sides does the polygon have? Explain.

_____

_____

Trendy Transformations

Provide a clear solution to the problem. Be sure to show all of your work, include all relevant drawings, and justify your answers.

Anna is printing shirts for her high school's math club. She starts with the basic design shown, centered on a 20 by 20 unit grid. Each grid measures 0.5 inch by 0.5 inch. She wants each club member to have a unique shirt, so she instructs them to choose three transformations to perform on the original design. Each transformation will be centered at the highest point of the preimage, and each preimage and each resulting image will be included as part of the design.

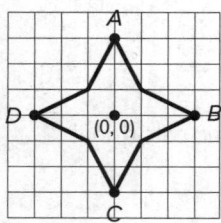

Part A

Jai chooses a translation vector of ⟨4, −3⟩, a 90° rotation about the highest point in the previous image, and then a reflection in the line $y = 3$. What are the coordinates of A''', B''', C''', and D'''? Justify your answer.

Part B

Amy chooses a reflection in the line $x = -2$, a translation of $\langle -1, 4 \rangle$, and a reflection through the line $x = -3$. What are the coordinates of the final image in Amy's design? Justify your answer.

Part C

On a medium or large shirt, the design can have a diagonal measure up to 10 inches. On a small shirt the design can have a diagonal measure up to 7.5 inches. Can Amy and Jai both order small shirts? Justify your answer.

Part D

After the shirts are ordered, Anna decides to sew a rectangular border snugly around each design. How much thread will she need to enclose Jai's design? How much thread will she need to enclose Amy's design? What assumptions did you make in your calculations?

1. *ABCD* is translated using the rule
   $(x, y) \rightarrow (x - 1, y + 3)$. State the
   coordinates of *A'*, *B'*, *C'*, and *D'*.

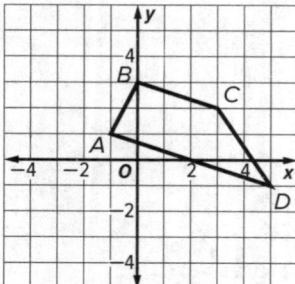

   $A' = $ ☐

   $B' = $ ☐

   $C' = $ ☐

   $D' = $ ☐

2. A triangle with vertices $X(-1, 1)$, $Y(-2, 3)$, $Z(4, 2)$
   is reflected. After the transformation, the image
   of the triangle has vertices $X'(-1, -1)$, $Y'(-2, -3)$,
   and $Z'(4, -2)$. The line of reflection is ☐ .

3. Consider the following diagram.

   Name three points shown in the diagram.

   ☐ , ☐ and ☐

   Give three names for the line shown in the
   diagram.

   ☐ , ☐ and ☐

4. Greta rotates △*DEF* 90° counter-clockwise
   about the origin to obtain △*D'E'F'*, with vertices
   $D'(-3, 1)$, $E'(-5, -3)$, $F'(-7, 4)$. The coordinates
   of △*DEF* are ☐ .

5. Tiana draws quadrilateral *DEFG*. Which
   transformations can Tiana apply to obtain
   a figure congruent to *DEFG*?

   Translation 5 units up and 1 unit left

   135° Rotation

   Dilation by a scale factor of 2

   Reflection in *x*-axis

6. Rectangle *PQRS* has vertices $P(9, 5)$,
   $Q(11, 1)$, $R(5, -2)$, $S(3, 2)$. The area of
   PQRS is ☐ units².

7. Ahmed connects points $S(5, -2)$ and $T(1, 9)$ to
   create $\overline{ST}$. He places point *M* at the midpoint
   of $\overline{ST}$. The coordinates of *M* are ☐ .

8. Circle the statements that are true about
   rotations, reflections, and translations.

   Right angles are preserved.

   Perpendicular lines are not preserved.

   Obtuse angles become acute angles.

   Parallel lines are preserved.

9. A transformation is performed on △*CDE* with
   vertices $C(0, 2)$, $D(-1, 7)$ and $E(2, 5)$ to obtain
   △*C'D'E'* with vertices $C'(0, -2)$, $D'(-1, 3)$, and
   $E'(2, 1)$. The type of transformation is a
   ☐ .

10. Consider each figure. Place a check mark under the rotations about the center of the figure that map the figure onto itself.

Figure	60° Rotation	90° Rotation	180° Rotation
Square			
Isosceles Trapezoid			
Parallelogram			
Regular Hexagon			
Rectangle			

11. Vanessa draws $\triangle XYZ$ with vertices $X(-4, -1)$, $Y(-1, -1)$ and $Z(0, -3)$. After two transformations, the vertices of the image are $X''(-6, 4)$, $Y''(-3, 4)$, and $Z''(-2, 6)$. Describe the transformations that Vanessa could have performed on $\triangle XYZ$ to obtain this image.

_____

12. Draw all lines of symmetry on the following figures.

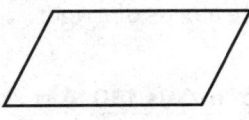　　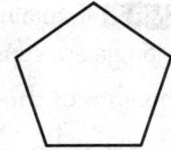　　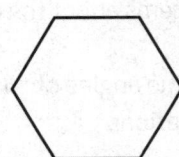

13. Two triangles, $\triangle FGH$ and $\triangle JKL$, are congruent. Aaron claims that the perimeter of $\triangle FGH$ is given by the expression $FG + KL + JL$. Is Aaron correct? Explain why or why not.

_____

14. Raul is constructing $\overline{YZ}$ to be congruent to $\overline{AB}$. Describe the steps that Raul should take.

_____

_____

_____

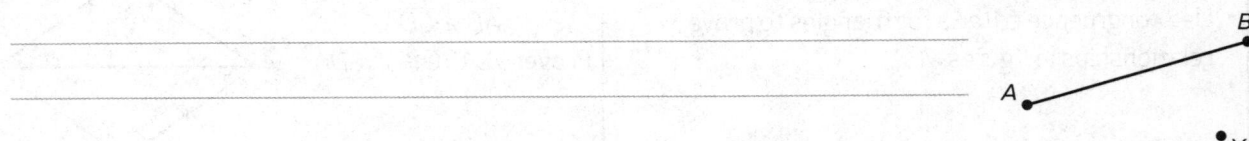

15. Quadrilateral $ABCD$ has vertices at $A(-2, 5)$, $B(-1, 12)$, $C(8, 3)$, and $D(14, -13)$.

   a. What is the perimeter of $ABCD$?

   _____

   _____

   b. If $ABCD$ is translated along $<-3, 4>$ and reflected in the x-axis, find the vertices of the image.

   _____

   c. Is the perimeter of $ABCD$ the same as the perimeter of its image?

   _____

   _____

4 Triangles and Congruence

CHAPTER FOCUS Learn about some of the objectives that you will explore in this chapter. Answer the preview questions. As you complete each lesson, return to these pages to check your work.

What You Will Learn	Preview Question
Congruence	
• Use the definition of congruence in terms of rigid motions to decide if two figures are congruent. • Given congruent triangles, show that corresponding parts are congruent. • Given triangles with corresponding congruent parts, show that the triangles are congruent.	**SMP 7** $\triangle ABC \cong \triangle DEF$. List all congruent corresponding parts. $\overline{AB} \cong \overline{DE}, \overline{BC} \cong \overline{EF}, \overline{AC} \cong \overline{DF}$ $\angle A \cong \angle D, \angle B \cong \angle E, \angle C \cong \angle F$
Proving Theorems About Triangles	
• Prove and apply theorems about the angles of triangles. • Use theorems about the angles of triangles to model real-world situations.	**SMP 2** Explain why the acute angles of a right triangle are supplementary. **The sum of the measures of a $\triangle$ is 180. A rt $\triangle$** **has 1 rt angle, and 180 − 90 = 90. The sum of** **the acute angles is 90, so the angles are suppl.**
Proving Triangles Congruent — SSS, SAS	
• Show that the SSS and SAS criteria for triangle congruence follow from the definition of congruence in terms of rigid motions. • Use congruence criteria for triangles to prove relationships in figures.	**SMP 7** Write a two-column proof to prove the following. **Given:** $\triangle ABC$ and $\triangle CDA$ $\qquad\overline{AB} \cong \overline{CD}$ **Prove:** $\triangle ABC \cong \triangle CDA$

Statements	Reasons
1. $\overline{AB} \cong \overline{CD}$	1. Given
2. $\overline{AB} \cong \overline{BC}, \overline{CD} \cong \overline{DA}$	2. Given
3. $\overline{BC} \cong \overline{DA}$	3. Substitution
4. $\overline{AC} \cong \overline{AC}$	4. Reflexive Prop. of $\cong$
5. $\triangle ABC \cong \triangle CDA$	5. SSS

What You Will Learn	Preview Question

Proving Triangles Congruent — ASA, AAS

- Show that the ASA criterion for triangle congruence follows from the definition of congruence in terms of rigid motions.
- Use congruence criteria for triangles to prove relationships in figures.

SMP 1 $\overline{YZ} \parallel \overline{QR}$. X is the midpoint of $\overline{YQ}$. Prove that $\triangle XYZ \cong \triangle XQR$.

Congruence in Right and Isosceles Triangles

- Use congruence criteria for right triangles.
- Prove that base angles of an isosceles triangle are congruent.
- Apply the Isosceles Triangle Theorem and its converse.

SMP 3 Jane says you can prove $\triangle ABC \cong \triangle DBC$ using SAS. Flynn says you can prove $\triangle ABC \cong \triangle DBC$ using LL. Who is correct? Explain.

Triangles and Coordinate Proof

- Use coordinates to prove simple geometric theorems algebraically.
- Prove that the segment joining the midpoints of two sides of a triangle is parallel to the third side and half the length of the third side.

SMP 1 Prove that the triangle with vertices $A(1, 6)$, $B(12, 3)$, and $C(2, 1)$ is a right triangle.

Objectives

- Use the definition of congruence in terms of rigid motions to decide if two figures are congruent.

- Given congruent triangles, show that corresponding parts are congruent.

- Given triangles with corresponding congruent parts, show that the triangles are congruent.

Recall that a rigid motion is a transformation that does not change the size or shape of a figure. You have seen that translations, reflections, and rotations are rigid motions. You can use rigid motions to define congruence.

In a plane, two figures are **congruent** if and only if there is a sequence of rigid motions that maps one figure to the other. This is called the **principle of superposition**.

EXAMPLE 1   Investigate Congruence

EXPLORE Ernesto is using tiles to make a mosaic. Some of the tiles are shown in the figure. He needs to determine which of the tiles are congruent.

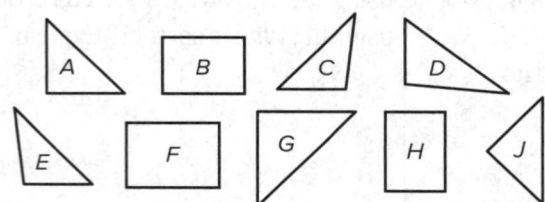

a. **USE TOOLS** Use a piece of tracing paper and a straightedge to trace tile A. Which tile is congruent to tile A? Explain how you can use the tracing paper to find a specific sequence of rigid motions that maps tile A to the other tile. Why does this show that the tiles are congruent?

_____

_____

_____

b. **CONSTRUCT ARGUMENTS** Use the tracing paper to help you find additional pairs of congruent tiles. Justify your choices.

_____

_____

_____

_____

EXAMPLE 2 **Determine Congruence**

Follow these steps to determine whether the given pairs of figures are congruent.

a. **CONSTRUCT ARGUMENTS** Use the definition of congruence in terms of rigid motions to show that $\triangle ABC$ is congruent to $\triangle PQR$.

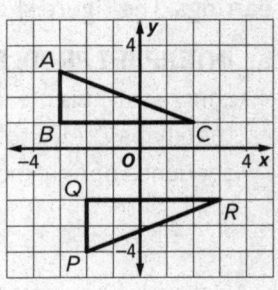

_____

_____

_____

_____

b. **COMMUNICATE PRECISELY** Is quadrilateral $DEFG$ congruent to quadrilateral $JKLM$? Use precise language to justify your answer.

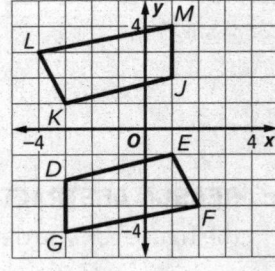

_____

_____

_____

_____

c. **COMMUNICATE PRECISELY** Is pentagon $PQRST$ congruent to pentagon $XWVZY$? Use precise language to justify your answer.

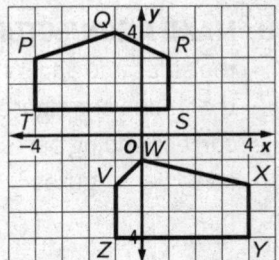

_____

_____

_____

_____

_____

d. **COMMUNICATE PRECISELY** You are shown two triangles, $\triangle ABC$ and $\triangle DEF$. You are told that $\angle A$ and $\angle D$ are congruent, $\angle B$ and $\angle E$ are congruent, and $\angle C$ and $\angle F$ are congruent. You are also told that $\overline{AB}$ and $\overline{DE}$ are congruent, $\overline{BC}$ and $\overline{EF}$ are congruent, and $\overline{AC}$ and $\overline{DF}$ are congruent. Use the definition of congruence in terms of rigid motions to show that $\triangle ABC$ is congruent to $\triangle DEF$.

_____

_____

_____

_____

_____

_____

EXAMPLE 3 Identify Congruent Corresponding Parts

Melinda cuts out two identical pieces of copper to make into earrings. The figure shows the dimensions that she knows.

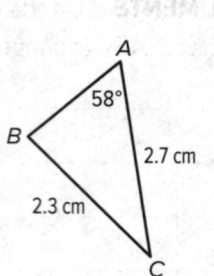
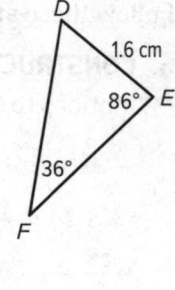

a. **INTERPRET PROBLEMS** Melinda knows that the two earrings are congruent. This means there must be a sequence of rigid motions that maps △ABC to △DEF. What happens to $\overline{AB}$ under this sequence of rigid motions?

_____

_____

b. **CONSTRUCT ARGUMENTS** What does your answer to **part a** tell you about the length of $\overline{AB}$? Explain.

_____

_____

_____

c. **REASON ABSTRACTLY** What other lengths or angle measures that are not marked in the figure can you determine by using the same reasoning as in **parts a** and **b**?

_____

_____

d. **MAKE A CONJECTURE** When two figures are congruent, the **corresponding parts** are the sides and angles that are images of each other under the sequence of rigid motions that maps one figure to the other. For example, in the above figure, $\overline{BC}$ and $\overline{EF}$ are corresponding parts. What conjecture can you make about the corresponding parts of congruent figures?

_____

_____

e. **CRITIQUE REASONING** Melinda said that if two figures are congruent, then they must have the same perimeter. Do you agree or disagree? Justify your answer.

_____

_____

_____

f. **CRITIQUE REASONING** Melinda said that if two figures are congruent, then they must have the same area. Do you agree or disagree? Justify your answer.

_____

_____

_____

The phrase *corresponding parts of congruent triangles are congruent* is frequently abbreviated CPCTC.

You use the symbol ≅ to indicate congruent figures. When you write a congruence statement, such as △ABC ≅ △DEF, you list corresponding vertices in the same order. For example, in the statement △ABC ≅ △DEF, you can conclude that ∠A corresponds to ∠B, and $\overline{AB}$ corresponds to $\overline{DE}$.

EXAMPLE 4 Identify Congruent Corresponding Parts

Use the fact that △GDQ ≅ △RYL and RL = 4.8 ft to answer the following questions.

a. **COMMUNICATE PRECISELY** List the three pairs of corresponding angles and three pairs of corresponding sides given that △GDQ ≅ △RYL.

_____

_____

b. **REASON ABSTRACTLY** Which angles and sides must be congruent? Explain.

_____

_____

_____

c. **CONSTRUCT ARGUMENTS** Can you determine any side lengths or angle measures in △GDQ? Justify your answer.

_____

_____

d. **CONSTRUCT ARGUMENTS** If it is also known that m∠L = 45 and that DQ = 5.2 feet, what other side lengths or angle measures can you determine? Justify your answer.

_____

_____

_____

The Key Concept box summarizes some properties of congruence.

KEY CONCEPT

Complete each example in the table.

Property	Example	
Reflexive Property of Congruence	△ABC ≅ _____	
Symmetric Property of Congruence	If △ABC ≅ △DEF, then △DEF ≅ _____ .	
Transitive Property of Congruence	If △ABC ≅ △DEF and △DEF ≅ △GHJ, then △ABC ≅ _____ .	

CONSTRUCT ARGUMENTS Determine whether the given figures are congruent. Explain.

1.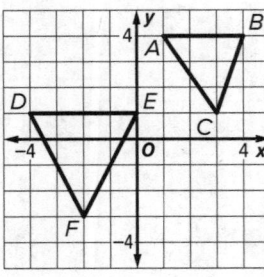

2.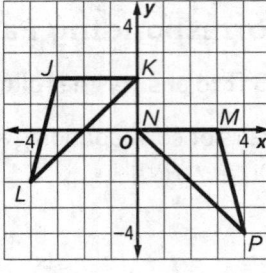

3.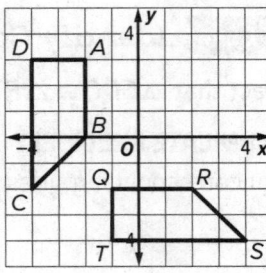

_____

_____

_____

_____

_____

4. **COMMUNICATE PRECISELY** Quadrilateral $PQRS \cong$ quadrilateral $TUVW$. Write congruence statements for all pairs of corresponding angles and corresponding sides.

REASON ABSTRACTLY In Exercises 5–8, use the given information about each figure.

5. $\triangle ABC \cong \triangle DEF$. Find DE and $m\angle B$.

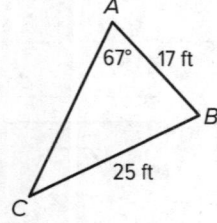

6. $\triangle JKL \cong \triangle MNP$. Find NP and $m\angle M$.

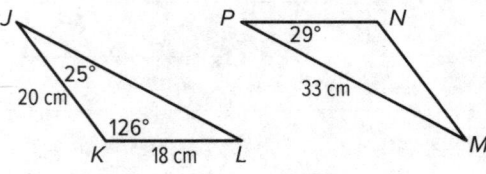

7. $\triangle STU \cong \triangle XYZ$. Find TU and $m\angle T$.

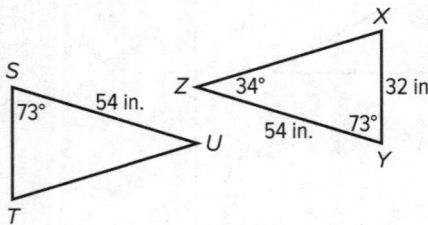

8. $ABCD \cong GHJK$. Find BC and $m\angle D$.

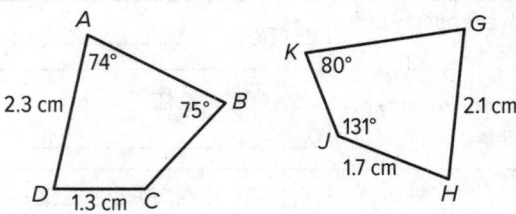

9. **USE A MODEL** Yasmina is building a skateboard ramp. She cuts out two congruent pieces of wood for the side panels of the ramp so that △MNP ≅ △QRS, as shown.

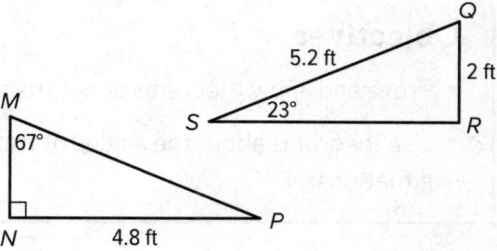

a. Yasmina wants to put duct tape around the perimeters of both panels to seal the edges. What is the total length of the duct tape she will need? Explain.

_____

_____

_____

b. Yasmina would like to know the total amount of wood she will need, but she says that she does not have enough information to find the area of the panels. Do you agree? If so, why. If not, show how she can find the total amount of wood she will need.

_____

_____

10. **INTERPRET PROBLEMS** Duncan is making wooden blocks by cutting out shapes from a one inch thick board. Outlines of the blocks are shown below with the labels A through J.

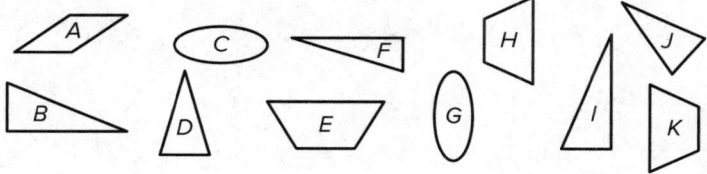

a. Which of the blocks are congruent to B? Explain your reasoning.

_____

_____

_____

b. Part of the board has a cut-out in the shape shown to the right. Which of the blocks could have been cut from this spot? Explain your reasoning.

_____

_____

c. Another part of the board has a cut-out in the shape shown at the right. Which of the blocks could have been cut from this spot? Explain your reasoning.

_____

_____

d. Based on your answer from **part b**, what can you conclude about H and K? Explain your reasoning.

_____

_____

Mathematical Practices
1, 2, 3, 4, 5, 6, 8

Objectives

- Prove and apply theorems about the angles of triangles.
- Use theorems about the angles of triangles to model real-world situations.

One important characteristic of all triangles is presented in the following theorem.

Triangle Angle-Sum Theorem The sum of the measures of the three interior angles of a triangle is 180. In the figure, $m\angle A + m\angle B + m\angle C = 180$.

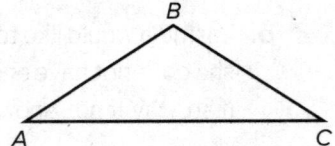

EXAMPLE 1 Prove the Triangle Angle-Sum Theorem

a. USE TOOLS Use tracing paper to verify the Triangle Angle-Sum Theorem. Describe your method and include a sketch in the space provided.

_____

_____

_____

_____

_____

b. CONSTRUCT ARGUMENTS Complete the paragraph proof.
Given: $\triangle ABC$.
Prove: $m\angle 1 + m\angle 2 + m\angle 3 = 180$

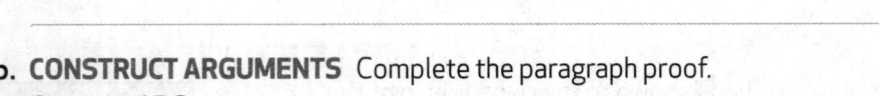

Draw $\overleftrightarrow{AD} \parallel \overline{BC}$ using the _____ Postulate. $\angle 4$ and $\angle BAD$ form a linear pair. By the

Supplement Theorem, $\angle 4$ and $\angle BAD$ are _____, so $m\angle 4 + m\angle BAD =$ _____

by the definition of supplementary angles. $m\angle BAD =$ _____

by the Angle Addition Postulate, so by the Substitution Property of Equality

$m\angle 4 + m\angle 2 + m\angle 5 = 180$. By _____,

$\angle 4 \approx \angle 1$ and $\angle 5 \approx \angle 3$, so _____ and _____ by Definition of

Congruent Angles. Therefore, $m\angle 1 + m\angle 2 + m\angle 3 = 180$ by the Substitution

Property of Equality.

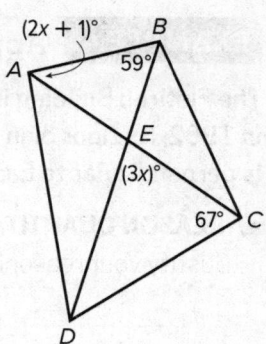

EXAMPLE 2 **Apply the Triangle Angle-Sum Theorem**

a. **REASON QUANTITATIVELY** Use the Vertical Angles Theorem to write an algebraic expression for $m\angle AEB$. Explain.

_____

_____

_____

b. **REASON ABSTRACTLY** Use the Triangle Angle-Sum Theorem to write and solve an equation to find the value of x. Justify each step of your solution.

_____

_____

_____

_____

_____

_____

c. **CALCULATE ACCURATELY** Use your answers to **parts a** and **b** to find $m\angle AEB$ and $m\angle CDE$. Use properties or theorems to support each step of your solution.

_____

_____

_____

_____

_____

Exterior Angle Theorem The measure of an exterior angle of a triangle equals the sum of the measures of the two remote interior angles. In the figure, $m\angle A + m\angle B = m\angle 1$.

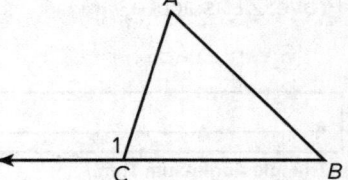

EXAMPLE 3 **Prove the Exterior Angle Theorem**

Use the figure above to prove the Exterior Angle Theorem.

Statements	Reasons
1. $m\angle A + m\angle B + m\angle ACB = 180$	Triangle Angle-Sum Thm.
2. _____ form a linear pair.	Def. of a linear pair
3. $m\angle 1 + m\angle ACB = 180$	_____
4. _____	Substitution
5. $m\angle 1 = m\angle A + m\angle B$	_____

EXAMPLE 4   Apply Theorems About Triangles

The Flatiron Building in New York City is one of America's oldest skyscrapers, completed in 1902. Its floor plan is approximately a right triangle. In the figure below, 5th Avenue is perpendicular to East 22nd Street, and $m\angle B$ is 10 less than 3 times $m\angle C$.

a. REASON QUANTITATIVELY Find the angle measures in the floor plan. Justify your reasoning.

_____

_____

_____

_____

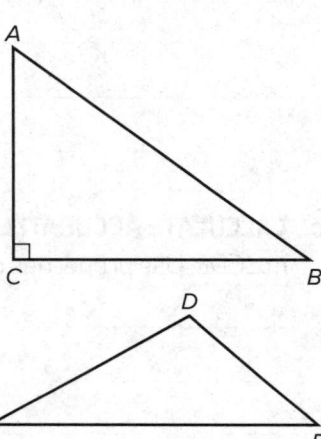

b. Find $m\angle BCD$ in two ways. Explain each method.

_____

_____

Corollary 1 to Triangle Angle-Sum Theorem The acute angles of a right triangle are complementary. In the figure, $\angle C$ is a right angle, so $\angle A$ and $\angle B$ are complementary. By the definition of complementary angles, $m\angle A + m\angle B = 90$.

Corollary 2 to Triangle Angle-Sum Theorem There can be at most one right or obtuse angle in a triangle. In the figure, if $\angle D$ is a right or obtuse angle, then $\angle E$ and $\angle F$ are acute angles. That is, if $m\angle D \geq 90$, then $m\angle E < 90$ and $m\angle F < 90$.

EXAMPLE 5   Prove Corollary 2 to the Triangle Angle-Sum Theorem

Fill in each missing statement or reason in the flow proof below.
Given: In $\triangle ABC$, $\angle A$ is a right or obtuse angle.
Prove: $\angle B$ is acute.

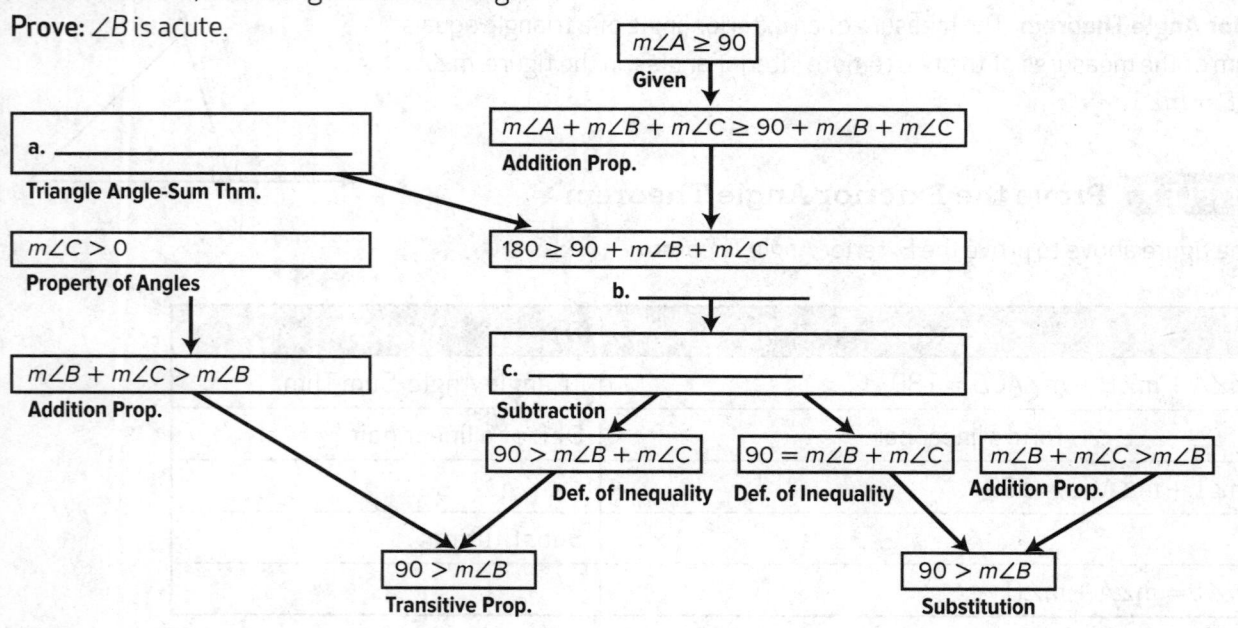

The Triangle Angle-Sum Theorem leads to a useful theorem relating the angles of two triangles.

Third Angles Theorem If two angles of one triangle are congruent to two angles of a second triangle, then the third angles of the triangles are congruent.

EXAMPLE 6 Prove the Third Angles Theorem

Fill in the reason for each statement in the following 2-column proof.

Given: $\angle P \cong \angle X$ and $\angle Q \cong \angle Y$
Prove: $\angle R \cong \angle Z$
Proof:

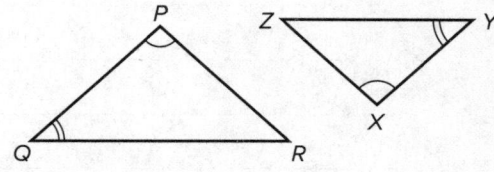

Statements	Reasons
1. $\angle P \cong \angle X, \angle Q \cong \angle Y$	1.
2. $m\angle P = m\angle X, m\angle Q = m\angle Y$	2.
3. $m\angle P + m\angle Q + m\angle R = 180$ $m\angle X + m\angle Y + m\angle Z = 180$	3.
4. $m\angle P + m\angle Q + m\angle R = m\angle X + m\angle Y + m\angle Z$	4.
5. $m\angle X + m\angle Y + m\angle R = m\angle X + m\angle Y + m\angle Z$	5.
6. $m\angle R = m\angle Z$	6.
7. $\angle R \cong \angle Z$	7.

PRACTICE

1. **REASON QUANTITATIVELY** Find the angle measures in $\triangle KLM$. Justify your calculations.

   a. Find $m\angle KML$.

   _____

   _____

   b. Find $m\angle L$.

   _____

   _____

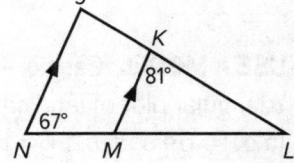

2. **REASON QUANTITATIVELY** Find the angle measures in $\triangle PQR$. Justify your calculations.

   a. Find $m\angle PRQ$.

   _____

   _____

   b. Find $m\angle QPR$.

   _____

   _____

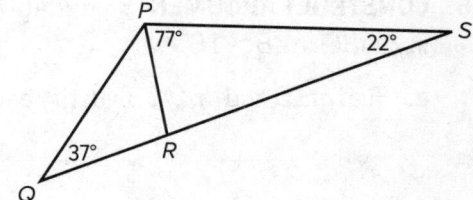

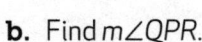

Copyright © McGraw-Hill Education

Proving Theorems About Triangles **123**

3. **PLAN A SOLUTION** Find the measure of the indicated angle. Show your work and state any theorems you use.

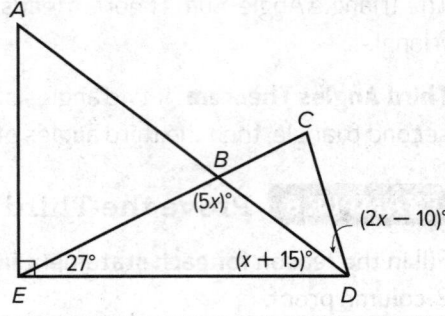

a. Find $m\angle C$.

_____

_____

_____

_____

b. Find $m\angle A$.

_____

_____

_____

_____

_____

4. **CONSTRUCT ARGUMENTS** Prove Corollary 1 to the Triangle Angle-Sum Theorem. Given: $\triangle ABC$ with $\angle C$ a right angle. Prove: $m\angle A + m\angle B = 90$

_____

_____

_____

5. **USE A MODEL** Cassie, a real estate agent, is assessing a triangular plot of land next to a ravine. She has determined that $m\angle 1 = 64$ and $m\angle 4 = 154$. Find $m\angle 2$. Which theorem did you use?

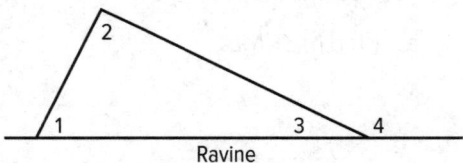

_____

_____

_____

6. **CONSTRUCT ARGUMENTS** In $\triangle ACD$, $m\angle DAC = 36$ and $\angle D \cong \angle ACD$. In $\triangle ABC$, $m\angle B > 18$.

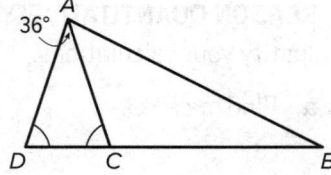

a. Find $m\angle D$ and $m\angle ACB$. Justify each step.

_____

_____

_____

_____

b. Give a reason why ∠B is acute. Then write and solve an inequality describing the measure of ∠B

7. **REASON QUANTITATIVELY** To navigate around a peninsula a ship sails from Port A at a bearing 58° west of due north and when it reaches point B, it changes course to the bearing 20° west of due south to reach Port C. The ship's route is shown in the figure.

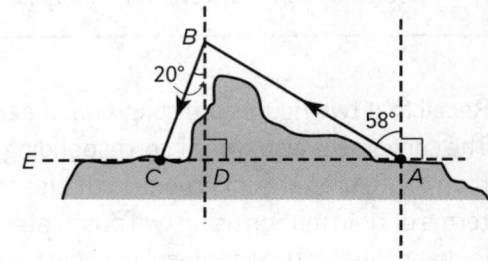

a. Find $m\angle DAB$, $\angle DBA$, and $m\angle BCD$. Explain your reasoning.

b. Find $m\angle BCE$. State any theorems you use to determine your answer.

8. **REASON QUANTITATIVELY** A wall panel needs to be cut to fill a space in transition to a staircase as shown in the figure. If $m\angle DFE = 60$, find $m\angle AFD$ and $m\angle FDC$, the angles at which the panel should be cut.

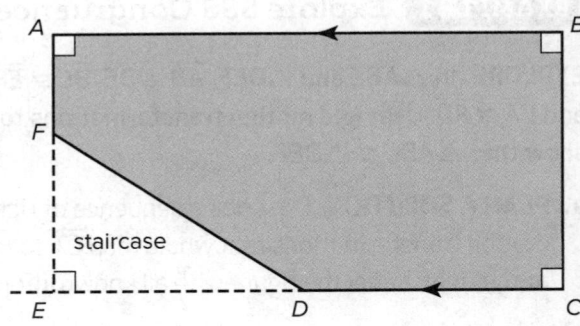

9. **REASON QUANTITATIVELY** In the two triangles △ABC and △DEF, ∠A ≅ ∠D and ∠C ≅ ∠F. Find the value of x if $m\angle B = 3x - 5$ and $m\angle E = x + 27$. Justify each step of your solution.

Objectives

- Show that the SSS and SAS criteria for triangle congruence follow from the definition of congruence in terms of rigid motions.

- Use congruence criteria for triangles to prove relationships in figures.

Mathematical Practices
1, 2, 3, 4, 6, 7

Recall that two figures are congruent if each pair of corresponding parts is congruent. The converse is also true. If corresponding parts of two different figures are congruent, then the figures are congruent. In this lesson you will explore triangle congruence in terms of rigid motion using two postulates: SSS (side-side-side) Congruence and SAS (side-angle-side) Congruence. To show that SSS is sufficient to show triangle congruence, the Perpendicular Bisector Theorem and its converse will be used. These theorems will be explored further in a later chapter.

KEY CONCEPT Perpendicular Bisector Theorem

Perpendicular Bisector Theorem If a point is on the perpendicular bisector of a segment, then it is equidistant from the endpoints of the segment.
Converse of the Perpendicular Bisector Theorem If a point is equidistant from the endpoints of a segment, then it is on the perpendicular bisector of the segment.

EXAMPLE 1 Explore SSS Congruence

EXPLORE In $\triangle ABC$ and $\triangle DEF$, $\overline{AB} \cong \overline{DE}$, $\overline{BC} \cong \overline{EF}$, and $\overline{CA} \cong \overline{FD}$. Use rigid motion transformations to show that $\triangle ABC \cong \triangle DEF$.

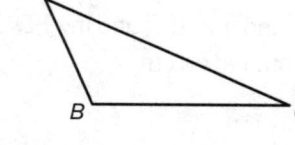 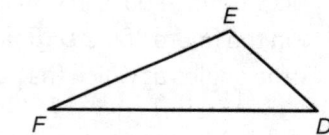

a. **PLAN A SOLUTION** Describe a sequence of right motion transformations that would map $\overline{CA}$ to $\overline{FD}$ as shown in the figure at the right. Label the figure with all known information.

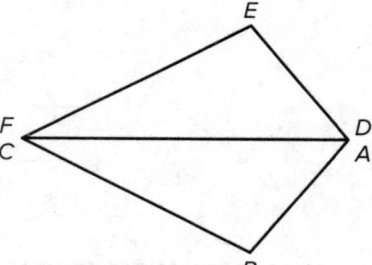

_____

_____

b. **REASON ABSTRACTLY** Given that $\overline{EF} \cong \overline{BC}$, what can you conclude about the relationship between $\overline{FD}$ and $\overline{EB}$? Explain.

_____

_____

_____

c. **REASON ABSTRACTLY** Which rigid motion transformation maps E to B?

_____

_____

d. CONSTRUCT ARGUMENTS Explain how your observations complete the argument that $\triangle ABC \cong \triangle DEF$.

_____

_____

_____

_____

EXAMPLE 2 **Use SSS to Determine Triangle Congruence**

The triangles $\triangle ABC$, $\triangle DEF$, and $\triangle GHI$ are placed on the coordinate grid shown to the right.

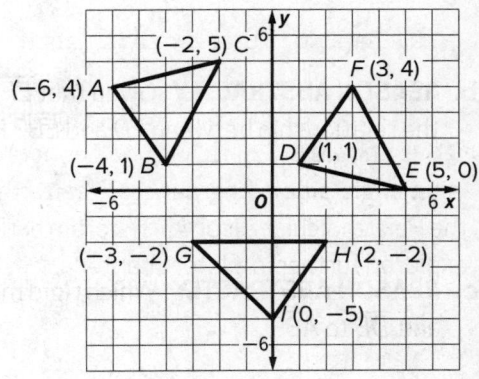

a. CALCULATE ACCURATELY Use the distance formula to find the lengths of the sides of $\triangle ABC$ and $\triangle DEF$. Show your work.

_____

_____

_____

_____

b. REASON QUANTITATIVELY Determine which sides of the two triangles are congruent. Explain your reasoning.

_____

_____

c. CONSTRUCT ARGUMENTS Use the results from **part b** to conclude that $\triangle ABC \cong \triangle DFE$.

_____

_____

d. USE STRUCTURE Use the distance formula to find the lengths of the sides of $\triangle GHI$. Is $\triangle ABC \cong \triangle GHI$? Explain your reasoning.

_____

_____

_____

In **Example 1**, rigid motion transformations are used to show that SSS is sufficient to prove triangle congruence. A similar argument can be used to show that SAS is a valid congruence criterion. This argument uses the Angle Bisector Theorem and its converse. These theorems will be explored further in a later chapter.

KEY CONCEPT **Angle Bisector Theorem**

Angle Bisector Theorem If a point is on the bisector of an angle, then it is equidistant from the sides of the angle.
Converse of the Angle Bisector Theorem If a point in the interior of an angle is equidistant from the sides of the angle, then it is on the bisector of the angle.

EXAMPLE 3 **Explore SAS Congruence**

In △ABC and △DEF, $\overline{AC} \cong \overline{DF}$, $\overline{BC} \cong \overline{EF}$, and
∠BCA ≅ ∠EFD. Use rigid motion transformations
to show that △ABC ≅ △DEF.

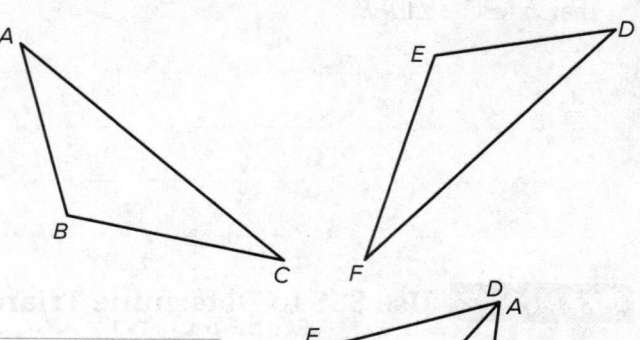

a. **PLAN A SOLUTION** Describe a sequence of rigid
motion transformations that would map $\overline{CA}$ to $\overline{FD}$
as shown in the figure at the right. Label the figure
with all known information.

_____

_____

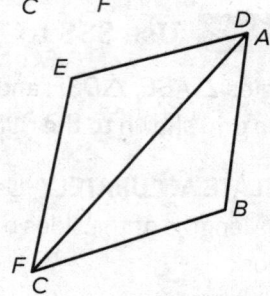

b. **REASON ABSTRACTLY** Given that $\overline{EF} \cong \overline{BC}$, what can you conclude about
the relationship between $\overline{FD}$ and $\overline{EB}$? Explain.

_____

_____

c. **REASON ABSTRACTLY** Which rigid motion transformation maps E to B
and $\overline{DE}$ to $\overline{AB}$?

_____

d. **REASON ABSTRACTLY** Since ∠BCA ≅ ∠EFD, what can be concluded about $\overline{FD}$?

_____

_____

e. **REASON ABSTRACTLY** What can you conclude about ∠EDF and ∠BAC? Explain.

_____

_____

_____

f. **CONSTRUCT ARGUMENTS** Explain how your observations complete the argument
that △ABC ≅ △DEF.

_____

_____

EXAMPLE 4 Determine Triangle Congruence

In the figure, $\overline{AC} \cong \overline{AD}$.

a. **INTERPRET PROBLEMS** Suppose you know ∠C ≅ ∠D. Can you prove
that △ABC ≅ △ABD? Why or why not?

_____

_____

_____

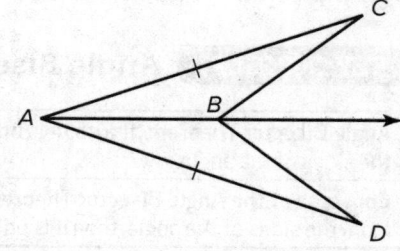

7. **USE A MODEL** An engineer is designing a new cell phone tower. Part of the tower is shown in the figure. The engineer makes sure that line m is parallel to line n and that $\overline{AB} \cong \overline{CD}$. Can she prove that $\triangle ABC \cong \triangle DCB$? Explain why or why not.

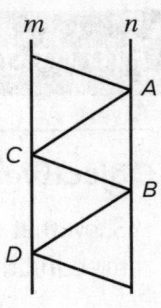

_____

_____

_____

8. **REASON ABSTRACTLY** For each pair of triangles, describe a sequence of rigid motions attempting to map $\triangle ABC$ to $\triangle DEF$ that could be used to determine whether or not the triangles are congruent.

a.

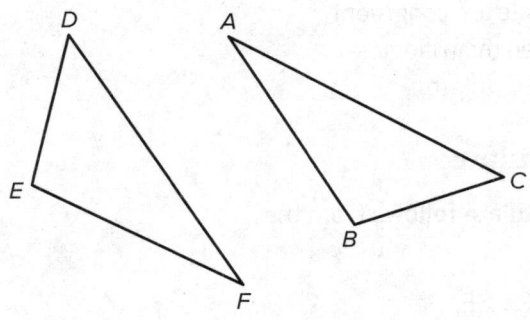

b.

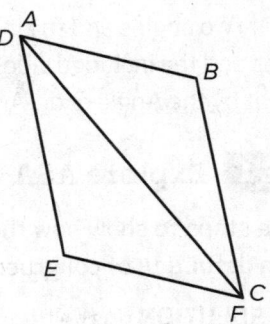

_____   _____

_____   _____

_____   _____

_____   _____

9. **CALCULATE ACCURATELY** Triangles $\triangle ABC$ and $\triangle DEF$ are placed on the coordinate grid with vertices $A(-5, 2)$, $B(-1, -1)$, $C(1, 3)$, $D(-3, -2)$, $E(1, 2)$, and $F(3, -3)$.

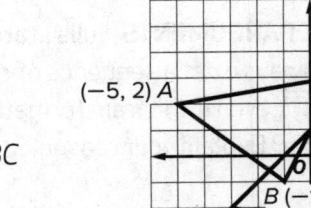

a. Use the distance formula to find the lengths of the sides of $\triangle ABC$ and $\triangle DEF$.

_____

b. Which sides of the two triangles are congruent?

_____

c. Are the two triangles congruent? Explain your reasoning.

_____

d. What can you conclude about $\triangle ABC$ and $\triangle DEF$ regarding rigid motions?

_____

_____

Proving Triangles Congruent—ASA, AAS

Objectives

- Show that the ASA criterion for triangle congruence follows from the definition of congruence in terms of rigid motions.
- Use congruence criteria for triangles to prove relationships in figures.

Mathematical Practices

1, 2, 3, 4, 5, 7

An **included side** is the side located between two consecutive angles of a polygon. In the triangle shown at the right, $\overline{PQ}$ is the included side between $\angle P$ and $\angle Q$. If two angles and the included side of one triangle are congruent to two angles and the included side of a second triangle, then the triangles are congruent by the Angle-Side-Angle (ASA) Congruence Postulate.

EXAMPLE 1 Explore ASA Congruence Postulate

Follow these steps to show how the ASA Congruence Postulate follows from the rigid-motion definition of congruence.

a. **PLAN A SOLUTION** In the figure at right, $\angle A \cong \angle D$, $\angle B \cong \angle E$, and $\overline{AB} \cong \overline{DE}$. Mark this information on the figure; then describe a sequence of rigid-motion transformations that would map $\overline{AB}$ to $\overline{DE}$ as shown in the figure below.

b. **CONSTRUCT ARGUMENTS** Julia states that if $\overline{AB} \cong \overline{DE}$, then there will always exist a sequence of rigid-motion transformations that map $\overline{AB}$ to $\overline{DE}$ even if the transformations are not easy to determine. Do you agree? Explain your reasoning.

c. **REASON ABSTRACTLY** In the figure at right, what is the relationship between $\overline{AB}$ and $\angle FEC$? $\overline{AB}$ and $\angle FDC$?

d. **CONSTRUCT ARGUMENTS** What rigid transformation maps C to F? Justify your answer.

e. USE STRUCTURE What can you say about the image of $\overrightarrow{BC}$ under this reflection? Why?

_____

f. USE REASONING Explain how your answer to **parts e** and **f** complete the argument that $\triangle ABC \cong \triangle DEF$.

_____

_____

_____

_____

_____

KEY CONCEPT ASA Congruence Postulate

Complete the congruence postulate. Then mark the figure to show an example of given information that would allow you to use the postulate to prove the triangles are congruent.

Postulate	Example
Angle-Side-Angle (ASA) Congruence If _____ _____ _____ , then the triangles are congruent.	

EXAMPLE 2 Use ASA to Determine Triangle Congruence

Pamela is studying Native American arrowheads. She has drawn a diagram to model aspects of the shape of a certain arrowhead and wants to know if $\overline{AB} \cong \overline{AD}$. She drew a dashed line $\overline{AC}$ through the diagram and found that $\overline{AC}$ bisects $\angle BAD$ and $\angle BCD$.

a. PLAN A SOLUTION Mark the congruent angles on the figure.

b. CONSTRUCT ARGUMENTS Write a two-column proof using the ASA Congruence Postulate.

Given: $\overline{AC}$ bisects $\angle BAD$ and $\angle BCD$.

Prove: $\overline{AB} \cong \overline{AD}$

Statements	Reasons
1.	
2.	
3.	
4.	
5.	

c. REASON ABSTRACTLY Describe a rigid motion that would map △ABC to △ADC.

_____

EXAMPLE 3 **Prove the Angle-Angle-Side (AAS) Congruence Theorem**

Follow these steps to prove the AAS Congruence Theorem.

Given: $\angle J \cong \angle M$, $\angle K \cong \angle N$, and $\overline{JL} \cong \overline{MP}$

Prove: $\triangle JKL \cong \triangle MNP$

a. PLAN A SOLUTION Mark the given information on the figure. Then explain how you can prove the AAS Congruence Theorem by using one of the three congruence postulates you have already established.

_____

_____

_____

b. CONSTRUCT ARGUMENTS Write a two-column proof of the AAS Congruence Theorem.

Statements	Reasons
1.	1.
2.	2.
3.	3.
4.	4.

KEY CONCEPT **AAS Congruence Theorem**

Complete the congruence theorem. Then mark the figure to show an example of given information that would allow you to use the theorem to prove that the triangles are congruent.

Theorem	Example
Angle-Angle-Side (AAS) Congruence If _____ _____ _____, then the triangles are congruent.	

EXAMPLE 4 Use the AAS Congruence Theorem

Althea used a kit to build a picnic table for her yard. The side view is shown in the figure. Althea made sure the tabletop is parallel to the ground and she checked that $\overline{BC} \cong \overline{DC}$. She wants to know if she can conclude that $\overline{AC} \cong \overline{EC}$.

a. **CONSTRUCT ARGUMENTS** Write a two-column proof using the AAS Congruence Theorem.

Given: $\overline{AB} \parallel \overline{DE}, \overline{BC} \cong \overline{DC}$

Prove: $\overline{AC} \cong \overline{EC}$

Statements	Reasons
1.	1.
2.	2.
3.	3.
4.	4.
5.	5.

b. **CRITIQUE REASONING** Althea said she proved that $\overline{AC} \cong \overline{EC}$ using a different congruence criterion. Do you think this is possible? If so, explain how. If not, explain why not.

_____

_____

_____

_____

_____

PRACTICE

1. In the figure, $\overline{JK}$ is perpendicular to $\overline{LM}$, and $\overline{JK}$ bisects $\angle LKM$.

   a. **CONSTRUCT ARGUMENTS** Write a paragraph proof to show that $\triangle LKJ \cong \triangle MKJ$.

   _____

   _____

   _____

   _____

   b. **REASON ABSTRACTLY** What rigid-motion transformation maps $\triangle LKJ$ to $\triangle MKJ$?

   _____

   _____

2. **CONSTRUCT ARGUMENTS** Write a two-column proof for the following.

Given: $\overline{AC}$ is parallel to $\overline{BD}$. Point D is the midpoint
of $\overline{CE}$. $\angle CAD \cong \angle DBE$

Prove: $\overline{AD} \cong \overline{BE}$

Statements	Reasons
1.	1.
2.	2.
3.	3.
4.	4.
5.	5.
6.	6.
7.	7.

3. **USE TOOLS** Use a compass and straightedge and the ASA
Congruence Postulate to construct a triangle congruent to $\triangle PQR$.
Show your work in the space at the right.

INTERPRET PROBLEMS In Exercises 4 and 5, explain whether there is enough
information given in the figure to prove that the triangles are congruent. If so, describe
a sequence of rigid motions mapping one triangle to the other.

4.   5.

6. **USE A MODEL** Dylan came to a river during a hike and he wanted to estimate the distance across it. He held his walking stick, $\overline{AB}$, vertically on the ground at the edge of the river and sighted along the top of the stick across the river to the base of a tree, T. Then he turned, without changing the angle of his head, and sighted along the top of the stick to a rock, R, located on his side of the river.

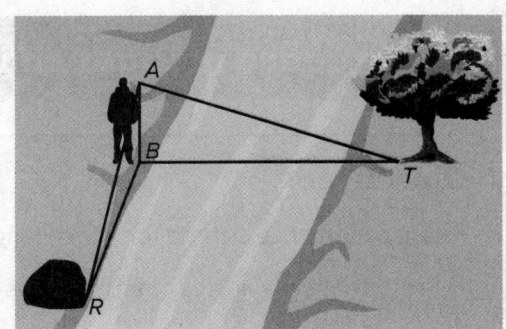

a. Explain why $\triangle ABT \cong \triangle ABR$.

_____

_____

_____

_____

b. Dylan finds that it takes 27 paces for him to walk from his current location to the rock. He also knows that each of his paces is 14 inches long. Explain how he can use this information to estimate the distance across the river.

_____

_____

_____

7. **CRITIQUE REASONING** Raj says that he can draw two triangles that have two sides and a nonincluded angle congruent and that the two triangles are congruent.

a. Raj says that there must be a SSA Congruence Theorem to justify the triangle he constructed. Do you agree? Explain.

_____

_____

_____

b. Provide a counterexample to disprove Raj's conjecture.

c. Two triangles have two sides and a nonincluded angle congruent. Prove that if any other pair of angles of the two triangles is congruent, then the two triangles are congruent.

_____

_____

_____

Objectives

- Use congruence criteria for right triangles.
- Prove that base angles of an isosceles triangle are congruent.
- Apply the Isosceles Triangle Theorem and its converse.

In Lessons 4.3 and 4.4, you developed and applied the SSS, SAS, ASA, and AAS Congruence Criteria. Now you will investigate SSA congruence.

EXAMPLE 1 Investigate SSA Congruence

EXPLORE Follow these steps to use a ruler, protractor, and compass to create triangles. Compare the triangles you draw with those of other students.

a. **USE TOOLS** In the space at the left below, draw a ray, $\overrightarrow{AX}$. Then use the protractor to draw a ray, $\overrightarrow{AY}$, so that $m\angle A = 30°$. Mark point B on $\overrightarrow{AY}$ so that $AB = 6$ cm. Finally, draw $\overline{BC}$ so that $BC = 4$ cm. To do this, open your compass to 4 cm, place the point on B, and draw an arc that intersects $\overrightarrow{AX}$. Label the triangle. Is there more than one way to draw $\triangle ABC$? If so, draw it a second way in the space at the right below.

b. **MAKE A CONJECTURE** Is there an SSA Congruence Criterion? Explain.

_____

_____

_____

_____

c. **USE TOOLS** Repeat **part a** in the space at the right. Use the same dimensions, but this time draw the triangle so that $BC = 3$ cm.

d. COMMUNICATE PRECISELY Describe how the situation in **part c** is different from the situation in **part a**.

_____

e. MAKE A CONJECTURE Is there ever a time when the SSA Congruence Criterion works? If so, when?

_____

The SSS, SAS, ASA, and AAS congruence criteria hold for right triangles and can be given special names using the parts of a right triangle. In addition, there is an SSA Congruence Theorem for right triangles.

KEY CONCEPT **Right Triangle Congruence**

Complete each congruence theorem. Then mark the figure to show an example of given information that would allow you to use the theorem to prove the right triangles are congruent.

Theorem	Example
Leg-Leg (LL) Congruence If _____ _____, then the triangles are congruent.	
Hypotenuse-Angle (HA) Congruence If _____ _____ _____, then the triangles are congruent.	
Leg-Angle (LA) Congruence If _____ _____ _____, then the triangles are congruent.	
Hypotenuse-Leg (HL) Congruence If _____ _____ _____, then the triangles are congruent.	

EXAMPLE 2 **Use Right Triangle Congruence**

In the figure, $\overline{AC} \cong \overline{AD}$ and $\overline{AB}$ is perpendicular to $\overline{CD}$.

a. **CONSTRUCT ARGUMENTS** Mark the given information on the figure. Then write a paragraph proof that $\triangle ABC \cong \triangle ABD$.

_____

_____

_____

_____

b. **COMMUNICATE PRECISELY** What type of triangle is $\triangle CAD$? What must be true about $\angle C$ and $\angle D$? Why?

_____

_____

_____

In an isosceles triangle, the congruent sides are called the **legs** of the triangle. The angle whose sides are the legs of the triangle is the **vertex angle**. The side opposite the vertex angle is the **base** of the triangle. The two angles formed by the base and the congruent sides are the **base angles**.

The Key Concept box summarizes a relationship that you may have discovered in **Example 2**.

KEY CONCEPT **Isosceles Triangles**

Complete each example.

Theorem	Example
Isosceles Triangle Theorem If two sides of a triangle are congruent, then the angles opposite those sides are congruent.	 Example: If $\overline{PQ} \cong \overline{PR}$, then _____.
Converse of the Isosceles Triangle Theorem If two angles of a triangles are congruent, then the sides opposite those angles are congruent.	 Example: If $\angle K \cong \angle L$, then _____.

EXAMPLE 3 **Prove the Isosceles Triangle Theorem**

Follow these steps to prove the Isosceles Triangle Theorem.

Given: $\overline{AB} \cong \overline{AC}$

Prove: $\angle B \cong \angle C$

a. USE STRUCTURE The first steps of the proof are to let P be the midpoint of $\overline{BC}$ and to draw the auxiliary line segment $\overline{AP}$. Why are these steps justified?

_____

_____

b. CONSTRUCT ARGUMENTS Write a two-column proof for the theorem.

Statements	Reasons
1.	1.
2.	2.
3.	3.
4.	4.
5.	5.
6.	6.
7.	7.

EXAMPLE 4 **Use the Isosceles Triangle Theorem**

Julia works for a company that makes lounge chairs. As shown in the figure, the back of each chair is an isosceles triangle that can be adjusted so the person sitting on the chair can recline.

a. CONSTRUCT ARGUMENTS Suppose the chair is adjusted so that $m\angle Q = 50$. What is $m\angle QRS$? Write a paragraph proof to justify your answer.

_____

_____

_____

_____

_____

b. DESCRIBE A METHOD Julia would like a general method that she can use to find $m\angle QRS$ if she knows $m\angle Q$. Write an expression for $m\angle QRS$ when $m\angle Q = x$. Explain.

_____

_____

_____

c. CRITIQUE REASONING Manuel says that he can use the Exterior Angle Theorem to get the result shown in **part b**. Is he correct? Explain.

_____

_____

_____

PRACTICE

1. CONSTRUCT ARGUMENTS In the figure, $\overline{BD}$ is the perpendicular bisector of $\overline{AC}$, and $\overline{AB} \cong \overline{CD}$. Write a paragraph proof to show that $\triangle AEB \cong \triangle CED$.

_____

_____

_____

_____

_____

2. CONSTRUCT ARGUMENTS Write a two-column proof of the converse of the Isosceles Triangle Theorem.

Given: $\angle N \cong \angle P$

Prove: $\overline{MN} \cong \overline{MP}$

Statements	Reasons
1.	1.
2.	2.
3.	3.
4.	4.
5.	5.
6.	6.
7.	7.

3. Each of the triangles shown below is isosceles.

a. **USE TOOLS** Use a ruler to find the midpoint of each side of each triangle. Then draw the triangle formed by connecting the midpoints of each side.

b. **MAKE A CONJECTURE** Look for patterns in your drawings. Make a conjecture about what you notice.

_____

_____

c. **CONSTRUCT ARGUMENTS** In the isosceles triangle at the right, $\overline{AB} \cong \overline{AC}$. Use the figure to help you explain why the conjecture you made in **part b** is true.

_____

_____

_____

_____

4. A boat is traveling at 25 mi/h parallel to a straight section of the shoreline, $\overline{XY}$, as shown. An observer in a lighthouse L spots the boat when the angle formed by the boat, the lighthouse, and the shoreline is 35° and again when this angle is 70°.

a. **USE STRUCTURE** Explain how you can prove that $\triangle BCL$ is isosceles.

_____

_____

b. **USE A MODEL** It takes the boat 15 minutes to travel from point B to point C. When the boat is at point C, what is its distance to the lighthouse?

_____

5. **CRITIQUE REASONING** Anisa says that if two exterior angles of a triangle are congruent, then the triangle is isosceles. Do you agree? Explain.

_____

_____

_____

Objectives

- Use coordinates to prove simple geometric theorems algebraically.
- Prove that the segment joining the midpoints of two sides of a triangle is parallel to the third side and half the length of the third side.

EXAMPLE 1 **Investigate a Triangle Property**

EXPLORE A town is preparing for a 5K run. The race will start at city hall, *C*. The course will take runners along straight streets to the library, *L*, to the science museum, *S*, and back to city hall for the finish. A city employee has been asked to develop a different route for runners who want a shorter race.

a. CALCULATE ACCURATELY The employee decides to create a shorter route by locating the midpoint *X* of $\overline{CL}$ and the midpoint *Y* of $\overline{CS}$. The runners will go from *C* to *X* to *Y* and back to *C*. Use the Midpoint Formula to locate the midpoints of $\overline{CL}$ and $\overline{CS}$. Draw $\overline{XY}$ on the figure.

_____

b. CALCULATE ACCURATELY Use the Slope Formula and Distance Formula to compare $\overline{XY}$ to $\overline{LS}$. What do you notice?

_____

_____

_____

c. MAKE A CONJECTURE Draw and label your own triangle, △*PQR*, in the space at the right. Then find the midpoints, *M* and *N*, of two sides and draw the segment joining these midpoints. Use the Midpoint Formula, Slope Formula, and Distance Formula to check if the relationship you noticed in **part b** holds for this figure. State your observations as a conjecture.

_____

_____

_____

d. USE A MODEL What is the length of the race course from *C* to *X* to *Y* and back to *C*? Explain how you know.

_____

_____

_____

e. DESCRIBE A METHOD Describe how the employee could find another route that is half as long as the original route. Explain your reasoning.

_____

_____

_____

_____

A **coordinate proof** uses figures in the coordinate plane and algebra to prove geometric relationships. You can use a coordinate proof to prove the conjecture you made in the previous exploration. The Key Concept box provides suggestions for placing triangles on the coordinate plane when writing a coordinate proof.

> **KEY CONCEPT** **Placing Triangles on the Coordinate Plane**
>
> **Step 1** Use the origin as a vertex or center of the triangle.
>
> **Step 2** Place at least one side of the triangle on an axis.
>
> **Step 3** Keep the triangle within the first quadrant if possible.
>
> **Step 4** Use coordinates that make computations as simple as possible.

EXAMPLE 2 **Write a Coordinate Proof**

Follow these steps to write a coordinate proof for the following.

Given: $\triangle PQR$, where M is the midpoint of $\overline{PQ}$ and N is the midpoint of $\overline{PR}$

Prove: $\overline{MN}$ is parallel to $\overline{QR}$, and $\overline{MN} = \frac{1}{2}QR$.

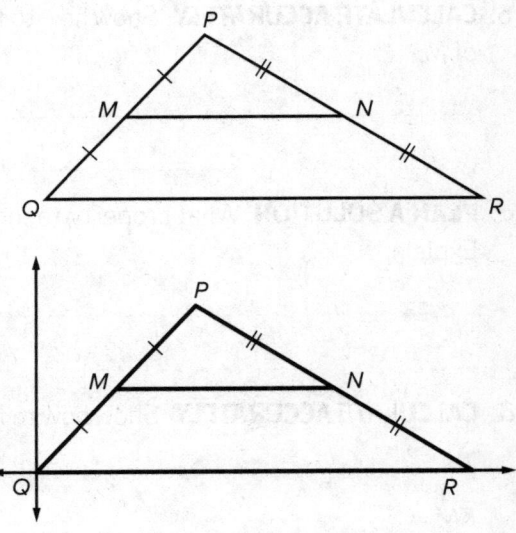

a. REASON ABSTRACTLY Place $\triangle PQR$ on the coordinate plane and assign coordinates to the vertices of the triangle. For convenience, place vertex Q at the origin and place $\overline{QR}$ along the positive x-axis. Since the proof will involve midpoints, it makes sense to assign coordinates that are multiples of two. What are appropriate coordinates for R in terms of a? What are appropriate coordinates for P in terms of b and c? Label the coordinates of the vertices P, Q, and R in the figure.

_____

_____

b. CALCULATE ACCURATELY Show how to find the coordinates of M and N.

_____

_____

c. PLAN A SOLUTION What theorem or postulate can you use to prove that $\overline{MN}$ is parallel to $\overline{QR}$?

_____

d. CONSTRUCT ARGUMENTS Prove that $\overline{MN}$ is parallel to $\overline{QR}$.

_____

_____

_____

e. CONSTRUCT ARGUMENTS Prove that $MN = \frac{1}{2}QR$.

_____

_____

_____

EXAMPLE 3   Write a Coordinate Proof

Follow these steps to prove that the midpoint of the hypotenuse of a right triangle is equidistant from the vertices of the triangle.

Given: $\triangle JKL$ is a right triangle with a right angle at $\angle K$, and M is the midpoint of $\overline{JL}$.

Prove: $JM = KM = LM$

a. REASON ABSTRACTLY Place the triangle above on the coordinate plane. Label the coordinates of the vertices J, K, and L.

_____

_____

b. CALCULATE ACCURATELY Show how to find the coordinates of M.

_____

_____

c. PLAN A SOLUTION What property, theorem, or formula can you use to complete the proof? Explain.

_____

_____

d. CALCULATE ACCURATELY Show how to find each of the following distances.

$JM = $ _____

$KM = $ _____

$LM = $ _____

e. USE STRUCTURE The segment that joins K and M divides $\triangle JKL$ into two smaller triangles. What types of triangles are these? Explain how you know.

_____

_____

_____

Some coordinate proofs are based on specific values for the coordinates of the figures. As in the more general proofs, you can use the Distance Formula, Midpoint Formula, Slope Formula, and/or congruence criteria to write the proof.

EXAMPLE 4 **Write a Coordinate Proof**

Andrew is using a coordinate plane to design a quilt. Two of the triangular patches for the quilt are shown in the figure. Andrew wants to be sure that $\angle A$ and $\angle D$ have the same measure.

Follow these steps to prove that $\angle A \cong \angle D$.

a. **PLAN A SOLUTION** Describe the main steps you can use to prove that $\angle A \cong \angle D$.

_____

_____

_____

_____

b. **CONSTRUCT ARGUMENTS** Write a paragraph proof that $\angle A \cong \angle D$.

_____

_____

_____

_____

_____

PRACTICE

1. **CONSTRUCT ARGUMENTS** $\triangle PQR$ is a right triangle with a right angle at $\angle Q$, and M is the midpoint of $\overline{PR}$. Draw a figure and assign coordinates to prove that the area of $\triangle QMR$ is one-half the area of $\triangle PQR$.

_____

_____

_____

_____

_____

_____

_____

2. Complete the following proof.

Given: △ABC is isosceles with $\overline{AB} \cong \overline{AC}$. D is the midpoint of $\overline{AB}$, E is the midpoint of $\overline{BC}$, and F is the midpoint of $\overline{AC}$.

Prove: △DEF is isosceles.

  a. USE STRUCTURE In the space at the right, show how to place △ABC on a coordinate plane. Show how to assign coordinates to the vertices of △ABC.

  b. REASON ABSTRACTLY What are the coordinates of D, E, and F?

_____

_____

  c. CONSTRUCT ARGUMENTS Explain how to complete the proof.

_____

_____

_____

_____

3. CRITIQUE REASONING A student was asked to prove that the segment joining the midpoints of two sides of a triangle is parallel to the third side. He set up the figure and coordinates shown at right. Then he gave the argument shown below the figure.

  a. Is the student's proof correct? If not, explain why not and explain what the student would need to do differently to write a correct proof.

_____

_____

_____

_____

_____

Let M be the midpoint of $\overline{RS}$. Then the coordinates of M are M(0, b). Let N be the midpoint of $\overline{ST}$. Then the coordinates of N are N(a, 0).

The slope of $\overline{MN}$ is $\frac{0-b}{a-0} = -\frac{b}{a}$.

The slope of $\overline{RT}$ is $\frac{0-2b}{2a-0} = -\frac{b}{a}$.

Since the slopes are equal, $\overline{MN} \parallel \overline{RT}$.

  b. Sharon argues that the proof in **Example 2** makes an assumption about the triangle and is therefore invalid. She claims that we assumed that the triangle lies on the x-axis with one of its vertices at the origin. In order to prove this theorem in general, we would have to assume that the triangle has three vertices at (a, b), (c, d), and (e, f). Is Sharon correct?

_____

_____

_____

_____

4. **USE A MODEL** A landscape architect is using a coordinate plane to design a triangular community garden. The fence that will surround the garden is modeled by $\triangle ABC$.

The architect wants to know if the any of the three angles in the fence will be congruent. Determine the answer for the architect and give a coordinate proof to justify your response.

_____

_____

_____

5. Complete the following proof.

Given: $\triangle JKL$ where P is the midpoint of $\overline{JK}$, Q is the midpoint of $\overline{KL}$, and R is the midpoint of $\overline{JL}$.

Prove: The area of $\triangle JKL$ is 4 times the area of $\triangle PQR$.

a. **USE STRUCTURE** In the space at the right, show how to place $\triangle JKL$ on a coordinate plane. Show how to assign coordinates to the vertices of $\triangle JKL$.

b. **CONSTRUCT ARGUMENTS** Write the proof.

_____

_____

_____

_____

c. **CRITIQUE REASONING** Jane approaches this proof differently. She argued that $\triangle KPQ$, $\triangle QRL$, $\triangle PJR$, and $\triangle RQP$ are congruent. Since they are all congruent, each of them is one-fourth of $\triangle JKL$. Is Jane correct? Outline how this proof would work making use of **Example 2**.

_____

_____

_____

6. **CONSTRUCT ARGUMENTS** Write the following proof.

Given: $\overline{AB} \cong \overline{AC}$

X is the midpoint of $\overline{AB}$, Y is the midpoint of $\overline{AC}$.

Prove: $\overline{BY} \cong \overline{CX}$

_____

_____

_____

Designing a Park

Provide a clear solution to the problem. Show all of your work, including relevant drawings. Justify your answers.

A town is building a skateboarding park within the region bounded by the pentagon shown on the map. Each unit on the grid represents 10 meters. The town wants the designer to do the following:

- Create the park in the shape of an equilateral triangle.
- Have one side be parallel to Jones Avenue.
- Make the perimeter of the park 180 meters.

Part A

You are told Jones Avenue makes a 120° angle with West Street. Use this information to construct the side of the park parallel to Jones Avenue from point *A* as accurately as possible. Explain your reasoning. Label the other endpoint as point *B*.

Part B

Construct the rest of the park. Explain what you did. Label the final vertex of the park as point *C*.

Part C

The mayor is impressed with your drawing, but would like to know the coordinates of points *B* and *C*. Give approximate values for each point based on your drawing. Then find the exact coordinate of point *B*.

Kites and Congruence

Provide a clear solution to the problem. Be sure to show all of your work, include all relevant drawings, and justify your answers.

In a kite design, four triangular pieces of fabric are sewn together to form a quadrilateral. Two rods are attached to the kite so that one rod bisects the angles, as shown.

Part A

The manufacturer wants to create templates for the triangular shapes that need to be cut. If $m\angle DCA = 50$, $AB = 20$ in., $EC = 10$ in., $m\angle ABC = 93$, and $\overline{AC}$ bisects $\overline{BD}$ where $BD = 24$ in., solve for all angles and side lengths for each triangle in the figure. Label the angle measures in the figure. Name all the congruent triangles shown within the design.

Part B

One worker notices that the rods intersect at right angles. Write a paragraph proof to prove that the rods will always intersect at right angles.

Part C

The manufacturer will use a bolt of fabric measuring 60 inches wide and 130 yards long. Draw a model and find how many kite patterns can be cut from this bolt of fabric.

1. The coordinates of the vertices of a triangle are (0, 0), (a, 0), and (b, c). The area of this triangle is ☐ square units.

2. Shelley has drawn two triangles, $\triangle XYZ$ and $\triangle TUV$. She knows that $\angle X \cong \angle T$ and $\angle Y \cong \angle U$. In order to prove that $\triangle XYZ \cong \triangle TUV$ using AAS, she also needs to know that ☐ or ☐.

3. Archaeologists have found two triangular building foundations. The diagram shows some measurements from the foundations. Which of the following additional measurements would be sufficient to prove that the two triangles are congruent?

$m\angle E = 36$ and $DE = 67$ ft

$BC = 65$ ft and $DE = 67$ ft

$m\angle E = 36$ and $m\angle A = 69$

$m\angle A = 69$ and $DE = 67$ ft

4. William draws triangle $\triangle PQR$. He rotates, translates, and reflects the triangle to create a new triangle with vertices S, T, and V. He finds that $\angle P \cong \angle T$, $\angle Q \cong \angle S$, and $\angle R \cong \angle V$. From this, William determines that he can write the congruence statement ☐.

5. The triangles below can be shown to be congruent by ☐.

6. Which of the following statements are true for the following diagram?

$\triangle MQN \cong \triangle MQP$ $\triangle PQO \cong \triangle OQN$

$\triangle MNP \cong \triangle OPN$ $\triangle MPO \cong \triangle MNO$

$\triangle ONQ \cong \triangle OPQ$ $\triangle QPM \cong \triangle QPO$

7. Consider the following diagram.

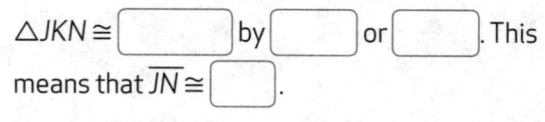

$\triangle JKN \cong$ ☐ by ☐ or ☐. This means that $\overline{JN} \cong$ ☐.

8. The HL Congruence Theorem states that in two ☐ triangles, if the ☐ are congruent and corresponding ☐ are congruent, the triangles are congruent.

9. Complete the following proof.

Given: $\overline{AB} \cong \overline{BC}$

Prove: $\angle A \cong \angle C$

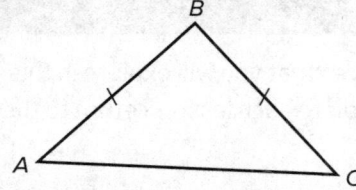

Statement	Reason
$\overline{AB} \cong \overline{BC}$	
Draw $\overline{BD}$ bisecting $\angle ABC$	
	Definition of angle bisector
$\overline{BD} \cong \overline{BD}$	
	SAS
$\angle A \cong \angle C$	

10. Consider the following diagram

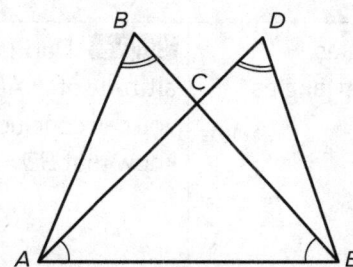

a. What can you conclude about $\triangle ABE$ and $\triangle EDA$? Justify your answer.

_____

b. What can you conclude about $\triangle BCA$ and $\triangle DCE$? Justify your answer.

_____

c. If $AE = 8$ and the perimeter of $\triangle EDA$ is 22, what is the perimeter of $\triangle DCE$? Explain how you know.

_____

_____

11. The vertices of $\triangle ABC$ are $A(-2, -4)$, $B(-1, 1)$ and $C(3, -2)$. The vertices of $\triangle DEF$ are $D(-3, 2)$, $E(2, 3)$ and $F(6, 0)$.

a. Are the triangles congruent? Justify your answer.

_____

_____

b. The vertices of $\triangle GHI$ are $G(-2, 0)$, $H(1, 4)$, and $I(3, -1)$. Is there a correspondence by which $\triangle GHI$ is congruent to $\triangle ABC$ or $\triangle DEF$? Justify your answer.

_____

_____

5 Relationships in Triangles

CHAPTER FOCUS Learn about some of the objectives that you will explore in this chapter. Answer the preview questions. As you complete each lesson, return to these pages to check your work.

What You Will Learn	Preview Question
Angle and Perpendicular Bisectors	
• Prove and apply the Perpendicular Bisector and Angle Bisector Theorems and their converses. • Prove and apply the Circumcenter and Incenter Theorems. • Construct the perpendicular bisector of a line segment.	**SMP 2** Can the circumcenter and the incenter of a triangle be the same point? Explain. **Yes; for an equilateral triangle; the angle** **bisectors are the perpendicular bisectors** **of the sides.**
Medians and Altitudes	
• Identify medians and altitudes in triangles. • Compare points of concurrency within triangles.	**SMP 3** Darius says you can conclude $\overline{BD}$ is an altitude of $\triangle ABC$ if you know that $BC = 10$. Lisa says you can conclude $\overline{BD}$ is an altitude of $\triangle ABC$ if you know that $BD = 8$. Who is correct? Explain. *[Triangle ABC with B at top, A bottom left, C bottom right, D on AC. Side AB labeled 10 in., AD = 6 in., DC = 6 in., BD drawn vertically.]* **They are both correct. If $BC = 10$, then by** **the Converse of the ⊥ Bisector Thm, $\overline{BD}$ is** **a ⊥ bisector of $\overline{AC}$ and by def. is an altitude.** **If $BD = 8$, then by the Converse of the** **Pythagorean Thm. $\triangle ABD$ is right. By def.** **$\overline{BD}$ is an altitude.** **SMP 2** When can a median of a triangle also be an altitude? Explain your reasoning. **The median to the base of an isosceles triangle is** **perpendicular to the base because of congruent** **triangles, so it is also an altitude.**

What You Will Learn	Preview Question
Inequalities in One Triangle	
• Apply properties of inequalities to the measure of angles and sides of a triangle. • Understand and apply the Exterior Angle Inequality Theorem. • Understand and apply theorems about the angle-side relationships in triangles.	**SMP 7** Write < or > so that each statement relating the sides of $\triangle ABC$ is true. AB ☐ AC AB ☐ BC BC ☐ AC
The Triangle Inequality	
• Prove the Triangle Inequality Theorem and use it to determine and prove facts about specific triangles and related figures.	**SMP 1** A triangle has side lengths of $3x - 2$, $x + 3$, and $x + 2$. What is the range of possible values for x? How do you know? _____ _____ _____
Inequalities in Two Triangles	
• Prove the Hinge Theorem • Apply the Hinge Theorem and its converse to real-world situations.	**SMP 2** How can you use the Hinge Theorem to determine which is greater, AB or EB? _____ _____ _____ _____

Objectives

- Prove and apply the Perpendicular Bisector and Angle Bisector Theorems and their converses.

- Prove and apply the Circumcenter and Incenter Theorems.

- Construct the perpendicular bisector of a line segment.

A **perpendicular bisector** is a segment, line, or plane that intersects a line segment at its midpoint and is also perpendicular to the line segment.

KEY CONCEPT

Perpendicular Bisector Theorem If a point is on the perpendicular bisector of a segment, then it is equidistant from the endpoints of the segment.

Converse of the Perpendicular Bisector Theorem If a point is equidistant from the endpoints of a segment, then it is on the perpendicular bisector of the segment.

EXAMPLE 1 Use the Perpendicular Bisector Theorem

Use the Perpendicular Bisector Theorem to solve a problem involving baseball.

On a baseball diamond, home plate and second base lie on the perpendicular bisector of the line segment joining first and third base. First base is 90 ft from home plate. How far is it from third base to home plate?

a. **USE A MODEL** In the space at the right, sketch a baseball diamond, labeling home plate as point A, first base as B, second base as C, and third base as D. Label the intersection of $\overline{AC}$ and $\overline{BD}$ as E.

b. **REASON ABSTRACTLY** Use the Perpendicular Bisector Theorem to make a conjecture regarding A. Describe your conclusion in the context of the situation.

_____

_____

_____

Three or more lines are **concurrent** if they intersect at a common point, the **point of concurrency**.

A triangle has three sides and, therefore, three perpendicular bisectors. The point of concurrency of the perpendicular bisectors of a triangle is called the **circumcenter** of the triangle.

KEY CONCEPT

Circumcenter Theorem The perpendicular bisectors of a triangle intersect at a point, called the *circumcenter*, that is equidistant from the vertices of the triangle.

EXAMPLE 2 Use the Circumcenter Theorem

Locate the center of the circle that passes through the vertices of the triangle shown.

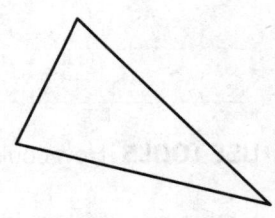

a. **USE TOOLS** Using a ruler and protractor, draw perpendicular bisectors for each side of the triangle shown at the right. Locate the point of intersection.

b. **USE TOOLS** Using a compass, construct a circle centered at the circumcenter that passes through each vertex.

c. **USE TOOLS** Describe briefly how you could solve a problem of this type using dynamic geometry software.

_____

_____

_____

_____

KEY CONCEPT

Angle Bisector Theorem If a point is on the bisector of an angle, then it is equidistant from the sides of the angle.

Converse of Angle Bisector Theorem If a point is equidistant from the sides of an angle, then it is on the bisector of the angle.

EXAMPLE 3 Prove the Converse of the Angle Bisector Theorem

Given: $\overline{DF} \perp \overline{BD}$, $\overline{EF} \perp \overline{BE}$, and $DF = FE$.
Prove: $\overrightarrow{BF}$ is the angle bisector of $\angle DBE$.

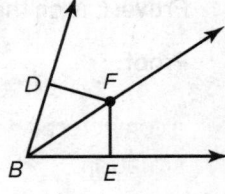

Statement	Reason
1. $DF = FE$	Given
2. $\overline{DF} \cong \overline{FE}$	a. _____
3. b. _____	Reflexive Property
4. $\overline{DF} \perp \overline{BD}, \overline{EF} \perp \overline{BE}$	Given
5. c. _____	HL Theorem
6. $\angle DBF \cong \angle EBF$	CPCTC
7. $\overrightarrow{BF}$ is the angle bisector of $\angle DBE$	Definition of angle bisector

The point of concurrency of the angle bisectors of a triangle is called the **incenter** of the triangle.

KEY CONCEPT

Incenter Theorem The angle bisectors of a triangle intersect at a point called the incenter, which is equidistant from each side of the triangle.

Copyright © McGraw-Hill Education

Angle and Perpendicular Bisectors **159**

EXAMPLE 4 **Construct the Incenter of a Triangle**

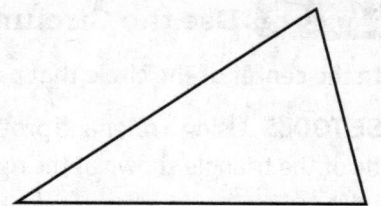

a. USE TOOLS Construct the incenter of the triangle. Explain your procedure.

_____

_____

b. USE TOOLS How could you use a ruler to verify your construction?

_____

_____

PRACTICE

1. **CONSTRUCT ARGUMENTS** Complete the proof of the Perpendicular Bisector Theorem.

   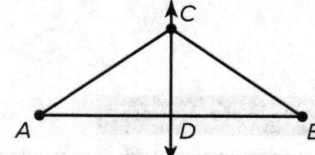

   Given: $AD = BD$ and $\overleftrightarrow{CD} \perp \overline{AB}$
   Prove: $AC = BC$

   It is given that $AB = BD$ so **a.** _____ by definition of congruence.
   By the Reflexive Property, $\overline{CD} \cong \overline{CD}$. Also, it is given that $\angle ADC$ and
   b. _____ are right angles, so $\triangle ADC$ and $\triangle BDC$ are right triangles.
   By the **c.** _____ Thm., $\triangle ADC \cong \triangle BDC$, so by CPCTC, **d.** _____.
   By the definition of congruence, _____.

2. **CONSTRUCT ARGUMENTS** Write a proof of the Converse of the Perpendicular Bisector Theorem.

   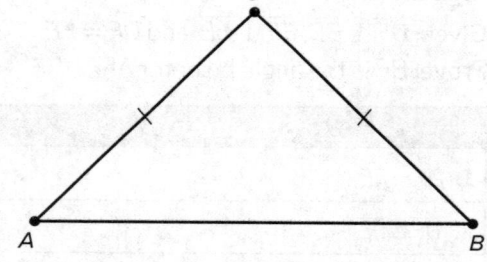

   Given: C is equidistant from A and B.
   Prove: C is on the perpendicular bisector of $\overline{AB}$.

   Proof:

   _____

   _____

   _____

   _____

   _____

   _____

   _____

   _____

   _____

   _____

   _____

   _____

3. **REASON QUANTITATIVELY** Use the Perpendicular Bisector Theorem or its converse to complete.

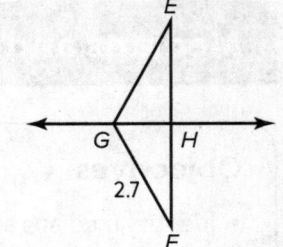

a. $\overleftrightarrow{GH}$ is the _____ of $\overline{EF}$. By the Perpendicular Bisector Thm., $EG =$ _____ , so $EG =$ _____ .

b. Find WZ. Justify your answer.

_____

_____

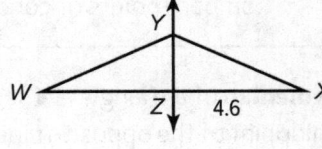

4. **REASON QUANTITATIVELY** Use the Angle Bisector Theorem or its converse to find each measure.

a. By the _____ Thm., $PQ =$ _____ , so $PQ =$ _____ .

b. Find $m\angle JKL$. Explain.

_____

_____

_____

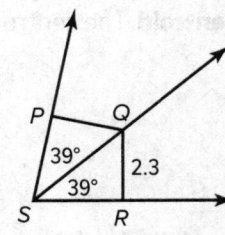

5. **CONSTRUCT ARGUMENTS** Prove the Incenter Theorem.

Given: $\overline{PA}$, $\overline{PB}$, and $\overline{PC}$ are the angle bisectors of $\angle BAC$, $\angle ABC$, and $\angle ACB$, respectively.
Prove: $PD = PE = PF$

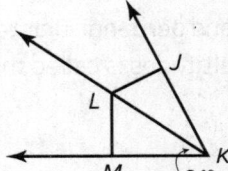

It is given that $\overline{PA}$ is the angle bisector of $\angle BAC$, so by the **a.** _____, P is equidistant from $\overline{AB}$ and $\overline{AC}$. Since $\overline{PD} \perp \overline{AB}$ and $\overline{PE} \perp \overline{AC}$, **b.** _____ by the definition of equidistant. Since it is given that $\overline{PB}$ is the angle bisector of $\angle ABC$, the same logic can be applied. P is equidistant from _____ by the Angle Bisector Theorem and therefore $PD = PF$. By the **d.** _____ of Equality, $PD = PE = PF$.

6. **CONSTRUCT ARGUMENTS** Write a proof of the Angle Bisector Theorem.

Given: BD is the angle bisector of $\angle ABC$.
Prove: D is equidistant from $\overline{AB}$ and $\overline{BC}$.

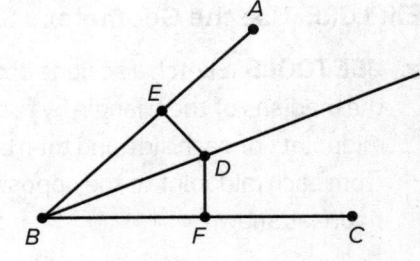

_____

_____

_____

_____

_____

_____

_____

_____

_____

_____

Objectives

- Identify medians and altitudes in triangles.
- Compare points of concurrency within triangles.

Mathematical Practices
1, 2, 3, 4, 5, 6, 7

A **median** of a triangle is a segment whose endpoints are a vertex of the triangle and the midpoint of the opposite side. The point of concurrency of the medians is called the **centroid**. The centroid is the balancing point or *center of gravity* for a triangular region.

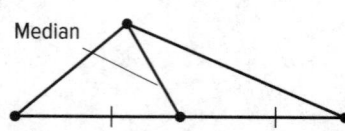

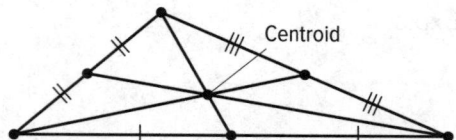

An **altitude** of a triangle is a segment from a vertex to the line containing the opposite side and perpendicular to the line containing that side. The point of concurrency of the altitudes is called the **orthocenter**.

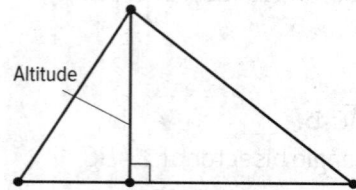

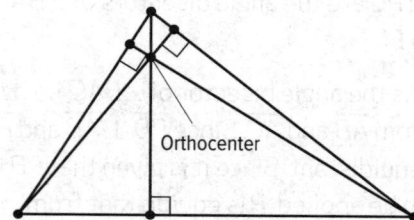

EXAMPLE 1 **Points of Concurrency**

EXPLORE **Use the Geometer's Sketchpad to investigate properties of medians.**

a. **USE TOOLS** Sketch a scalene obtuse triangle. Construct the medians of the triangle by first constructing the midpoints of each side and then constructing a segment from each midpoint to the opposite vertex. Label the figure as shown.

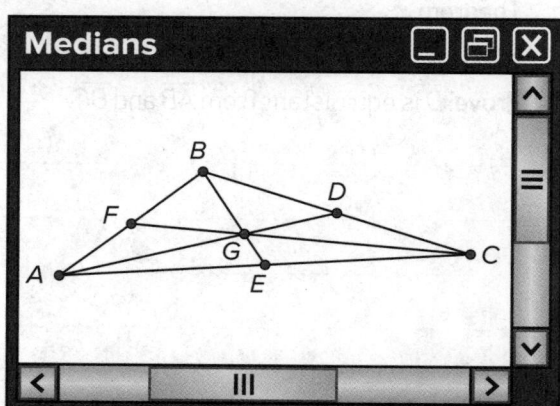

b. **FIND A PATTERN** Use the Measure tool to determine the length of each median.

$AD = $ _____ $BE = $ _____ $CF = $ _____

Next measure the distance from the vertex endpoint of each median to the point of concurrency.

$AG =$ _____ $BG =$ _____ $CG =$ _____

Multiply AD, BE, and CF by $\frac{2}{3}$. What do you find?

_____

c. **MAKE A CONJECTURE** Move one or more of the vertices on your triangle. Repeat **part b**.
Make a conjecture about the relationship between a median of a triangle and the line
segment that connects the centroid to the vertex that lies on the median.

_____

_____

KEY CONCEPT

Complete the statement of the theorem.

The Centroid Theorem
The _____ of a triangle intersect at a point called the _____ that is _____ of the distance from each vertex to the midpoint of the opposite side.

EXAMPLE 2 **Points of Concurrency**

EXPLORE Investigate properties of altitudes using the Geometer's Sketchpad.

a. **USE TOOLS** Draw an acute triangle. Construct the
altitudes of the triangle.

b. **USE TOOLS** Drag one vertex and observe how the
orthocenter moves. When does the orthocenter lie
outside of the triangle?

_____

_____

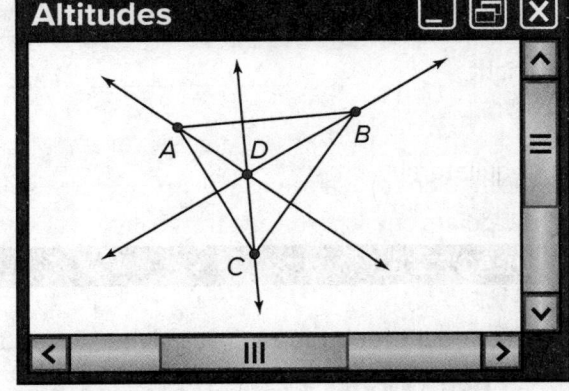

c. **USE TOOLS** Can the orthocenter coincide with
one of the vertices of the triangle? Justify
your answer.

_____

_____

_____

_____

_____

EXAMPLE 3 Compare Points of Concurrency

MAKE A CONJECTURE Draw △ABC and construct its circumcenter. Label the point Circumcenter. Construct the other three points of concurrency and label the points accordingly. Hide the lines and segments used to construct the points of concurrency, but do not hide the midpoints of the segments that form △ABC. Drag the vertices of △ABC to form each triangle listed below. What relationships do you notice among the points of concurrency for each triangle?

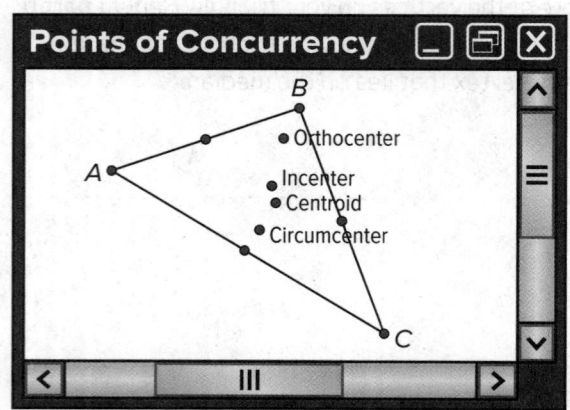

a. isosceles acute _____

b. isosceles obtuse _____
_____

c. scalene obtuse _____

d. scalene acute _____
_____

e. right _____
_____

f. equilateral _____

PRACTICE

1. Construct the altitudes of a triangle using a compass and straightedge.

   a. **USE A MODEL** Construct a triangle using only a straightedge. Label the points A, B, and C.

b. PLAN A SOLUTION Explain how to use a compass and straightedge to construct the altitude from B that intersects $\overleftrightarrow{AC}$. Label all points.

_____

_____

_____

_____

_____

c. USE TOOLS Construct the altitude from B that intersects $\overleftrightarrow{AC}$. Label the point of intersection of $\overleftrightarrow{AC}$ and the altitude point G.

d. Construct the other two altitudes and identify the point of concurrency.

e. COMMUNICATE PRECISELY Compare your altitude $\overline{BG}$ to the ones drawn by your classmates. Explain what conditions allow $\overline{BG}$ to lie within $\triangle ABC$, what conditions allow $\overline{BG}$ to lie outside $\triangle ABC$, and what conditions allow $\overline{BG}$ to lie along one of the sides of $\triangle ABC$.

_____

_____

_____

2. **COMMUNICATE PRECISELY** Explain why it is sometimes necessary to extend the sides of the triangle to construct the altitudes when it was not necessary for the construction of the medians.

_____

_____

_____

_____

3. **CRITIQUE REASONING** Kim drew right triangle ABC with hypotenuse $\overline{AB}$. She then drew altitude $\overline{CD}$. Kim says that $\triangle ACD$ has the same angle measures as $\triangle BCD$. Is she correct? Prove your answer.

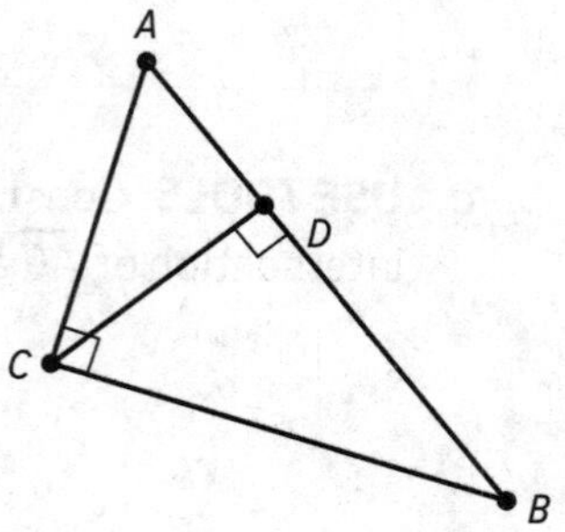

_____

_____

_____

_____

_____

_____

_____

_____

_____

_____

_____

_____

_____

_____

_____

_____

4. **REASON ABSTRACTLY** In $\triangle ABC$, point D is the midpoint of $\overline{BC}$. $\overline{AD}$ is both an altitude and a median of $\triangle ABC$. For each vertex B and C, the median and altitude are distinct line segments. How could you best classify $\triangle ABC$? Draw a sketch and explain your reasoning.

_____

_____

_____

_____

5. **PLAN A SOLUTION** The three sides of a triangular park have lengths 60 yards, 50 yards, and 50 yards. The community wants to construct a fountain in the center of the park.

a. Which point of concurrency would provide the best approximation for the center of the park? Justify your answer.

_____

_____

_____

b. Use a model of the park to recommend an exact location of the fountain. Describe the method you used.

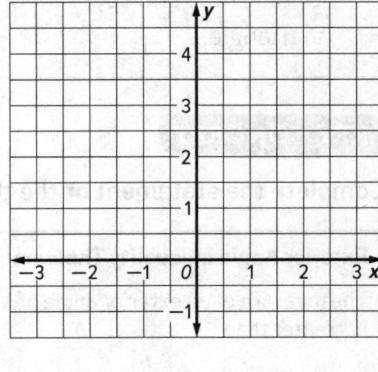

_____

_____

_____

_____

_____

_____

_____

_____

6. Alan challenges Elli to balance a cardboard triangle on the tip of each index finger. The figures in **part b** represent the triangles that she is asked to balance.

a. **INTERPRET PROBLEMS** Which point of concurrency should Elli uses to find the placement for her index finger on the triangles? Explain.

_____

_____

_____

b. **CALCULATE ACCURATELY** On each triangle below, find the point that Elli should use to balance the triangle.

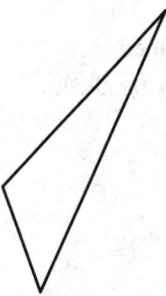

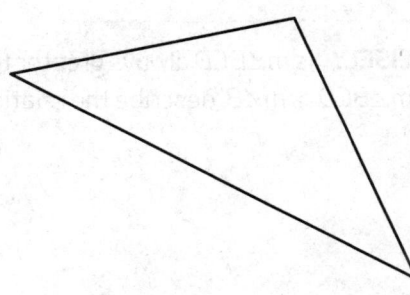

Objectives

- Apply properties of inequalities to the measure of angles and sides of a triangle.
- Understand and apply the Exterior Angle Inequality Theorem.
- Understand and apply theorems about the angle-side relationships in triangles.

KEY CONCEPT

Complete the statement of the theorem.

Exterior Angle Inequality Theorem

The measure of an exterior angle of a triangle is greater than

_____

_____

$m\angle 3 > m\angle 1$ and $m\angle 3 > m\angle 2$

EXAMPLE 1 **Explore the Exterior Angle Inequality Theorem**

EXPLORE Use the Geometer's Sketchpad to create a figure similar to the one below and label the points as shown.

a. Record the measures of $\angle A$, $\angle B$, and $\angle BCD$ below.

$m\angle A =$ _____ $m\angle B =$ _____ $m\angle BCD =$ _____

Move point C. Record the angle measures below.
$m\angle A$ _____ $m\angle B$ _____ $m\angle BCD$ _____

Move point C again and record the angle measures below.
$m\angle A$ _____ $m\angle B$ _____ $m\angle BCD$ _____

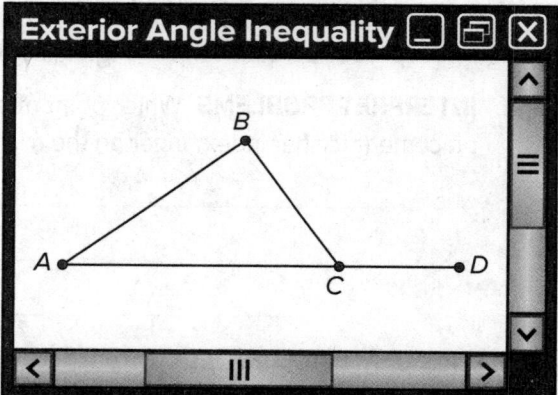

b. **MAKE A CONJECTURE** What can you conclude about the relationship between $m\angle A$ and $m\angle BCD$, and $m\angle B$, and $m\angle BCD$?

_____

c. **COMMUNICATE PRECISELY** Is $m\angle BCD$ always greater than $m\angle BCA$? Since $m\angle BCD > m\angle A$ and $m\angle BCD > m\angle B$, describe the relationship between $m\angle BCA$, $m\angle A$, and $m\angle B$.

_____

_____

_____

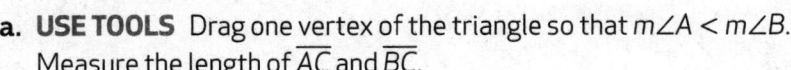 **EXAMPLE 2** **Explore Angle-Side Relationships in Triangles**

EXPLORE Use the Geometer's Sketchpad to create a triangle with vertices A, B, and C, similar to the one the below. Analyze the side and angle relationships.

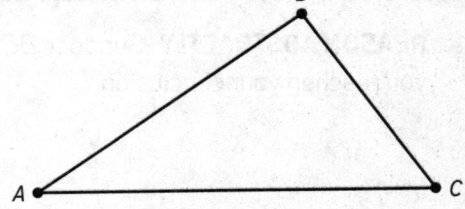

a. USE TOOLS Drag one vertex of the triangle so that $m\angle A < m\angle B$. Measure the length of $\overline{AC}$ and $\overline{BC}$.

$\overline{AC}$ _____ $\overline{BC}$ _____

b. USE TOOLS Drag one vertex of the triangle so that $m\angle A > \angle B$. Measure the length of $\overline{AC}$ and $\overline{BC}$.

$\overline{AC}$ _____ $\overline{BC}$ _____

c. CONSTRUCT ARGUMENTS If $m\angle A > m\angle B$, make a conjecture about the length of $\overline{BC}$ in relation to $\overline{AC}$.

_____

_____

d. USE TOOLS Draw points D and E, and connect them. Measure $\overline{DE}$. Then draw $\overline{DF}$ so that $\overline{DE} > \overline{DF}$. Draw $\overline{EF}$. Measure $m\angle E$ and $m\angle F$ and state the relationship between $m\angle E$ and $m\angle F$ in terms of an inequality.

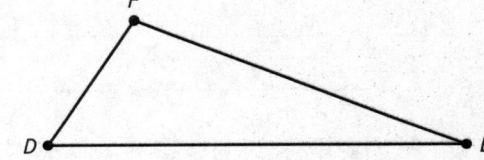

$m\angle E$ _____ $m\angle F$ _____

_____

e. CONSTRUCT ARGUMENTS If $\overline{DE} > \overline{DF}$, make a conjecture about the measure of $\angle E$ in relation to $\angle F$.

_____

_____

KEY CONCEPT

Angle-Side Relationships in Triangles
Complete the statement of each theorem.

Theorem 5.1
If one side of a triangle is longer than another side, then

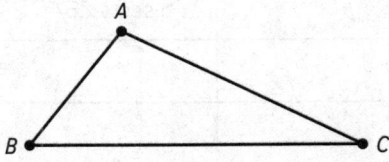

_____

$AC > AB$. Therefore $\angle B > \angle C$.

Theorem 5.2
If one angle of a triangle has a greater measure than another angle, then

_____

_____

$m\angle A > m\angle B$. Therefore $BC > AC$.

EXAMPLE 3 **Relationships within Triangles**

In △ABC, $\overline{AY}$ bisects ∠A, $\overline{BX}$ bisects ∠B, and Q is the intersection of $\overline{AY}$ and $\overline{BX}$.

a. REASON ABSTRACTLY Suppose BC > AC. Compare m∠BAY and m∠ABX. Explain how you reached your conclusion.

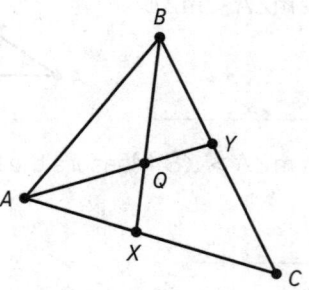

_____

_____

b. COMMUNICATE PRECISELY Claire claims that $\overline{XC} \cong \overline{YC}$ if BC > AC. Is she correct? Explain why or why not.

_____

_____

c. CONSTRUCT ARGUMENTS Complete the following proof.

Given: BC > AC
Prove: BQ > AQ

Statement	Reason
BC > AC	Given
m∠BAC > m∠ABC	
	$\overline{AY}$ bisects ∠BAC
$\frac{1}{2}$m∠BAC > $\frac{1}{2}$m∠ABC	
m∠BAY > m∠ABX	

d. FIND A PATTERN If m∠AQB > m∠BQY, which side of △ABQ has the greatest length? Explain.

_____

_____

_____

1. **CONSTRUCT ARGUMENTS** Prove that if one angle of a triangle has a greater measure than another angle, then the side opposite the greater angle is longer than the side opposite the lesser angle by indirect proof.

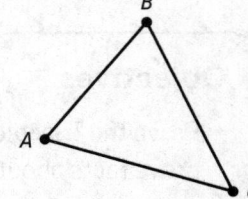

Given: $m\angle B > m\angle C$

Prove: $AC > AB$

Suppose AC is *not* greater than AB. In other words, suppose $AC = AB$ or $AC < AB$.

a. Show that the given relationship $m\angle B > m\angle C$ makes it impossible for $AC = AB$.

_____

_____

_____

b. Show that if $AC < AB$, $m\angle B$ would have to be less than $m\angle C$.

_____

_____

_____

c. What do **parts a** and **b** tell you about $m\angle B$ and $m\angle C$? Explain why this leads to a contradiction and completes the proof.

_____

_____

2. **COMMUNICATE PRECISELY** Suppose $WY > YX$. Show that $m\angle ZWY > m\angle YWX$. Explain each step of your reasoning.

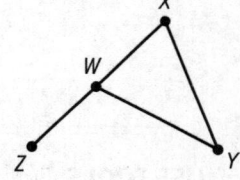

Given: $WY > YX$

Prove: $m\angle ZWY > m\angle YWX$.

_____

_____

_____

3. **CRITIQUE REASONING** Fatima and Aaron are looking at a map of Aaron's farm, which has corners A, B, and C. The edge of his property, marked by $\overline{BC}$, meets his neighbor's property line $\overline{AD}$ so that $\angle BCD$ is acute. Fatima suggests that $\overline{BC}$ is longer than $\overline{AB}$. Is she correct? Explain.

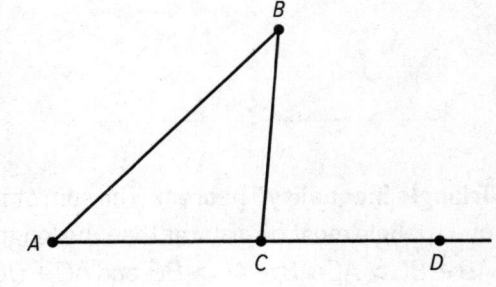

_____

_____

_____

_____

The Triangle Inequality

Objectives

- Prove the Triangle Inequality Theorem and use it to determine and prove facts about specific triangles and related figures.

EXAMPLE 1 **Investigate Triangle Side Relationships**

EXPLORE Use dynamic geometry software to investigate side relationships in a triangle.

a. **USE TOOLS** Plot any three points and label them *A*, *B*, and *C*. Join them to form a triangle. Measure each side of the triangle. Complete the table.

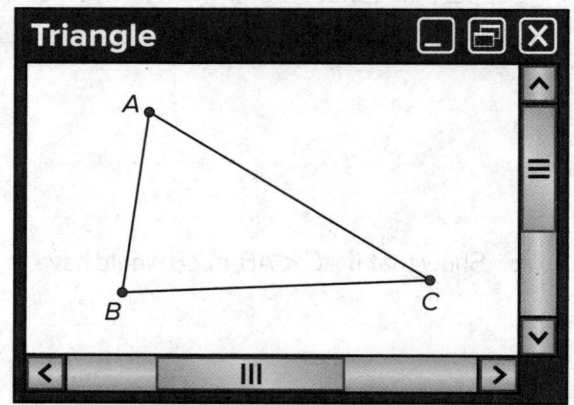

First Triangle (part a)				Second Triangle (part b)			
AB		AC + BC		AB		AC + BC	
BC		AB + AC		BC		AB + AC	
AC		AB + BC		AC		AB + BC	

b. **MAKE A CONJECTURE** Move the three points to form a new triangle. Record your results in the table. What do you notice about the sums of pairs of side lengths?

_____

_____

_____

c. **USE TOOLS** Drag point *A* toward the segment $\overline{BC}$. What do you notice about the sum of the segments and their lengths? What does the figure look like when the sum of two of the segments equals the length of the third segment?

_____

_____

_____

_____

Triangle Inequality Theorem The sum of the lengths of any two sides of a triangle must be greater than the length of the third side. In $\triangle ABC$, $AB + BC > AC$, $AB + AC > BC$, and $AC + BC > AB$.

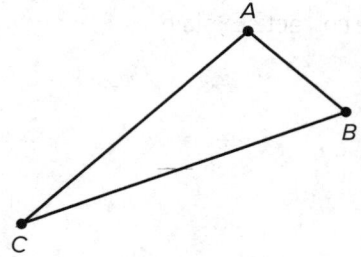

EXAMPLE 2 **Prove the Triangle Inequality Theorem**

Given: $\triangle ABC$
Prove: $AB + BC > AC$

a. **USE TOOLS** Copy $\triangle ABC$ from the bottom of the previous page. Use a compass and straightedge to extend CB to point D, so that $BD = AB$. Connect A and D to form $\triangle ABD$.

b. Complete the paragraph proof.

_____

_____

_____

_____

_____

_____

c. **COMMUNICATE PRECISELY** A classmate mentions that a separate proof is needed for obtuse triangles. Use a sketch to help explain why or why not.

_____

_____

_____

_____

EXAMPLE 3 **Apply the Triangle Inequality Theorem**

Carol is planning to fence a triangular plot of land. Two of the sides of the plot measure 230 yard and 490 yard. What are the maximum and minimum lengths of fencing Carol will need?

a. **PLAN A SOLUTION** To find the maximum or minimum lengths of fencing, we need to find the maximum and minimum values for which side? Explain.

_____

_____

b. **REASON QUANTITATIVELY** Find the maximum length of fencing.

_____

_____

c. COMMUNICATE PRECISELY Write an inequality to model the length of fencing. Justify the lower value.

_____

_____

_____

d. FIND A PATTERN What would have to be true about the two known side lengths of the plot of land for the lower value in **part c** to be zero? Explain.

_____

_____

_____

EXAMPLE 4 **Use the Triangle Inequality Theorem to Prove a Relationship for Rectangles**

Prove that the perimeter of any rectangle is greater than the sum of its diagonals.

a. PLAN A SOLUTION Given rectangle *JKLM*, state the conclusion of this result as an inequality, and briefly describe a strategy to prove the result.

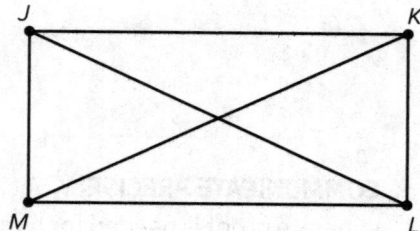

_____

_____

_____

b. CONSTRUCT ARGUMENTS Follow your strategy to prove the result.

Statement	Reason
1.	
2.	
3.	
4.	
5.	
6.	

PRACTICE

1. Enrique and Kelly are discussing the fastest way to get to school. If they choose to use the sidewalk, they would need to walk south for 1 mile, and then east for half a mile. Or they can cut through a field in a straight line from their location to the school. They know that they can walk at a rate of 5 miles per hour on the sidewalk and 4 miles per hour through the field.

   a. CRITIQUE REASONING Kelly explains to Enrique that they would need to know the distance of the path through the field in order to decide which route is longer. Is this true? Explain.

   _____

   _____

   _____

b. USE STRUCTURE Use the Triangle Inequality Theorem to determine the maximum and minimum amounts of time it would take to travel across the field. Compare these times to the time it would take to use the sidewalk.

_____

_____

_____

_____

_____

2. **CONSTRUCT ARGUMENTS** Recall that the shortest distance from a point to a line is the length of the perpendicular from the point to the line.

   a. Given: $\triangle ABC$ is acute. D is a point on $\overline{BC}$ such that $\overline{AD} \perp \overline{BC}$.
   Prove: Side $\overline{AB}$ is longer than $\overline{BD}$.

_____

_____

   b. Use **part a** to prove the Triangle Inequality Theorem for an acute triangle.

_____

_____

3. **REASON QUANTITATIVELY** Based on the information in $\triangle JKL$, determine the range of values of x.

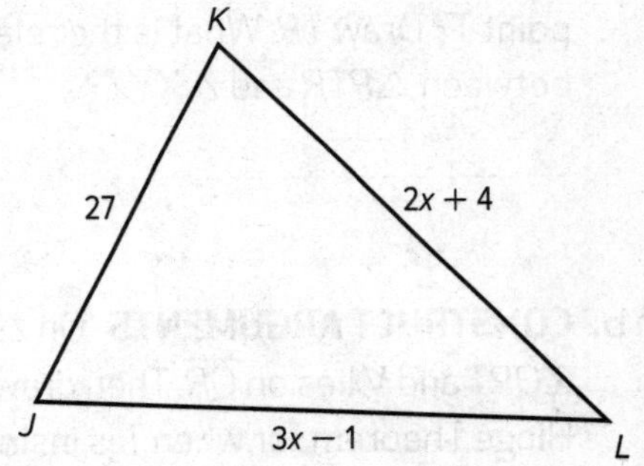

   a. Write three inequalities for x.

_____

_____

_____

   b. CALCULATE ACCURATELY Solve each inequality.

_____   _____

_____   _____

_____   _____

   c. REASON QUANTITATIVELY Use the inequalities to determine the range of values for x.

_____

   d. USE STRUCTURE Was each inequality from **part b** stated in your answer to **part c**? Why or why not?

_____

_____

Inequalities in Two Triangles

Use with Lesson 5-6

Objectives

- Prove the Hinge Theorem
- Apply the Hinge Theorem and its converse to real-world situations.

Mathematical Practices
1, 2, 3, 4, 7

Hinge Theorem If two sides of a triangle are congruent to two sides of another triangle, and the included angle of the first is greater than the included angle of the second triangle, then the third side of the first triangle is longer than the third side of the second triangle.

Converse of the Hinge Theorem If two sides of a triangle are congruent to two sides of another triangle, and the third side of the first triangle is longer than the third side of the second triangle, then the included angle of the first is greater than the included angle of the second.

EXAMPLE 1 Prove the Hinge Theorem

Given: $\triangle PQR$ and $\triangle XYZ$, $\overline{PQ} \cong \overline{XY}$, $\overline{PR} \cong \overline{XZ}$, $m\angle QPR > m\angle XYZ$
Prove: $QR > YZ$

a. PLAN A SOLUTION On $\triangle PQR$ draw $\overrightarrow{PT}$ such that $\angle TPR \cong \angle YXZ$ and $\overline{PT} \cong \overline{YZ}$. Label the point of intersection of $\overrightarrow{PT}$ and $\overline{QR}$ as S. Depending on the shape of the triangles, what conclusion can you make about the location of point T? Draw $\overline{TR}$. What is the relationship between $\triangle PTR$ and $\triangle XYZ$?

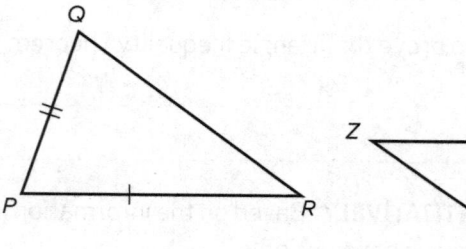
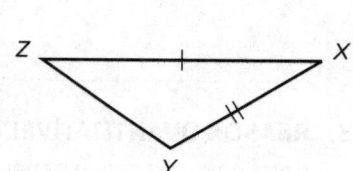

_____

_____

b. CONSTRUCT ARGUMENTS On $\triangle PQR$ in the diagram above, draw $\overrightarrow{PV}$ so that it bisects $\angle QPT$ and V lies on $\overline{QR}$. Then draw $\overline{VT}$ and $\overline{TR}$. Complete the two-column proof of the Hinge Theorem for when T is inside $\triangle PQR$.

Statements	Reasons
1. $\overline{PQ} \cong \overline{XY}$, $\overline{XY} \cong \overline{PT}$	1. Given
2. $\overline{PQ} \cong \overline{PT}$	2.
3.	3. Def. of $\angle$ bisector
4. $\overline{PV} \cong \overline{PV}$	4.
5.	5. SAS
6. $\overline{QV} \cong \overline{TV}$	6.
7. $VR + TV > TR$	7.
8.	8. Substitution
9. $QR > TR$	9.
10. $\triangle PTR \cong \triangle XYZ$	10. SAS (see part a above)
11. $\overline{TR} \cong \overline{YZ}$	11. Given
12. $QR > YZ$	12.

Copyright © McGraw-Hill Education

Copyright © McGraw-Hill Education

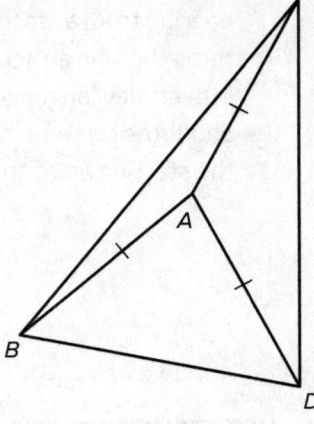

EXAMPLE 2 **Use the Hinge Theorem and Its Converse to Prove Relationships**

$\overline{AB}$, $\overline{AC}$, and $\overline{AD}$ are congruent, and $m\angle BAD < m\angle CAD < m\angle BAC$.

a. CONSTRUCT ARGUMENTS Order $\angle BCD$, $\angle CDB$, and $\angle DBC$ from smallest to largest measure. Prove the result.

_____

_____

_____

_____

_____

_____

_____

b. MAKE A CONJECTURE Suppose an arbitrary scalene triangle PQR has circumcenter S. If $m\angle QPR < m\angle PQR < m\angle PRQ$, what is the relationship between $\angle QSR$, $\angle PSR$, and $\angle PSQ$?

_____

_____

_____

PRACTICE

1. Two airplanes take off from the same airstrip. The first plane flies west for 150 miles, and then flies 30° southwest for 220 miles. The second plane flies east for 220 miles and then flies $\theta°$ southeast for 150 miles.

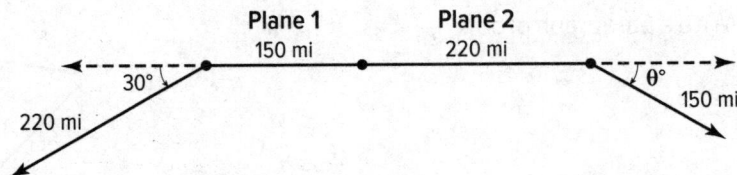

a. REASON QUANTITATIVELY If $\theta < 30$, which plane is farther from the airstrip after the second leg? Justify your answer.

_____

_____

_____

b. USE A MODEL Suppose the first plane is farther from the airstrip after both planes complete their second leg. Write an inequality for θ and justify your answer.

_____

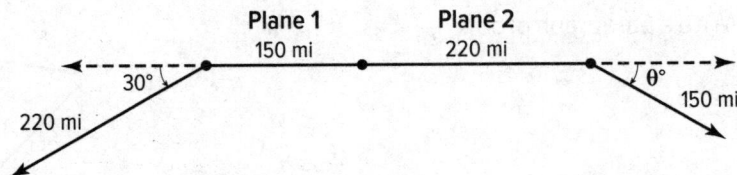

c. USE STRUCTURE Suppose a plane flies on one bearing from a starting point to point x and then turns through an acute angle and flies to point y. In everyday language, state a general conclusion about the relationship between the distance from the starting point to y and the angle of the turn.

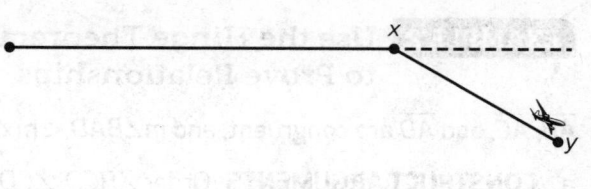

_____

_____

_____

2. USE STRUCTURE Under each figure, fill in the blank to complete the inequality. Justify your reasoning.

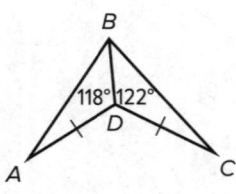

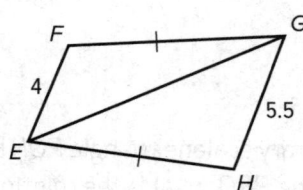

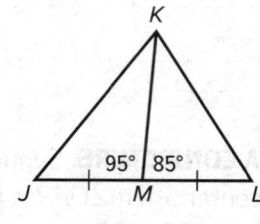

AB _____ BC

m∠GEH _____ m∠FGE

KL _____ JK

_____

_____

_____

3. Consider the quadrilateral *EFGH*.

a. CONSTRUCT ARGUMENTS Write a paragraph proof.

Given: $\overline{EF} \cong \overline{GH}$, $m\angle F > m\angle G$.
Prove: $EG > FH$

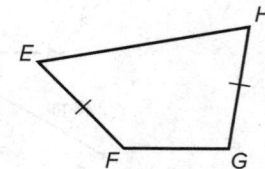

_____

_____

_____

b. USE STRUCTURE Suppose $\overline{JM} \cong \overline{KL}$ and $\overline{JK} > \overline{ML}$. What can you say about the relationship between ∠JMK and ∠MKL? Explain.

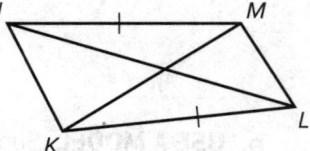

_____

_____

_____

_____

4. **INTERPRET PROBLEMS** Submarine A is moving toward the most recent position it has for submarine B. It sails due east for 38 kilometers and then sails 52° north of east for 25 kilometers, arriving at submarine B's starting position. Meanwhile, submarine B sails 38° south of west for 25 kilometers and then due west for 38 kilometers.

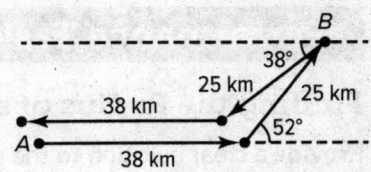

a. Does Submarine B end up northeast, northwest, southeast, or southwest of submarine A's starting point? Explain.

_____

_____

_____

_____

b. If the lengths of sub B's legs were switched, how much could you say about sub B's final position?

_____

_____

5. Jim and Ari start hiking from the same point. Jim hikes 5 miles due east and turns to hike 4.5 miles 30° south of east. Ari hikes 5 miles due west and turns to hike 4.5 miles 15° north of west.

a. **USE A MODEL** Draw a model to represent the situation.

b. **CONSTRUCT ARGUMENTS** Show that Jim is closer to the starting point than Ari is.

_____

_____

_____

_____

_____

c. **CRITIQUE REASONING** Ari notices that he hiked further than Jim. He suggests that if he would have hiked 4 miles instead of 5 miles at the start, he could use the Hinge theorem to show that he hiked a shorter distance than Jim. Critique Ari's reasoning.

_____

_____

_____

Finding the Radius of a Chariot Wheel

Provide a clear solution to the problem. Be sure to show all of your work, include all relevant drawings, and justify your answers.

A group of archeologists found a section of what they believe to be an ancient chariot wheel. They want to find the length of its radius to compare the wheel to others found in the area. They put the section on a centimeter grid to get the coordinates of three points as shown.

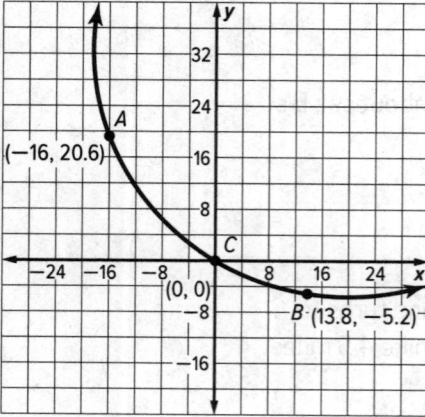

Part A

Which of the following is going to help the archeologists find the radius of the wheel: centroid, incenter, circumcenter, or orthocenter? Explain your answer.

Part B

Use △ACB to find the center of the wheel. Explain your solution process. Round your answer to the nearest tenth of a centimeter. Add to the drawing on the previous page to show your solution.

Part C

What is the length of the radius of the wheel? Explain. Round your answer to the nearest tenth of a centimeter.

Part D

Another wheel section found nearby had a radius of 45 cm. Do you think this wheel section came from the same type of chariot as the first section? Find the difference of the two lengths and their percent difference. Explain your reasoning.

Finding a Shortest Distance

Provide a clear solution to the problem. **Be sure to show all of your work,**
include all relevant drawings, and justify your answers.

A park is in the shape of a right triangle as shown below. The city council wants to
build paths from point *P* to points *A*, *B*, and *C* such that the total length of the three
paths is as short as possible. Where should point *P* be placed?

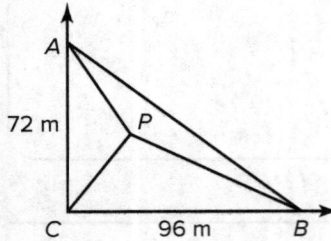

Part A
What would be the total length *PA* + *PB* + *PC* if point *P* were at *C*? Do you think the town
would have a shorter total distance by placing point *P* at the circumcenter of triangle
ABC? Explain your answer.

Part B

Position the diagram of the park on a coordinate grid so point *C* is at the origin and each square represents a length of 1 meter. Place point *P* at (20, 20). Does this give a shorter total distance? Explain your answer. Round all distances to the nearest tenth.

Part C

Place point *P* at the centroid of the triangle. Does this give a shorter total distance? Explain your answer. Round all distances to the nearest tenth.

Part D

Based on your results, can you conclude that the shortest distance you have found is the actual shortest possible distance? Explain your answer.

1. △DEF has vertices D(2, 6), E(−3, 2), and F(−3, 9). What are the coordinates of the orthocenter of △DEF?

( , ☐)

2. Select the side lengths that could be used to form a triangle.

4 mm, 8 mm, 5 cm

12 in., 25 in., 15 in.

45 m, 22 m, 20 m

15 in., 2 ft, 9 in.

3 yd, 4 ft, 7 ft

3. Consider the following diagram.

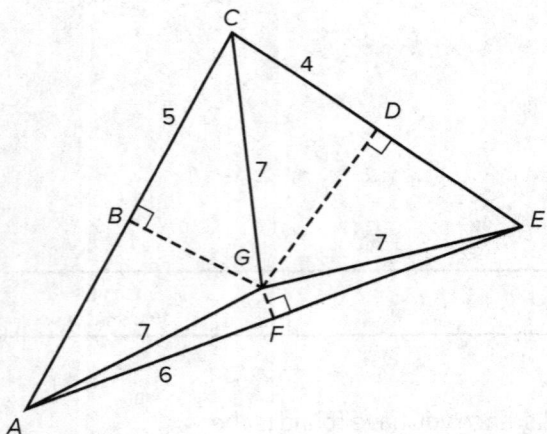

The perimeter of △ACE is ☐ units.

4. If XY = 31.8 inches, XZ = 40.6 inches, and YZ = 37.1 inches, then the angles of △XYZ from smallest to largest are ☐ .

5. If x is an integer, and 5 < x < 11, what are the possible values of x that would allow a triangle to be constructed with side lengths of 5, 11, and x?

☐

6. Which of the following statements are true for the following diagram?

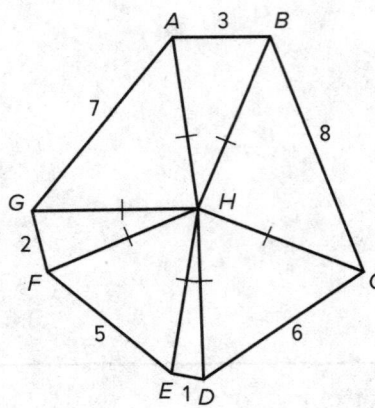

∠AHG ≅ ∠GHE   m∠AHB > m∠FHE

m∠BHC > m∠CHD   m∠DHC = 2m∠AHB

m∠GHF > m∠AHB   m∠AHB < m∠FHE

7. Complete the following.

To construct the centroid of △JKL, first construct the ☐ of $\overline{JK}$, $\overline{KL}$, and $\overline{JL}$. Draw line segments connecting each ☐ to the ☐ vertex. The point of ☐ is the centroid of the triangle.

8. Consider the following diagram.

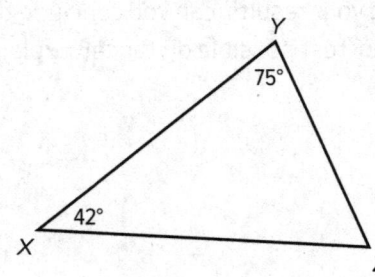

Complete each statement with >, <, or =.

XY ☐ YZ   XY ☐ XZ

XZ ☐ XY + YZ   YX + XZ ☐ XY + YZ

9. Consider the following diagram. Is $\triangle ABD \cong \triangle BCD$? Explain.

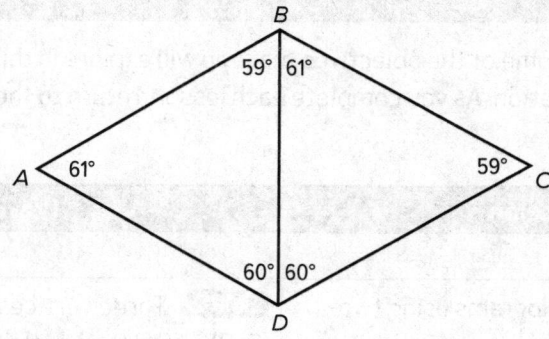

_____

_____

10. In the diagram below, $\overline{KN}$, $\overline{JM}$, and $\overline{LO}$ are the medians of $\triangle JLN$. What is the perimeter of $\triangle JKP$? Justify your answer.

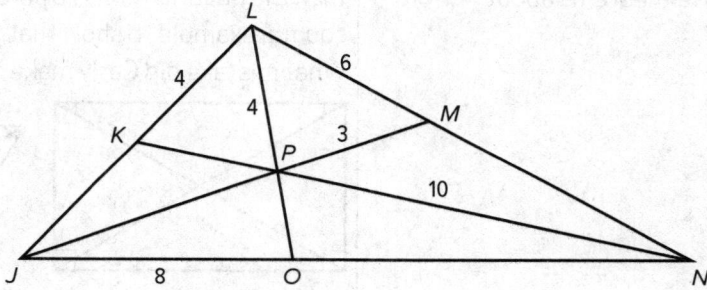

_____

_____

11. In the diagram below, $\overline{AD}$ is the bisector of $\angle BAC$, and $\overline{CD}$ is the bisector of $\angle BCA$. What is $m\angle ADC$? Show your work.

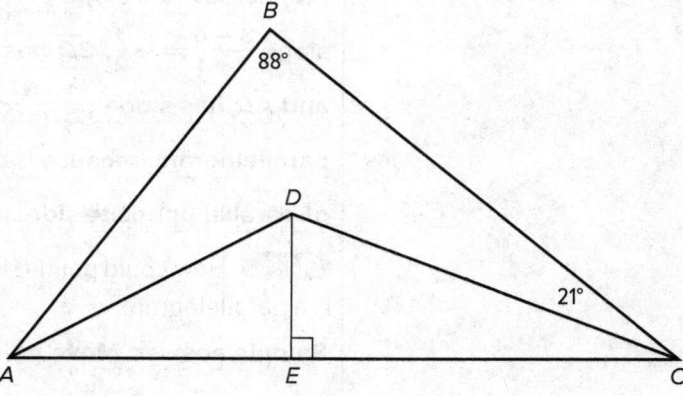

_____

_____

CHAPTER FOCUS Learn about some of the objectives that you will explore in this chapter. Answer the preview question. As you complete each lesson, return to these pages to check your work.

What You Will Learn	Preview Question
Parallelograms	
• Prove theorems about parallelograms using two-column and paragraph proofs. • Use coordinates to prove theorems about parallelograms.	SMP 2 Three vertices of a parallelogram are $(0, 4)$, $(5, 0)$, and $(10, 4)$. List all possible locations of the fourth vertex. **(15, 0), (−5, 0), and (5, 8)**
Proving Theorems About Parallelograms	
• Prove theorems about parallelograms by making formal geometric constructions. • Use coordinates to prove theorems about parallelograms.	SMP 3 Carly drew the following figure to prove that if the diagonals of a quadrilateral are congruent, then the quadrilateral is a parallelogram. Draw a counterexample to show that Carly is incorrect. What mistake did Carly make? 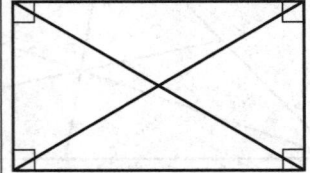  **You cannot prove a general statement with an example.** SMP 2 The vertices of quadrilateral $ABCD$ are $A(-2, 3)$, $B(1, 6)$, $C(7, 3)$, and $D(5, 1)$. Find the slope of each side and determine if $ABCD$ is a parallelogram. Explain your reasoning. **Slopes:** $\overline{AB}$ **has slope** $\frac{6-3}{1-(-2)} = 1$, $\overline{BC}$ **has slope** $\frac{3-6}{7-1} = -\frac{1}{2}$, $\overline{CD}$ **has slope** $\frac{3-1}{7-5} = 1$, **and** $\overline{AD}$ **has slope** $\frac{1-3}{5-(-2)} = -\frac{2}{7}$. $ABCD$ **is not a parallelogram because it does not have two pairs of parallel opposite sides.** SMP 2 How could point C be moved so that $ABCD$ is a parallelogram? **Sample answer: Move point C to $C(8, 4)$; then the slope of** $\overline{BC}$ **is** $\frac{4-6}{8-1} = -\frac{2}{7}$, **and the slope of** $\overline{CD}$ **is unchanged. Then $ABCD$ has two pairs of parallel opposite sides, so it is a parallelogram.**

What You Will Learn	Preview Question

Rectangles

- Prove theorems about rectangles using two-column proofs.
- Use coordinates to prove theorems about rectangles.
- Make formal geometric constructions to understand theorems about rectangles.

SMP 2 *ABGH* is a rectangle and *CDEF* is a parallelogram. Can you conclude *BCFG* is a rectangle? Explain.

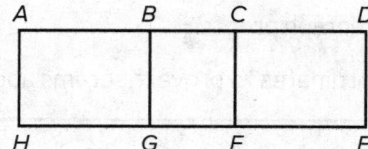

Rhombi and Squares

- Determine whether a figure defined by four points on a coordinate plane is a rhombus or square.
- Prove theorems about rhombi and squares.
- Construct rhombi and squares.

SMP 6 Classify the quadrilateral shown on the coordinate grid. Explain.

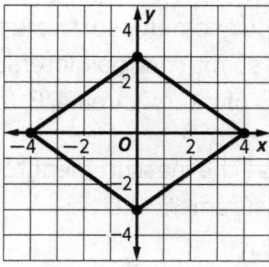

Trapezoids and Kites

- Determine whether a figure defined by four points is a trapezoid or kite.
- Prove theorems about trapezoids and kites using coordinates.

SMP 2 What are the coordinates of the endpoints of the midsegment of the trapezoid?

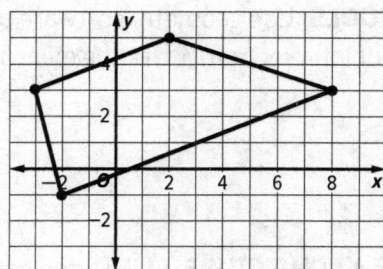

Objectives

Mathematical Practices
1, 2, 3, 5, 6, 7, 8

- Prove theorems about parallelograms using two-column and paragraph proofs.
- Use coordinates to prove theorems about parallelograms.

A **parallelogram** is a quadrilateral with both pairs of opposite sides parallel to one another.

EXAMPLE 1 Investigate the Properties of Parallelograms

EXPLORE Use geometry software to explore parallelograms. As you explore, think about what relationships are true for all parallelograms.

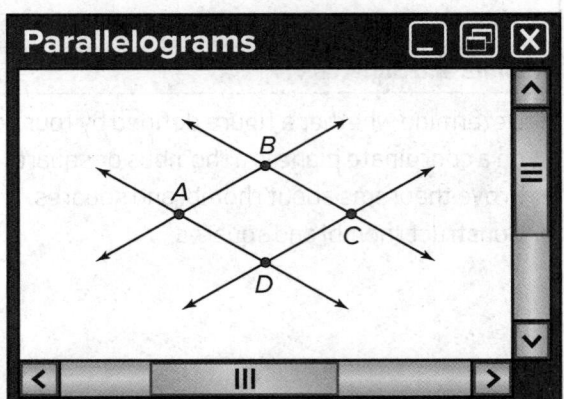

a. **USE TOOLS** Use geometry software to draw two pairs of parallel lines so that one pair intersects the other. Label the points of intersection A, B, C, and D.

b. **USE TOOLS** Use the measurement tools in the software to find the measurements listed.

AB _____ BC _____ CD _____ DA _____

∠ABC _____ ∠BCD _____ ∠CDA _____ ∠DAB _____

c. **MAKE A CONJECTURE** Make a conjecture about opposite angles and opposite sides in a parallelogram.

_____

_____

d. **USE TOOLS** Use geometry software to construct the diagonals of ABCD. Label the point of intersection M. Use the measurement tools to find the measurements listed.

AM _____ MC _____ DM _____ MB _____

e. **MAKE A CONJECTURE** Make a conjecture about the diagonals of a parallelogram.

_____

_____

f. FIND A PATTERN Manipulate the parallelogram you constructed in **part a**. Are the relationships that you noticed the same?

_____

_____

_____

Several properties are true for all parallelograms. All of these properties can be proved using definitions, properties, and theorems that you already know.

KEY CONCEPT

Complete the table by writing the complete theorem that corresponds to each abbreviation.

Theorem	Statement	Abbreviation
6.3		Opp. sides of a ▱ are ≅.
6.4		Opp. ∠s of a ▱ are ≅.
6.5		Cons. ∠s in a ▱ are supplementary.
6.6		If a ▱ has 1 rt. ∠, it has 4 rt. ∠s.
6.7		Diag. of a ▱ bisect each other.
6.8		Diag. separates a ▱ into 2 ≅ △s.

EXAMPLE 2 **Prove That Opposite Angles of a Parallelogram Are Congruent**

Plan and complete a two-column proof of Theorem 6.4: *If a quadrilateral is a parallelogram, then its opposite angles are congruent.*

a. PLAN A SOLUTION If you wanted to prove that $\angle P \cong \angle R$ using CPCTC, how could you alter the diagram at the right to assist in your proof? What fact about points and lines justifies your alteration?

_____

_____

_____

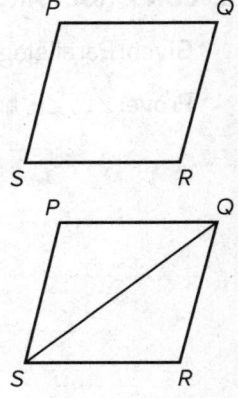

Parallelograms **189**

b. CONSTRUCT ARGUMENTS Fill in the missing statements and reasons to complete the proof.

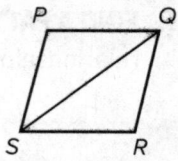

Given: Parallelogram *PQRS*
Prove: ∠*P* ≅ ∠*R*

Statements	Reasons
1. *PQRS* is a parallelogram.	**1.**
2. $\overline{PQ} \parallel \overline{RS}$ and $\overline{QR} \parallel \overline{SP}$	**2.**
3. ∠*PSQ* ≅ ∠*RQS* and ∠*PQS* ≅ ∠*RSQ*	**3.**
4.	**4.** Reflexive Property of Congruence
5. △*PQS* ≅ △*RSQ*	**5.**
6. ∠*P* ≅ ∠*R*	**6.**

c. DESCRIBE A METHOD How could you continue the proof to prove ∠*Q* ≅ ∠*S*?

_____

_____

EXAMPLE 3 **Prove that Consecutive Angles of a Parallelogram are Supplementary**

Plan and write a paragraph proof of Theorem 6.5: *If a quadrilateral is a parallelogram, then its consecutive angles are supplementary.*

a. CONSTRUCT ARGUMENTS Complete the paragraph proof.

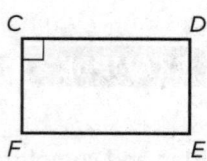

Given: *JKLM* is a parallelogram.
Prove: ∠*J* and ∠*K*, ∠*K* and ∠*L*, ∠*L* and ∠*M*, and ∠*M* and ∠*J* are supplementary.

It is given that *JKLM* is a parallelogram, so ____ ‖ ____ and ____ ‖ ____. When two parallel lines are cut by a transversal, _____ are supplementary. Therefore ∠*J* and ____, ∠*K* and ____, ∠*L* and ____, and ∠*M* and ____ are supplementary.

EXAMPLE 4 **Prove Right Angles in Parallelograms**

Write a paragraph proof of Theorem 6.6: *If a parallelogram has one right angle, then it has four right angles.*

a. CONSTRUCT ARGUMENTS Write a paragraph proof.

Given: Parallelogram *CDEF*; ∠*C* is a right angle.

Prove: ∠*D*, ∠*E*, and ∠*F* are right angles

_____

_____

_____

_____

_____

EXAMPLE 5	**Prove That Diagonals of a Parallelogram Bisect Each Other**

Use algebra to prove Theorem 6.7: *If a quadrilateral is a parallelogram, then its diagonals bisect each other.*

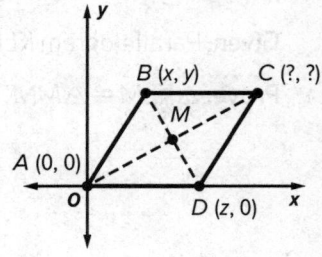

a. REASON ABSTRACTLY You know that opposite sides of a parallelogram are parallel and that parallel lines have equal slopes. How can this information help you find the coordinates of point *C* in parallelogram *ABCD*?

_____

_____

_____

b. CALCULATE ACCURATELY What are the midpoints of $\overline{AC}$ and $\overline{BD}$?

_____

c. COMMUNICATE PRECISELY If $\overline{AC}$ and $\overline{BD}$ have the same midpoint, how does that show that the diagonals bisect each other?

_____

_____

_____

PRACTICE

1. **CONSTRUCT ARGUMENTS** Prove Theorem 6.3: *If a quadrilateral is a parallelogram, then its opposite sides are congruent.*

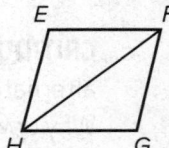

a. Fill in the missing statements and reasons.

Given: Parallelogram *EFGH*
Prove: $\overline{EF} \cong \overline{GH}$ and $\overline{HE} \cong \overline{FG}$

Statements	Reasons
1.	1. Given
2. $\overline{EF} \parallel \overline{GH}$ and $\overline{HE} \parallel \overline{FG}$	2.
3.	3. Alt. Int. ∠s Thm.
4. $\overline{FH} \cong \overline{FH}$	4.
5.	5. ASA
6. $\overline{EF} \cong \overline{GH}$ and $\overline{HE} \cong \overline{FG}$	6.

b. Explain why this proof is true for all parallelograms.

_____

_____

2. **CONSTRUCT ARGUMENTS** Write a paragraph proof of Theorem 6.8:
If a quadrilateral is a parallelogram, then each diagonal separates the parallelogram into two congruent triangles.

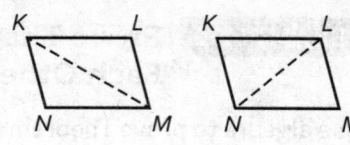

Given: Parallelogram *KLMN*

Prove: △*KLM* ≅ △*MNK* and △*NKL* ≅ △*LMN*

_____

_____

_____

3. Jasmine sketched a parallelogram on a coordinate plane as shown in the diagram.

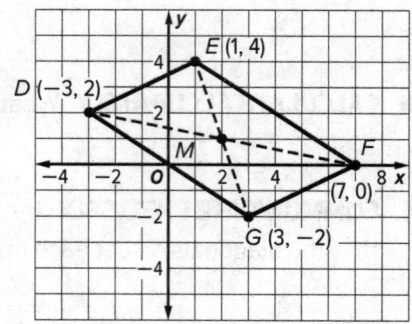

a. **USE STRUCTURE** Show how she can use algebra to show that the opposite sides of the parallelogram are congruent.

_____

_____

_____

b. **USE STRUCTURE** Show how she can use algebra to show that the diagonals bisect each other.

_____

_____

_____

c. **CRITIQUE REASONING** Raj suggests to Jasmine that she has found alternative proofs to Theorems 6.3 and 6.7 using algebra. Is he correct? Why or why not?

_____

_____

_____

d. **PLAN A SOLUTION** How could Jasmine alter her *DEFG* so that **parts a** and **b** are valid proofs of Theorems 6.3 and 6.7?

_____

_____

_____

_____

4. Below is a two-column proof of Theorem 6.7: *If a quadrilateral is a parallelogram, then its diagonals bisect each other.*

Given: Parallelogram WXZY

Prove: $\overline{WM} \cong \overline{MZ}$ and $\overline{XM} \cong \overline{MY}$

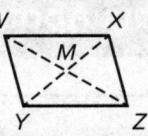

Statements	Reasons
1. WXZY is a parallelogram.	**1.** Given
2. $\overline{WX} \parallel \overline{ZY}$ and $\overline{WY} \parallel \overline{XZ}$	**2.** Definition of parallelogram
3. $\angle XWZ \cong \angle YZW$ and $\angle XYZ \cong \angle WXM$	**3.** Alternate Interior Angles Theorem
4. $\angle WMX \cong \angle YMZ$	**4.** Vertical angles are congruent.
5. $\triangle WXM \cong \triangle ZYM$	**5.** AAA
6. $\overline{WM} \cong \overline{MZ}$ and $\overline{XM} \cong \overline{MY}$	**6.** CPCTC

a. CRITIQUE REASONING What is the error in the proof?

b. CONSTRUCT ARGUMENTS How would you correct the error?

c. Rewrite the proof with your edits.

Statements	Reasons
1.	**1.** Given
2.	**2.** Definition of parallelogram
3.	**3.** Alternate Interior Angles Theorem
4.	**4.** Theorem 6.3
5.	**5.**
6.	**6.** CPCTC

5. FIND A PATTERN Maple Street and Elm Street are parallel. Jefferson Avenue and McKinley Avenue are parallel. Jon works at a pizza restaurant on the corner of Jefferson Avenue and Maple Street. He needs to deliver a pizza to a house on the corner of McKinley Avenue and Elm Street. Jon is deciding whether to travel on Maple Street and McKinley Avenue or on Jefferson Avenue and Elm Street. If he wants to travel the shortest distance, which route should he choose? Explain your reasoning.

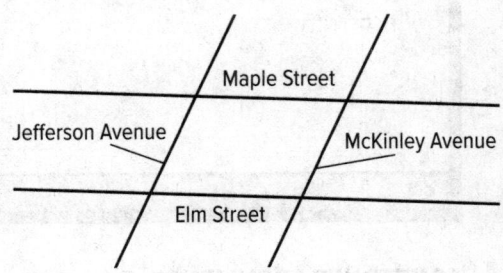

Proving Theorems About Parallelograms

Use with Lesson 6-3

Objectives

- Prove theorems about parallelograms by making formal geometric constructions.
- Use coordinates to prove theorems about parallelograms.

By the definition of a parallelogram, if both pairs of opposite sides of a quadrilateral are parallel, then it is a parallelogram. So, to prove that a quadrilateral is a parallelogram, show that both pairs of opposite sides are parallel.

EXAMPLE 1   Investigate Conditions of Parallelograms

EXPLORE Use dynamic geometry software to explore parallelograms. As you do so, think about different ways to use opposite sides to prove that a quadrilateral is a parallelogram.

a. USE TOOLS Use The Geometer's Sketchpad to draw two segments that share an endpoint. Label these as $\overline{EF}$ and $\overline{FG}$, as shown below on the left.

b. USE TOOLS Draw a circle using the "Circle by Center + Radius" tool with center E and radius FG. Draw a second circle in the same way with center G and radius EF, as shown below on the right. Label the point of intersection of the circles H. Construct $\overline{GH}$ and $\overline{HE}$. Then select and hide the circles.

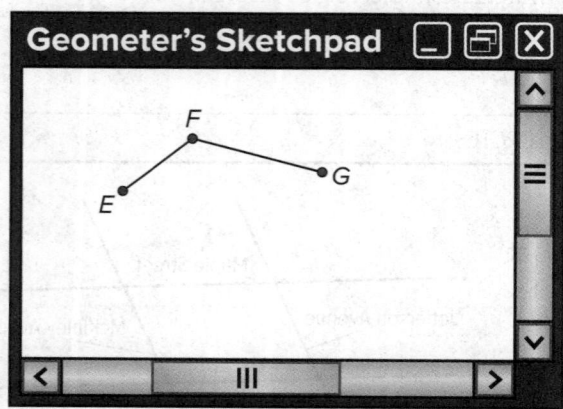

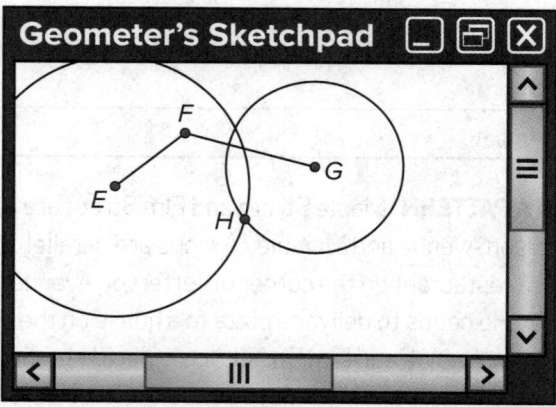

c. CONSTRUCT ARGUMENTS Explain why $\overline{EF}$ and $\overline{GH}$ are congruent and why $\overline{FG}$ and $\overline{HE}$ are congruent.

_____

_____

d. USE TOOLS Use the slope tool to find the slopes of $\overline{EF}$, $\overline{FG}$, $\overline{GH}$, and $\overline{HE}$. What can you conclude about the opposite sides of $EFGH$?

_____

e. **MAKE A CONJECTURE** What is a reasonable conjecture about parallelograms based on your exploration of quadrilateral *EFGH*?

_____

_____

Showing that opposite sides are parallel is just one way to prove that a quadrilateral is a parallelogram. There are other conditions that ensure a quadrilateral is a parallelogram as well. Remember that only one condition needs to be satisfied to complete a proof.

KEY CONCEPT

Complete the table by writing the complete theorem that corresponds to each abbreviation.

Theorem	Statement	Abbreviation
6.9		If both pairs of opp. sides are ≅, then quad. is a ▱.
6.10		If both pairs of opp. ∠s are ≅, then quad. is a ▱.
6.11		If diag. bisect each other, then quad. is a ▱.
6.12		If one pair of opp. sides is ≅ and ∥, then quad. is a ▱.

EXAMPLE 2 **Prove That a Quadrilateral Is a Parallelogram**

Complete the two-column proof to show that if both pairs of opposite sides are congruent, then a quadrilateral is a parallelogram.

a. **CONSTRUCT ARGUMENTS** Fill in the missing statements and reasons to complete the proof.

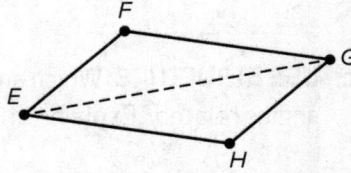

Given: $\overline{EF} \cong \overline{GH}$, $\overline{FG} \cong \overline{EH}$
Prove: *EFGH* is a parallelogram.

Statements	Reasons
1. Draw $\overline{EG}$.	1. Through any two points there is exactly one line.
2. $\overline{EF} \cong \overline{GH}$, $\overline{FG} \cong \overline{EH}$	2. Given
3.	3. Reflexive Property of Congruence
4. $\triangle EFG \cong \triangle GHE$	4.
5.	5. CPCTC
6. $\overline{EF} \parallel \overline{GH}$, $\overline{FG} \parallel \overline{EH}$	6.
7. *EFGH* is a parallelogram.	7. A parallelogram is a quadrilateral with both pairs of opposite sides parallel.

b. **CRITIQUE REASONING** A student said that because $\angle F \cong \angle H$, it can also be shown that $\triangle EFG \cong \triangle GHE$ by SAS. Do you agree? Justify your answer.

_____

c. CONSTRUCT ARGUMENTS Explain why Statement 3 is necessary.

_____

_____

_____

d. CONSTRUCT ARGUMENTS Describe a general strategy for proving that opposite sides are parallel once it has been shown that the triangles formed by drawing the diagonal are congruent.

_____

_____

_____

You can solve real-world problems by proving that quadrilaterals are paralllograms.

EXAMPLE 3 **Solve a Real-World Problem**

Jane is assembling an accordian drying rack that can be folded flat or opened up to various heights as shown. In the figure, E is the midpoint of $\overline{AC}$ and $\overline{BD}$. Jane wants to show that figure $ABCD$ is a parallelogram.

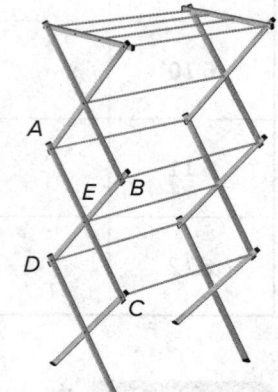

a. PLAN A SOLUTION What must Jane show in order to prove that $ABCD$ is a parallelogram? Explain.

_____

_____

b. USE STRUCTURE Which segments in the figure have point E as an endpoint? How are these segments related? Explain.

_____

_____

c. USE STRUCTURE Which angles in the figure have point E as a vertex? How are these angles related? Explain.

_____

_____

You can also prove that a quadrilateral in the coordinate plane is a parallelogram.

EXAMPLE 4 **Use Coordinates to Prove a Parallelogram**

Coordinates for three of the four vertices of parallelogram $ABCD$ are given in the table. Note that the coordinates of point A are missing.

Point	A	B	C	D
x	?	−2	−1	1
y	?	−2	4	5

a. CONSTRUCT ARGUMENTS What strategy could you use to identify the coordinates of point A? Explain.

_____

_____

b. USE STRUCTURE Draw parallelogram *ABCD* in the coordinate plane at the right. What are the coordinates of point *A*?

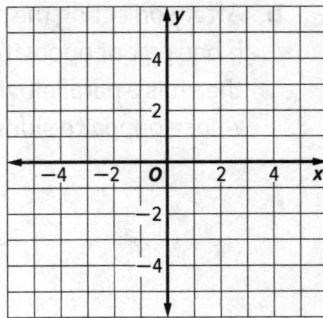

_____

c. REASON QUANTITATIVELY Find the slope of each side. The first has been done for you. What does this tell you about the quadrilateral?

slope of $\overline{AB} = \dfrac{-2-(-1)}{-2-0}$ or $\dfrac{1}{2}$ slope of $\overline{DC} =$

slope of $\overline{AD} =$ slope of $\overline{CB} =$

_____

_____

PRACTICE

1. USE TOOLS Use dynamic geometry software to construct parallelogram *PQRS* as shown. Remember to select and hide the parallel lines when the construction is complete.

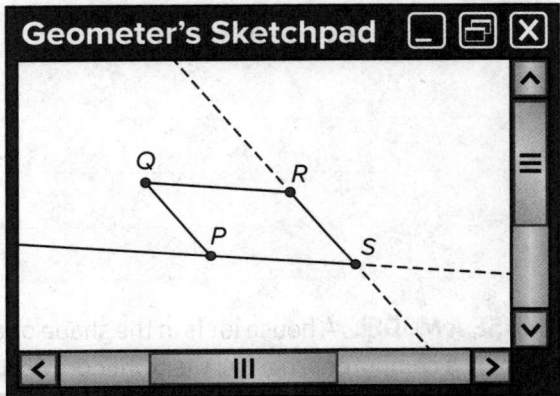

Geometer's Sketchpad

a. USE TOOLS Use the measurement tools in the software to measure ∠*P*, ∠*Q*, ∠*R*, and ∠*S*. What do you notice? Change the shape or location of quadrilateral *PQRS*. Does this relationship remain the same?

_____

_____

b. MAKE A CONJECTURE What can you conclude about quadrilateral *PQRS*?

_____

2. CRITIQUE REASONING A student wrote the paragraph proof below to prove that *PQRS* is a parallelogram. The proof contains a critical error. Find that error and correct it. Explain.

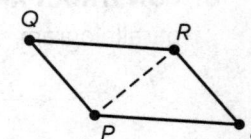

a. Given: ∠*P* ≅ ∠*R* , ∠*Q* ≅ ∠*S*
Prove: *PQRS* is a parallelogram.

Draw $\overline{PR}$ to form two triangles. Because the sum of the angles of one triangle is 180, the sum is 360 for two triangles. So, $m\angle P + m\angle Q + m\angle R + m\angle S = 360$. Since ∠*P* ≅ ∠*R* and ∠*Q* ≅ ∠*S*, $m\angle P = m\angle R$ and $m\angle Q = m\angle S$. By substitution, $m\angle P + m\angle Q + m\angle Q = 360$, $2(m\angle P) + 2(m\angle Q) = 360$, and dividing by 2 gives $m\angle P + m\angle Q = 180$. Likewise, $2(m\angle P) + 2(m\angle S) = 360$, and dividing by 2 gives $m\angle P + m\angle S = 180$. Consecutive angles are congruent, so $\overline{PS} \parallel \overline{QR}$ and $\overline{PQ} \parallel \overline{SR}$. Opposite sides are parallel, so PQRS is a parallelogram.

_____

_____

_____

b. After correcting the proof, the student suggests that you prove Theorem 6.12: If one pair of opposite sides of a quadrilateral is both congruent and parallel, then it is a parallelogram. Complete this proof by drawing quadrilateral *ABCD* whose opposite sides are congruent and parallel.

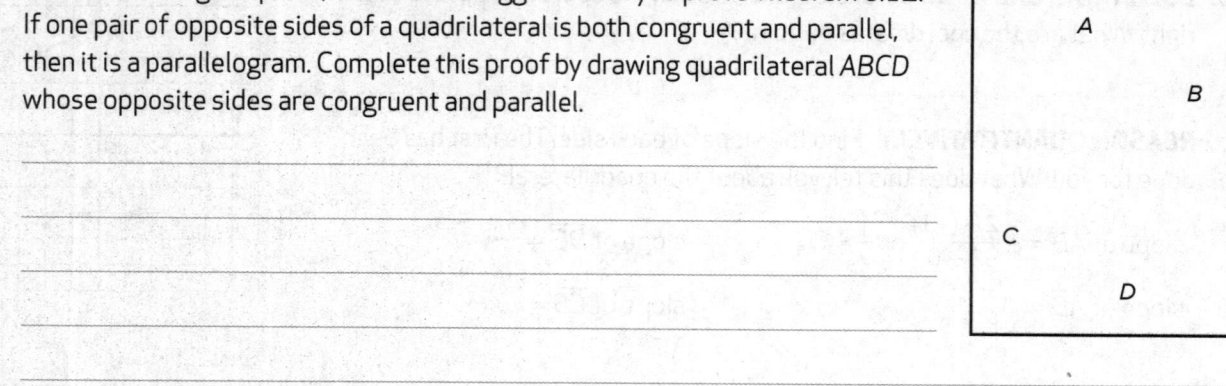

_____

_____

_____

_____

3. CONSTRUCT ARGUMENTS Write a paragraph proof of Theorem 6.11: If the diagonals of a quadrilateral bisect each other, then it is a parallelogram

Given: *ABCD* is a quadrilateral with diagonals that bisect each other.

Prove: *ABCD* is a parallelogram.

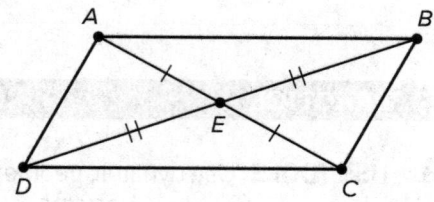

_____

_____

_____

_____

4. USE A MODEL A house lot is in the shape of a parallelogram. To represent the lot in a computer program, the owner draws a quadrilateral in the coordinate plane with vertices at $J(-1, 3)$, $K(2, 3)$, $L(1, -1)$, and $M(-3, -1)$. The owner later discovers that the coordinates for point *K* were entered incorrectly.

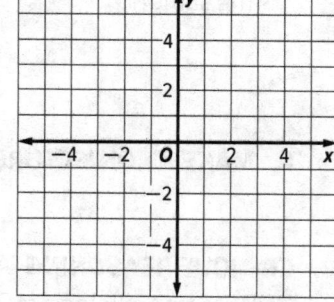

a. USE STRUCTURE Identify the correct coordinates for point *K*. Draw the corresponding parallelogram in the coordinate plane.

_____

b. CONSTRUCT ARGUMENTS Use the Slope Formula to prove that *JKLM* is a parallelogram.

_____

_____

c. CRITIQUE REASONING Jamal suggests that it would easier to prove that *JKLM* is a parallelogram by using Theorem 6.12: *If one pair of opposite sides of a quadrilateral is both parallel and congruent, then the quadrilateral is a parallelogram.* Do you agree? Explain.

_____

_____

_____

_____

5. USE STRUCTURE Using your answers from **Example 3 parts b** and **c**, mark the diagram at right for congruency. How are $\triangle AEB$, $\triangle BEC$, $\triangle CED$, and $\triangle DEA$ related? Explain.

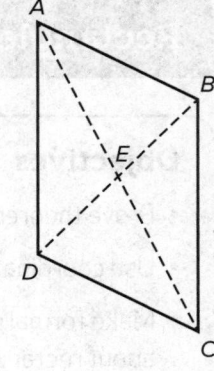

_____

   a. USE REASONING How can congruent triangles be used to show that opposite sides of *ABCD* are parallel? Explain.

   _____

   _____

   b. CONSTRUCT AN ARGUMENT Write a paragraph proof showing that *ABCD* is a parallelogram.

   _____

   _____

   _____

   _____

6. CRITIQUE REASONING A student said that another way to prove that the quadrilateral *ABCD* from **Example 4** is a parallelogram is to use the Distance Formula. Do you agree? Justify your answer. If you agree, complete the proof.

_____

_____

_____

_____

7. REASON QUANTITATIVELY A kite manufacturer is experimenting with different designs. The designer wants to modify a current design layout.

   a. A current kite design is represented in the coordinate plane with vertices at *A*(4, 20), *B*(20, 34), *C*(36, 20), and *D*(20, 0). The designer wants to modify the design by shortening the length of the kite. Draw the kite design in the coordinate plane and determine which point should be moved to modify the kite. What are the new coordinates if the kite is to be in the shape of a parallelogram?

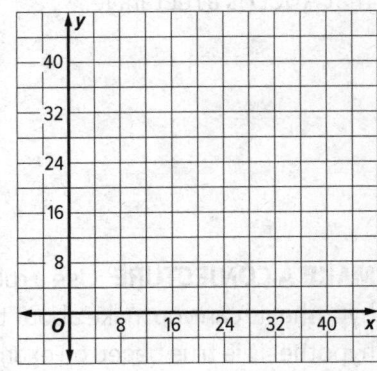

   _____

   b. Prove that the new kite design is in fact in the shape of a parallelogram.

   _____

   _____

   _____

Objectives

- Prove theorems about rectangles using two-column proofs.

- Use coordinates to prove theorems about rectangles.

- Make formal geometric constructions to understand theorems about rectangles.

Mathematical Practices
1, 2, 3, 5

A **rectangle** is a parallelogram with four right angles. Because a rectangle is a parallelogram, all the properties of parallelograms apply to rectangles.

EXAMPLE 1 Investigate Properties of Rectangles

EXPLORE Use a compass and straightedge to explore rectangles and their properties.

a. USE TOOLS Construct rectangle *ABCD* using the constructions of parallel and perpendicular lines.

b. CONSTRUCT ARGUMENTS Use the definition of a rectangle to explain how you know that *ABCD* is a rectangle.

_____

_____

_____

_____

c. MAKE A CONJECTURE Use a ruler to find *AC* and *BD*. What do you notice? What hypothesis can you make about the diagonals of a rectangle? Can you assume your hypothesis is true based on examples?

_____

_____

_____

Theorem 6.13: If a parallelogram is a rectangle, then its diagonals are congruent. Because Theorem 6.13 holds for all rectangles, we may add congruent diagonals to the list of properties of a rectangle.

EXAMPLE 2 **Prove that the Diagonals of a Rectangle Are Congruent**

a. **CONSTRUCT ARGUMENTS** Fill in the missing reasons to complete the proof.
Given: *RSTU* is a rectangle.

Prove: $\overline{RT} \cong \overline{SU}$

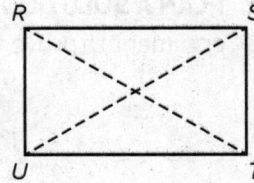

Statement	Reason
1. *RSTU* is a rectangle.	1.
2. *RSTU* is a parallelogram.	2.
3. $\overline{RU} \cong \overline{ST}$	3.
4. $\overline{UT} \cong \overline{UT}$	4.
5. ∠*RUT* and ∠*STU* are right angles.	5.
6. ∠*RUT* ≅ ∠*STU*	6.
7. △*RUT* ≅ △*STU*	7.
8. $\overline{RT} \cong \overline{SU}$	8.

b. **REASON ABSTRACTLY** Explain why this proof is true for all rectangles.

_____

_____

The converse of Theorem 6.13 is true as well.
Theorem 6.14: If the diagonals of a parallelogram are congruent, then the parallelogram is a rectangle.
Finding congruent diagonals is a valuable tool for proving that a parallelogram is a rectangle.

EXAMPLE 3 **Apply Properties of Rectangles**

PLAN A SOLUTION Nia was asked to prove that the figure at the right is a rectangle. She has a ruler but no protractor or other tool to measure angles. How can she prove that the figure is a rectangle?

a. State the theorem that can be used to prove that the figure above is a parallelogram using only a ruler.

_____

_____

b. State the theorem that can be used to prove that a parallelogram is a rectangle using only a ruler.

_____

_____

c. Using the theorems found in **part a** and **b**, describe how Nia could show that the figure is a rectangle.

_____

_____

EXAMPLE 4 Proving Rectangles on a Coordinate Plane

The coordinates of a quadrilateral are shown. Use algebra to prove that it is a rectangle.

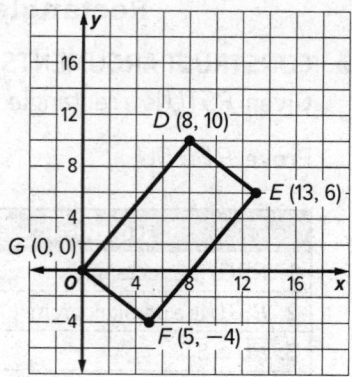

a. **PLAN A SOLUTION** Describe how you could construct an argument to prove that *DEFG* is a rectangle.

_____

_____

_____

b. **REASON QUANTITATIVELY** Show that *DEFG* is a rectangle. Explain.

_____

_____

_____

_____

_____

_____

_____

_____

PRACTICE

1. **CRITIQUE REASONING** David argues that to prove a quadrilateral is a rectangle, it is sufficient to prove that its diagonals are congruent. Do you agree? If so, explain why. If not, explain and draw a counterexample.

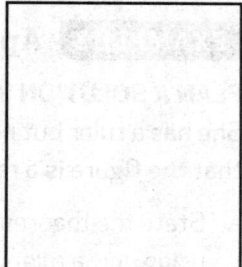

_____

_____

a. How can you alter David's argument to make it correct?

_____

_____

_____

b. David also says that to show that two diagonals of a quadrilateral are congruent, it is sufficient to show that all four angles of the quadrilateral are right angles. Is he correct? Explain.

_____

_____

_____

2. CONSTRUCT ARGUMENTS Fill in the missing parts to complete the proof.

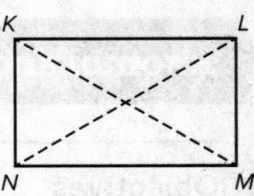

Given: Parallelogram $KLMN$, $\overline{KM} \cong \overline{LN}$

Prove: $KLMN$ is a rectangle.

Statement	Reason
1. $KLMN$ is a parallelogram; $\overline{KM} \cong \overline{LN}$.	1.
2. $\overline{KN} \cong \overline{LM}$	2.
3. $\overline{NM} \cong \overline{NM}$	3.
4.	4. SSS Theorem
5. $\angle KNM \cong \angle LMN$	5.
6. $\angle KNM$ and $\angle LMN$ are supplementary.	6.
7. $\angle KNM$ and $\angle LMN$ are right angles.	7.
8.	8. If a parallelogram has 1 rt. $\angle$, then it has 4 rt. $\angle$s.
9. $KLMN$ is a rectangle	9.

3. Students were asked to find whether the quadrilateral formed by connecting $J(0, 2)$, $K(2, 6)$, $L(10, 2)$, and $M(2, -2)$ is a rectangle. Two students solutions are shown below.

Samantha	Juan
I found the lengths of the sides: $JK = MJ = 2\sqrt{5}$ and $KL = LM = 4\sqrt{5}$, so pairs of sides are equal, and it is a parallelogram. I found the slope of each side: $\overline{JK} = 2$, $\overline{KL} = \frac{-1}{2}$, $\overline{LM} = \frac{-1}{2}$, $\overline{MJ} = 2$. That shows that $\overline{JK} \perp \overline{KL}$ and $\overline{LM} \perp \overline{MJ}$. If 1 $\angle$ of a parallelogram is a rt. $\angle$, then all 4 angles are rt. $\angle$s. That makes it a rectangle.	I graphed it, and I could see that it is not a rectangle. 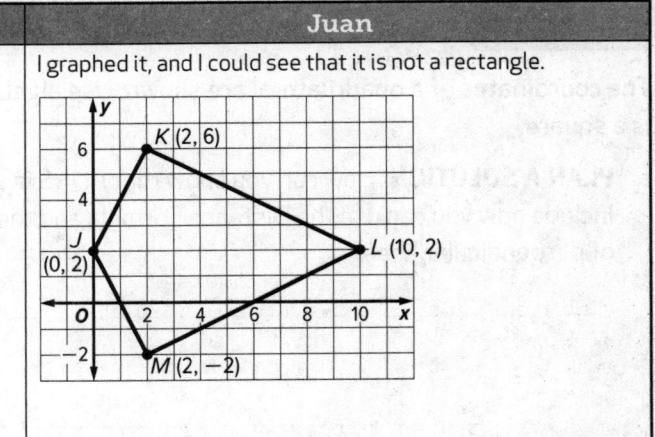

a. **REASON ABSTRACTLY** Evaluate each students' solution.

_____

_____

_____

b. **CONSTRUCT ARGUMENTS** Explain how you would solve the problem.

_____

_____

_____

Objectives

- Determine whether a figure defined by four points on a coordinate plane is a rhombus or square.
- Prove theorems about rhombi and squares.
- Construct rhombi and squares.

Mathematical Practices
1, 2, 3, 5, 6, 7

A rhombus is a quadrilateral with four congruent sides. Since the opposite sides are congruent, a rhombus is also a parallelogram and has all of the properties of a parallelogram. Additionally, the diagonals of rhombi have the following properties:

If a parallelogram is a rhombus, then its diagonals are perpendicular

If a parallelogram is a rhombus, then each diagonal bisects a pair of opposite angles.

A square is a parallelogram with four congruent sides and four congruent angles. This makes a square a rectangle and a rhombus. All of the properties of parallelograms, rectangles, and rhombi also apply to squares.

EXAMPLE 1   Classify a Quadrilateral

The coordinates of a quadrilateral are shown. Use algebra to prove that *PQRS* is a square.

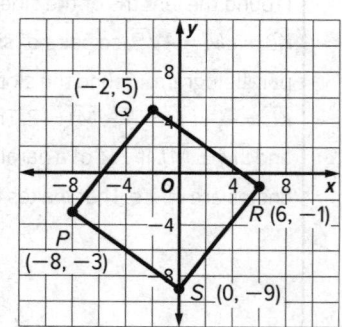

a. **PLAN A SOLUTION** How can you show that *PQRS* is a square? Include how you can use the Distance Formula and the slopes of perpendicular lines.

_____

_____

_____

_____

_____

b. **REASON QUANTITATIVELY** Prove that *PQRS* is a square. Explain.

_____

_____

_____

_____

_____

_____

_____

c. CRITIQUE REASONING Dylan believes quadrilateral *PQRS* is a square if the diagonals are congruent and perpendicular, and a rhombus if the diagonals are perpendicular but not congruent. Brittany believes this information is not sufficient to classify the quadrilateral. Who is correct? Explain your answer.

_____

_____

EXAMPLE 2 Proving a Parallelogram is a Rhombus

CONSTRUCT ARGUMENTS Prove that if the diagonals of a parallelogram are perpendicular, then the parallelogram is a rhombus.

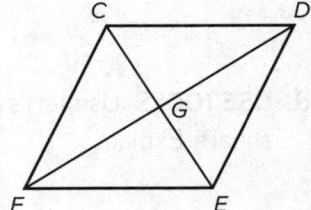

Given: *CDEF* is a parallelogram; $\overline{CE} \perp \overline{DF}$

Prove: *CDEF* is a rhombus

Statements	Reasons
1. *CDEF* is a parallelogram, $\overline{CE} \perp \overline{DF}$	1.
2. $\overline{DG} \cong \overline{FG}$	2.
3. ∠*CGF* and ∠*CGD* are right angles.	3.
4.	4. All right angles are congruent.
5. △*CGF* ≅ △*CGD*	5.
6. $\overline{CF} \cong \overline{CD}$	6.
7.	7. Opposite sides of a parallelogram are congruent.
8.	8.
9.	9.

EXAMPLE 3 Constructing a Rhombus

a. USE TOOLS Follow these steps to construct rhombus *WXYZ*.

- In the space to the right, use your compass to construct circle *W* containing point *Y*.

- With the compass at point *Y*, construct circle *Y* containing point *W*.

- Label the points of intersection *X* and *Z*.

- Draw $\overline{WX}, \overline{XY}, \overline{YZ},$ and $\overline{WZ}$.

b. COMMUNICATE PRECISELY Write a paragraph proof to prove *WXYZ* is a rhombus.

_____

_____

_____

c. CRITIQUE REASONING Jeff says that he can construct a rhombus using a circle drawn on patty paper. He constructs circle W through point X and then draws chord $\overline{XY}$, where Y is a point on the circle. He then folds the paper to reflect W across $\overline{XY}$. Is $WXW'Y$ a rhombus? Explain.

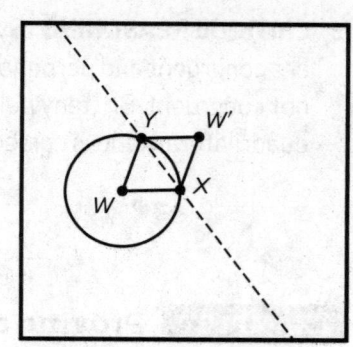

d. USE TOOLS Use Jeff's process to construct a square. Explain.

1. a. CRITIQUE REASONING Jillian is using coordinate geometry to classify quadrilateral $ABCD$. She finds $AB = BC = CD = AD = \sqrt{12}$ and decides $ABCD$ is a rhombus, but not a square. Do you agree with her conclusion? Explain your answer.

b. CRITIQUE REASONING Jillian is attempting to classify another quadrilateral, $EFGH$. She finds that the diagonals $EG = FH = 5$. Is it possible for a quadrilateral to be both a rectangle and a rhombus?

2. COMMUNICATE PRECISELY The vertices of parallelogram $QRST$ are $Q(-4, 7)$, $R(1, 9)$, $S(6, 7)$, and $T(1, 5)$. Determine whether $QRST$ is rectangle, rhombus, or square. List all that apply and explain your answer.

3. **COMMUNICATE PRECISELY** Mitul drew segment $\overline{LN}$, and constructed its perpendicular bisector. He labeled the intersection P. Then he constructed circle PM, where M is on the perpendicular line. He labeled the other intersection of the circle with the line point K. Write a paragraph proof that quadrilateral $KLMN$ is a rhombus. Following Mitul's method, make your own construction using a compass and straightedge.

_____

_____

_____

_____

_____

4. **CONSTRUCT ARGUMENTS** Prove that if the triangle formed by the diagonals and a side of a parallelogram is isosceles, then the parallelogram is a rectangle.

Given: $ACDE$ is a parallelogram, and $\triangle ACB$ is an isosceles triangle with base $\overline{AC}$.

Prove: $ACDE$ is a rectangle

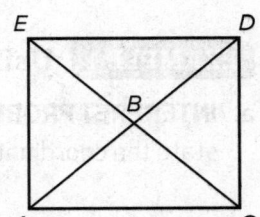

Statements	Reasons
1.	1. Given
2. $\overline{AB} \cong \overline{CB}$	2. Def. of Isosceles
3.	3. Diags. of a parallelogram bisect each other.
4. $AB = CB = EB = BD$	4. Definition of congruence and Transitive Prop.
5.	5. Addition Property of Equality
6. $ACDE$ is a rectangle.	6. If diagonals of par. are equal, it is a rectangle.

5. **USE STRUCTURE** If the diagonals of quadrilateral $LMNP$ form congruent triangles, prove that the quadrilateral is a square. Draw and label a figure and write a paragraph proof.

_____

_____

_____

_____

_____

_____

_____

Objectives

- Determine whether a figure defined by four points is a trapezoid or kite.

- Prove theorems about trapezoids and kites using coordinates.

A trapezoid is a quadrilateral with exactly one pair of parallel sides called bases. The nonparallel sides are called legs. The midsegment of a trapezoid is the segment that connects the midpoints of the legs of a trapezoid.

If the legs of a trapezoid are congruent, then it is an isosceles trapezoid.

A kite is a quadrilateral with exactly two pairs of consecutive congruent sides.

EXAMPLE 1   Using Coordinate Geometry to Explore Kites

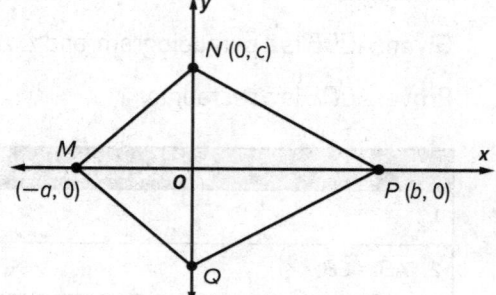

a. INTERPRET PROBLEMS Without introducing new variables, state the coordinates of point Q, assuming that $MNPQ$ is a kite.

Q _____

b. USE STRUCTURE Talia notices that the figure may be analyzed as two triangles, $\triangle MNP$ and $\triangle MQP$. What can we reason about opposite angles $\angle N$ and $\angle Q$? Explain.

_____

_____

c. CONSTRUCT ARGUMENTS Given kite $MNPQ$, show that $\angle NMQ \cong \angle NPQ$.

_____

_____

_____

d. CONSTRUCT ARGUMENTS Given kite $MNPQ$, show that $\overline{MP}$ is perpendicular to $\overline{NQ}$.

_____

_____

_____

e. REASON ABSTRACTLY If $a = b = c$, is $MNPQ$ still a kite? Justify your answer. Categorize the quadrilateral as specifically as you can.

_____

_____

_____

Kites

6.25	If a quadrilateral is a kite, then its diagonals are perpendicular. **Example** If quadrilateral *ABCD* is a kite, then $\overline{AC} \perp \overline{BD}$.	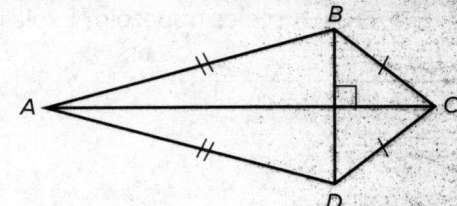
6.26	If a quadrilateral is a kite, then exactly one pair of opposite angles is congruent. **Example** If quadrilateral *JKLM* is a kite, $\overline{JK} \cong \overline{KL}$, and $\overline{JM} \cong \overline{LM}$, then $\angle J \cong \angle L$ and $\angle K \not\cong \angle M$.	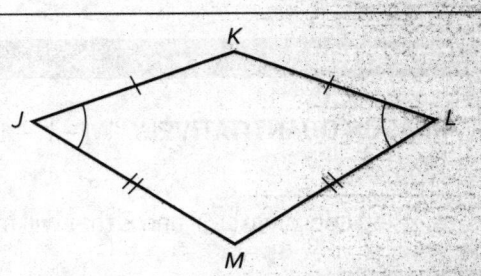

EXAMPLE 2   Using Coordinate Geometry to Classify and Prove Theorems About Trapezoids

a. **PLAN A SOLUTION** Plot quadrilateral *PQRS* with vertices *P*(0, 0), *Q*(a, b), *R*(c, b), and *S*(a + c, 0), where a > 0, b > 0, and c > 0, on the axes to the right.

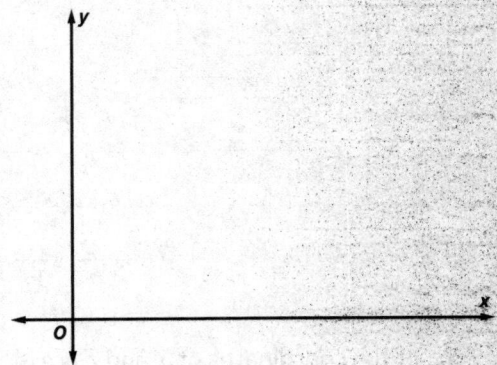

b. **CALCULATE ACCURATELY** Ryan says *PQRS* is an isosceles trapezoid with bases $\overline{QR}$ and $\overline{PS}$. Do you agree with Ryan? Justify your answer.

_____

_____

_____

_____

_____

c. **CONSTRUCT ARGUMENTS** Show that if a trapezoid is isosceles, then its diagonals are congruent.

Given: *PQRS* is an isosceles trapezoid with bases $\overline{QR}$ and $\overline{PS}$.

Prove: $\overline{QS} \cong \overline{PR}$.

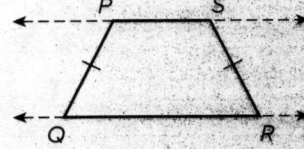

_____

_____

_____

d. CRITIQUE REASONING Makayla says quadrilateral *PQRS* is an isosceles trapezoid because the diagonals are congruent. Would this be enough information to classify *PQRS* as an isosceles trapezoid? Explain.

1. **REASON QUANTITATIVELY** $W(-1, -11)$ and $X(1, 1)$ are two vertices of quadrilateral *WXYZ*.

   a. Find coordinates *Y* and *Z* that will make *WXYZ* a kite. Justify your answer.

   b. Find coordinates *Y* and *Z* that will make *WXYZ* an isosceles trapezoid. Justify your answer.

   c. If the coordinates of *Y* and *Z* are $(4, 1)$ and $(4, -11)$, identify the shape of *WXYZ*.

2. **CONSTRUCT ARGUMENTS** *EFGH* is shown to the right, with $a \neq b \neq c \neq 0$. Use coordinate geometry to prove that *EFGH* is a kite.

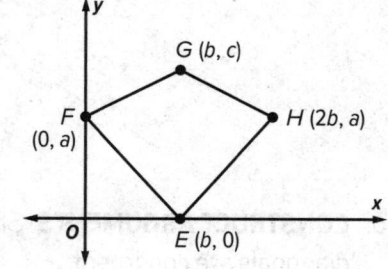

3. **USE STRUCTURE** Isosceles trapezoid *JKLM* is shown to the right.

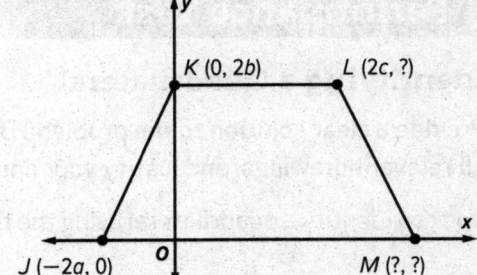

a. Without introducing new variables, state the coordinates of points *L* and *M*.

L _____   M _____

b. Let point *P* be the midpoint of $\overline{JK}$ and *Q* be the midpoint of $\overline{LM}$. Use coordinate geometry to show that the midsegment of *JKLM* is parallel to the bases of *JKLM* and equal to half the sum of their lengths.

_____

_____

_____

_____

_____

c. Let point *R* be the midpoint of $\overline{JM}$ and *S* be the midpoint of $\overline{KL}$. Use coordinate geometry to show that *PSQR*, the quadrilateral connecting the midpoints of *JKLM*, is a rhombus.

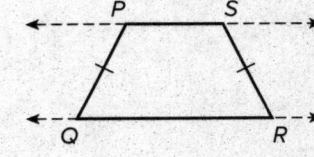

_____

_____

_____

_____

_____

_____

4. **CONSTRUCT ARGUMENTS** Quadrilateral *ABCD* is shown to the right.

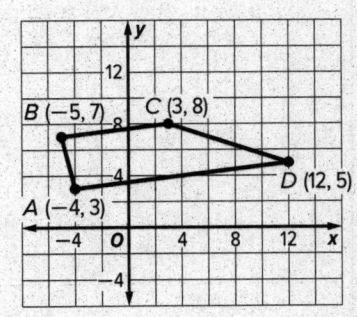

a. Show that *ABCD* is a trapezoid.

_____

_____

_____

b. Prove that *ABCD* is not an isosceles trapezoid.

_____

_____

_____

Identifying a Quadrilateral

Provide a clear solution to the problem. Be sure to show all of your work, include all relevant drawings, and justify your answers.

You can identify a quadrilateral using the theorems you have learned.

Part A
Construct a parallelogram *ABCD* using a compass and straightedge. Explain your construction and prove why the construction resulted in a parallelogram.

Part B

Is your construction a rhombus? Why or why not? If not, how could you alter your construction so that it is a rhombus?

Part C

Quadrilateral *PQRS* is shown. Prove that if △*PTQ* ≅ △*STR*, then *PQRS* is a parallelogram.

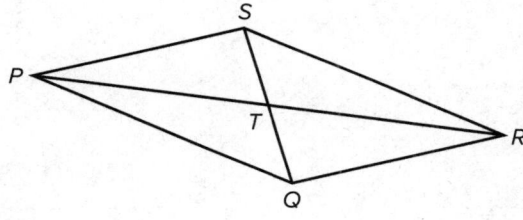

Part D

Using the same figure from **Part C**, prove that if *PQRS* is a parallelogram and △*PTQ* ≅ △*PTS*, then *PQRS* is a rhombus.

Using Triangles to Make a Quadrilateral

Provide a clear solution to the problem. Be sure to show all of your work, include all relevant drawings, and justify your answers.

You can make a quadrilateral by combining two triangles.

Part A
Fold a piece of paper, and cut out the corner to create two triangles. Then cut two more triangles from the same corner, starting the cut from the same point. Describe the triangles you made.

Part B
In the space below, trace both triangles. Try to make as many different quadrilaterals as you can, without folding or overlapping the triangles. Indicate whether any sides or angles are congruent with appropriate symbols.

Part C

Classify each quadrilateral that you made. Justify each classification with theorems or definitions. Explain your reasoning.

Part D

Do you think the type of triangle you start with limits the types of quadrilaterals you can make? Justify your answer with drawings or sentences.

1. In the diagram below, ABCD is a parallelogram.

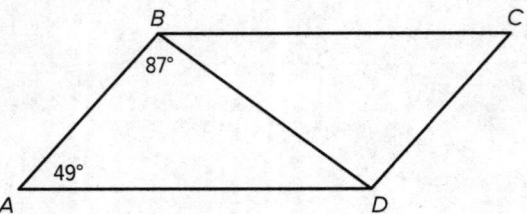

Complete the following.

$m\angle CDA = $ ☐

2. Rhombus JKLM has vertices $J(-1, -4)$, $K(1, 1)$, and $L(6, 3)$. The coordinates of M are (☐ , ☐).

3. Circle the figures that are parallelograms.

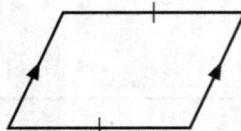   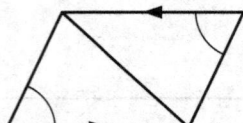

   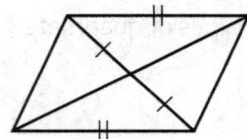

4. The area of the kite shown below is ☐ square units.

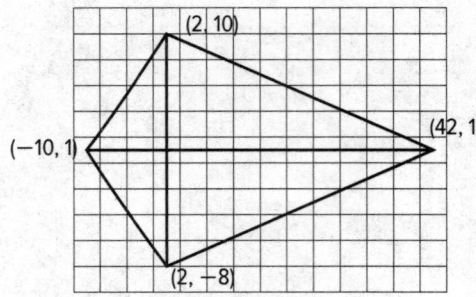

5. The diagonals and sides of quadrilateral WXYZ form four congruent isosceles triangles. The most specific name that can be given to quadrilateral WXYZ is ☐.

6. Rectangle DEFG has a length that is 2 centimeters longer than its width.

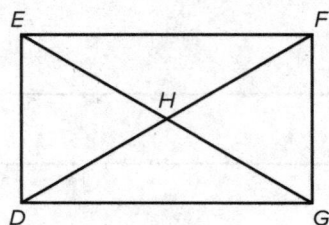

If $FG < EF$, $DF = 58$ centimeters and the perimeter of $\triangle DEF$ is 140 centimeters, the perimeter of $\triangle FHG$ is ☐ centimeters.

7. In the table below, the column on the left gives a characteristic of a quadrilateral. Check the columns corresponding to the types of quadrilaterals that have that characteristic.

	Parallelogram	Rectangle	Square	Rhombus	Trapezoid	Kite
Diagonals are congruent.						
Exactly one pair of opposite angles is congruent.						
Opposite sides are parallel.						
Diagonals are perpendicular.						
Has less than two pairs of opposite congruent sides.						
All sides are congruent.						
All angles are right angles.						
Exactly one set of opposite sides are parallel.						

8. Complete the steps and reasons in the following proof.

Given: ∠A is supplementary to ∠B

∠A is supplementary to ∠D

Prove: ABCD is a parallelogram

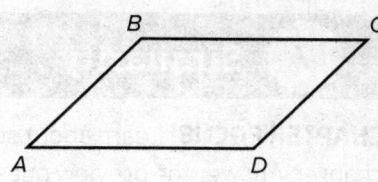

Statement	Reason
∠A is supplementary to ∠B.	
$\overline{AD} \parallel \overline{BC}$	Converse of the Consecutive Interior Angles Theorem
	Given
$\overline{AB} \parallel \overline{CD}$	
ABCD is a parallelogram.	

9. In the diagram below, MNOP is a trapezoid, Q is the midpoint of $\overline{MN}$, and R is the midpoint of $\overline{OP}$.

   a. What are the coordinates of Q? Show your work.

   _____

   _____

   b. What are the coordinates of R? Show your work.

   _____

   _____

   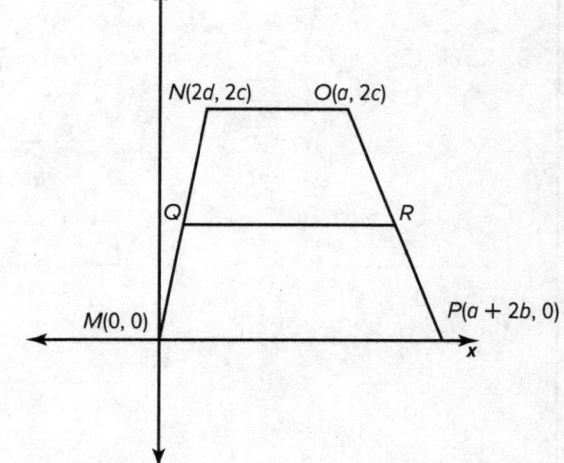

   c. Show that $QR = \dfrac{NO + MP}{2}$.

   _____

   _____

   _____

10. The coordinates of the vertices of quadrilateral LMNP are L(1, 7), M(4, 3), N(3, 1), and P(1, 2).

   a. Find the length and slope of each side of LMNP.

   _____

   _____

   b. Classify LMNP as a parallelogram, rhombus, trapezoid, kite, or square. Explain your reasoning.

   _____

   _____

   c. Verify that the diagonals of LMNP intersect at right angles by finding the slope of each diagonal and confirming that the two diagonals are perpendicular.

   _____

   _____

CHAPTER FOCUS Learn about some of the objectives that you will explore in this chapter. Answer the preview questions. As you complete each lesson, return to these pages to check your work.

What You Will Learn	Preview Question
Dilations	
• Compare transformations that preserve distance and angle measures to those that do not. • Verify the properties of dilations. • Represent and describe dilations as functions.	**SMP 1** What is the rule that describes the transformation of *ABCD* to *A′B′C′D′*? What is the center of dilation? What composition of functions would produce the same image? 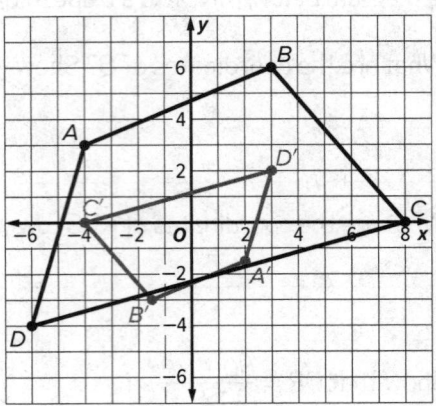 $(x, y) \rightarrow (-0.5x, -0.5y)$; **The center of dilation is** **(0, 0); A dilation by a scale factor of 0.5 followed** **by a rotation of 180° about (0, 0).**
Similar Triangles	
• Use the definition of similarity in terms of similarity transformations. • Establish the AA Similarity Postulate for triangles. • Use similarity to prove theorems and solve problems about triangles.	**SMP 6** △*DEF* is a dilation of △*ABC*. What is the scale factor of the dilation? Explain your reasoning. 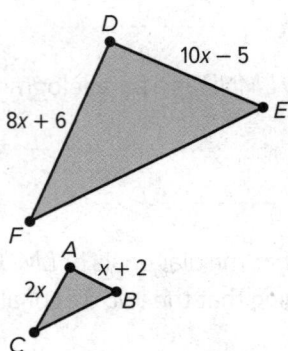 **5; The triangles are similar, so** $\dfrac{10x-5}{x+2} = \dfrac{8x+6}{2x}$ **and** $x = 3$; $\dfrac{10(3)-5}{3+2} = \dfrac{25}{5}$, **or 5.**

What You Will Learn	Preview Question
Perpendicular and Parallel Lines	
• Prove the slope criteria for perpendicular lines. • Prove the slope criteria for parallel lines.	**SMP 2** What is the slope-intercept form of the equation for the line through (a, b) that is perpendicular to $y = mx + b$? Justify your answer.
Parallel Lines and Proportional Parts	
• Prove the Triangle Proportionality Theorem and its converse. • Prove the Triangle Midsegment Theorem. • Use a compass and straightedge to partition a line segment in a given ratio.	**SMP 5** Use a compass and straightedge to locate point P on $\overline{XY}$ such that $XP = 2PY$. $X \bullet$ ——————— $\bullet Y$
Parts of Similar Triangles	
• Prove theorems about parts of similar triangles. • Use geometric shapes and their measures to describe objects and solve problems related to similarity.	**SMP 8** A diagram of a shuffleboard is shown. Triangle ADG is isosceles. $\overline{DK}$ is the perpendicular bisector of $\triangle ADG$. If $DK = 108$ inches, $DC = 38$ inches, $CH = 12$ inches, $BJ = 24$ inches, and $AK = 36$ inches, find AD, DH, DJ, and DB. Round to the nearest whole number.

Dilations

Objectives

- Compare transformations that preserve distance and angle measures to those that do not.

- Verify the properties of dilations.

- Represent and describe dilations as functions.

Mathematical Practices
1, 2, 3, 5, 6, 7, 8

EXAMPLE 1 **Compare Transformations**

EXPLORE A special effects designer uses function rules to change the size, shape, and/or location of images in a movie. The designer is experimenting with different ways to change the appearance of a tower in a science fiction movie.

a. CALCULATE ACCURATELY Each figure shows the starting position of the tower and the function that describes the transformation that the designer applies to it. Draw the image of the tower under each transformation.

i. $(x, y) \rightarrow (4x, y)$

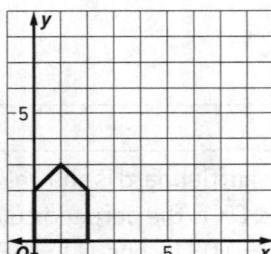

ii. $(x, y) \rightarrow (x + 5, y + 4)$

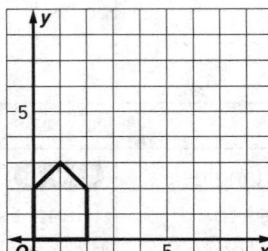

iii. $(x, y) \rightarrow (2x, 2y)$

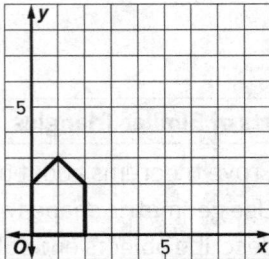

iv. $(x, y) \rightarrow (x, 2y)$

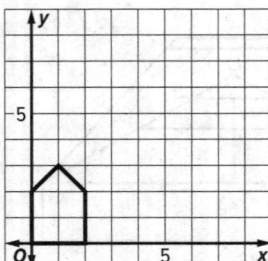

v. $(x, y) \rightarrow (0.5x, 0.5y)$

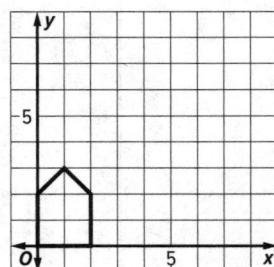

vi. $(x, y) \rightarrow (x + 3, -y + 7)$

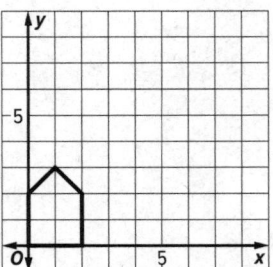

b. USE STRUCTURE Which transformations result in congruent figures? How do you know?

_____

_____

c. **COMMUNICATE PRECISELY** Which of the transformations seem to preserve angle measure? How could you be sure?

_____

_____

d. **COMMUNICATE PRECISELY** The designer wants to change the size but not the shape of the tower. Which transformations can the designer use? How would you describe these transformations in everyday language?

_____

e. **CALCULATE ACCURATELY** In figure iii, what is the relationship of the area of the image to the area of the preimage? Explain how to calculate this.

_____

_____

A **dilation** with center C and scale factor k ($k > 0$) is a function that maps a point P to its image P' as follows.

- If P coincides with C, then $P' = P$.

- If P does not coincide with C, then P' lies on $\overrightarrow{CP}$ such that $CP' = k(CP)$.

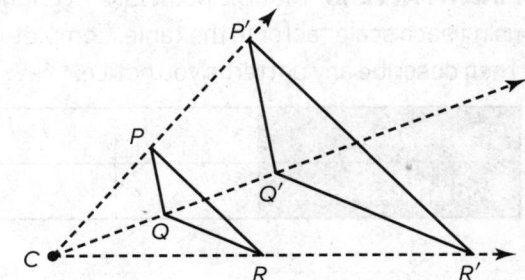

EXAMPLE 2   Investigate Properties of Dilations

EXPLORE Use the Geometer's Sketchpad to investigate properties of dilations as follows.

a. **USE TOOLS** Draw a triangle and label the vertices A, B, and C. Then draw a point P. Draw the image of $\triangle ABC$ after a dilation with center P and scale factor $k = \frac{3}{2}$, as shown. Change the shape and location of $\triangle ABC$ and notice how $\triangle A'B'C'$ is related to $\triangle ABC$.

b. **FIND A PATTERN** Repeat the process in Step a, but use different values of k, including $\frac{1}{4}$, $\frac{1}{2}$, 1, 2, and 3. What can you say about a dilation when the scale factor, k, is less than 1? equal to 1? greater than 1?

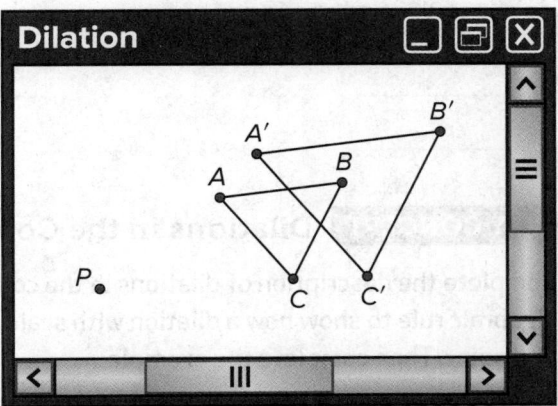

_____

_____

c. **USE TOOLS** Measure the length of each side of $\triangle ABC$ and the length of each side of $\triangle A'B'C'$. Change the scale factor and notice how the lengths compare. What conclusions can you make?

_____

_____

d. MAKE A CONJECTURE Draw a point *P* and a line *j*. Then draw the image of line *j* after a dilation with center *P* and scale factor 2, as shown. Label the image *j'*. Change the location of line *j* and the scale factor. Make a conjecture that relates lines *j* and *j'*.

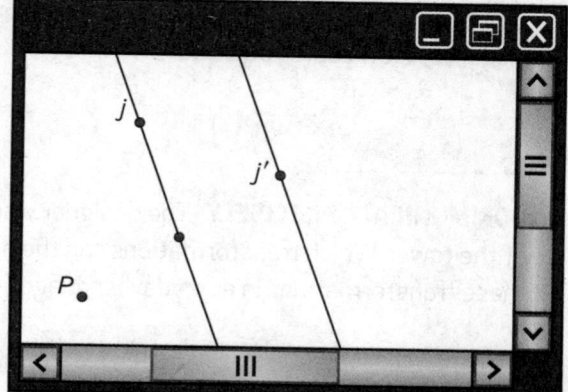

_____

_____

e. MAKE A CONJECTURE Is there ever a situation in which your conjecture in **part d** does not hold? What happens to a line under a dilation in this case?

_____

_____

f. FIND A PATTERN Plot the point $E(-4, 8)$. Find the image of point *E* under a dilation using each scale factor in the table. Complete the table by writing the coordinates of *E'*. Then describe any patterns you notice.

Scale Factor	$\frac{1}{4}$	$\frac{1}{2}$	2	$\frac{5}{2}$	3
Coordinates of *E'*					

_____

_____

g. FIND A PATTERN Calculate the coordinates of *E'* in **part f** when the scale factors are increasing integers such as 4, 5, and 6. Describe any patterns you notice in the value of the *x*- and *y*-coordinates.

_____

_____

_____

KEY CONCEPT   Dilations in the Coordinate Plane

Complete the description of dilations in the coordinate plane and complete the algebraic rule to show how a dilation with scale factor *k* changes the coordinates of a point. Then complete the example.

Description	Example
To find the coordinates of a point after a dilation centered at the origin with scale factor *k*, _____ _____ $(x, y) \rightarrow$ _____	Dilation with scale factor $\frac{3}{2}$. *A*(4, 2) *A'* $A(4, 2) \rightarrow A'(\underline{\quad}, \underline{\quad})$

EXAMPLE 3 **Use Dilations to Solve a Problem**

An architect is using a coordinate plane to design a studio apartment in the shape of a trapezoid. Each unit of the coordinate plane represents one meter. In the original plan, the vertices of the apartment are $J(-3, 3)$, $K(3, 3)$, $L(0, -3)$, and $M(-3, -3)$. The architect wants to enlarge the apartment so that $J'K' = 10$ meters.

a. **INTERPRET PROBLEMS** What scale factor should the architect use to dilate the original apartment so that the image has a side with the required length? Explain.

_____

_____

b. **COMMUNICATE PRECISELY** Write a function that shows how a point, (x, y), is mapped under this dilation if the center of dilation is the origin.

_____

c. **CALCULATE ACCURATELY** Draw and label $JKLM$ and its image, $J'K'L'M'$, on the coordinate plane at the right.

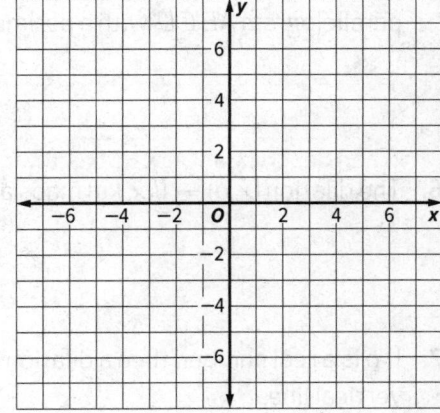

d. **CRITIQUE REASONING** The architect claims that the enlarged apartment will have a perimeter that is more than twice as large as the original perimeter. Do you agree or disagree? Explain.

_____

_____

PRACTICE

CALCULATE ACCURATELY Draw and label the image of the figure after a dilation with the given scale factor and center of dilation at the origin. Write a function to describe the transformation.

1. $k = \frac{2}{3}(x, y) \rightarrow \left(\frac{2}{3}x, \frac{2}{3}y\right)$

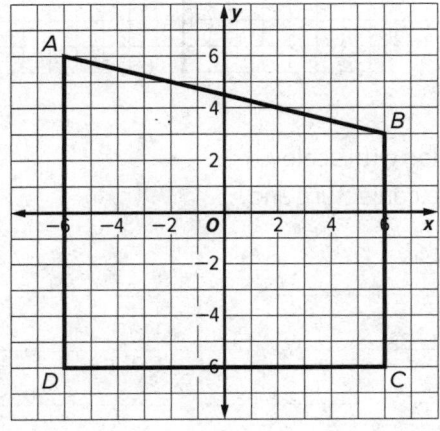

2. $k = 1.5(x, y) \rightarrow (1.5x, 1.5y)$

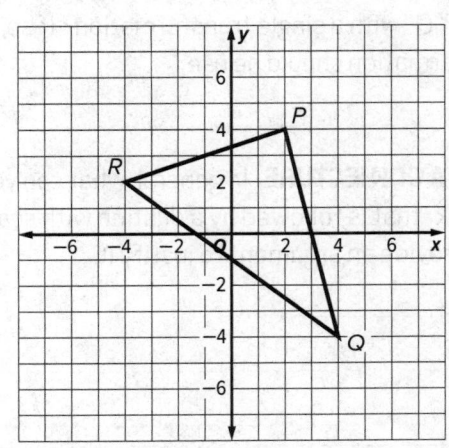

EVALUATE REASONABLENESS Determine whether each statement is *always,* *sometimes,* or *never* true. Explain.

3. If c is a real number, then a dilation centered at the origin maps the line $y = cx$ to itself.

_____

4. If $k > 1$, then a dilation with scale factor k maps $\overline{AB}$ to a segment that is congruent to $\overline{AB}$.

_____

_____

5. If $0 < k < 1$, then a dilation of parallelogram $ABCD$ with a scale factor of k will result in parallelogram $A'B'C'D'$ with a perimeter equal to that of the original parallelogram.

_____

_____

6. The dilation $(x, y) \rightarrow (kx, ky)$ maps a point to a point that is farther from the origin.

_____

_____

7. If a is a real number, then a dilation centered at the origin maps the line $x = a$ to a vertical line.

_____

_____

8. David is using a coordinate plane to experiment with quadrilaterals, as shown in the figure.

   a. **CALCULATE ACCURATELY** David creates $M'N'P'Q'$ by enlarging $MNPQ$ with a dilation with scale factor 2. Then he creates $M''N''P''Q''$ by dilating $M'N'P'Q'$ with a scale factor of $\frac{1}{3}$. The center of dilation for each dilation is the origin. Draw and label the final image, $M''N''P''Q''$.

   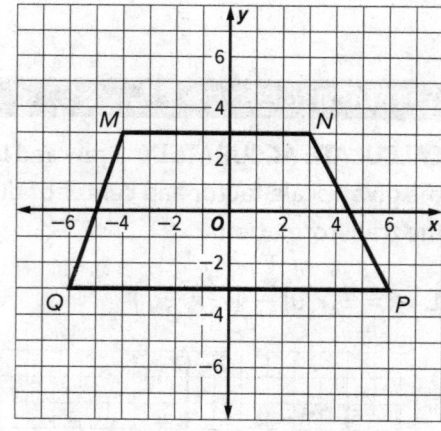

   b. **REASON ABSTRACTLY** Can David map $MNPQ$ directly to $M''N''P''Q''$ with a single transformation? If so, what transformation should he use?

   _____

   c. **MAKE A CONJECTURE** In general, what can you say about a dilation with scale factor k_1 that is followed by a dilation with scale factor k_2? State a conjecture and then provide an argument to justify it.

   _____

   _____

   _____

9. **REASON QUANTITATIVELY** The figure shows an architect's plan for a bedroom. Each unit of the coordinate plane represents one foot.

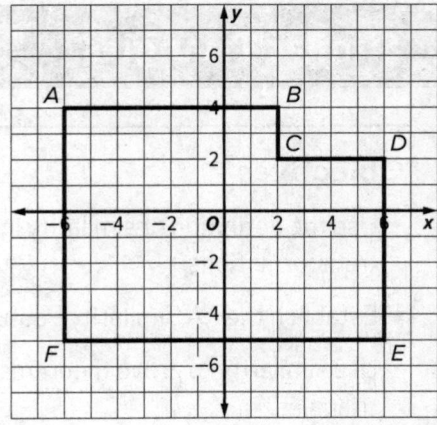

a. The architect would like to enlarge or reduce the bedroom as needed so that its perimeter is 63 feet. What transformation should the architect use? Justify your answer.

_____

_____

_____

b. The architect would like to enlarge or reduce the bedroom as needed so that its perimeter is 31.5 feet. What scale factor should the architect use in his dilation? Explain.

_____

_____

c. Based on **parts a** and **b**, write an equation that can be used to find the scale factor (x) of a dilation given any perimeter (y).

_____

_____

10. The point P' is the image of point $P(a, b)$ under a dilation centered at the origin with scale factor $k \neq 1$.

a. **REASON ABSTRACTLY** Assuming that point P does not lie on the y-axis, what is the slope of $\overleftrightarrow{PP'}$? Explain how you know.

_____

_____

b. **USE STRUCTURE** In **part a**, why is it important that P does not lie on the y-axis?

_____

_____

11. A city planner is designing the streets in a new shopping district. She has already planned Palm Street, as shown on the coordinate plane.

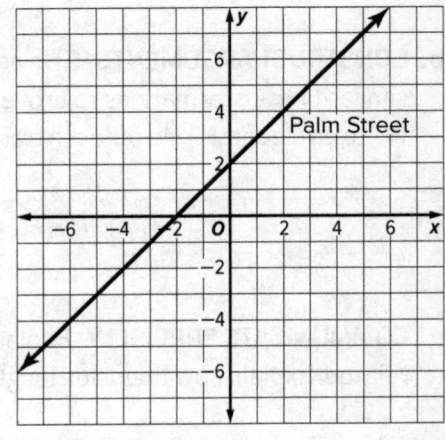

a. **REASON ABSTRACTLY** The city planner will apply a dilation to Palm Street to create Cedar Street. Describe how the two streets will be related.

_____

b. **USE STRUCTURE** The city planner uses a dilation with scale factor 2 and center of dilation at the origin to create Cedar Street. Draw Cedar Street on the coordinate plane and write its equation.

_____

Objectives

- Use the definition of similarity in terms of similarity transformations.

- Establish the AA Similarity Postulate for triangles.

- Use similarity to prove theorems and solve problems about triangles.

A **similarity transformation** is a transformation that preserves the shape (but not necessarily the size) of a figure. Similarity transformations include the rigid motion transformations (translations, reflections, and rotations) as well as dilations.

Two figures are **similar** if there is a sequence of similarity transformations that maps one figure onto the other. In the figure at the right, △JKL is dilated by a factor of 1.5 about point L and reflected in line ℓ.

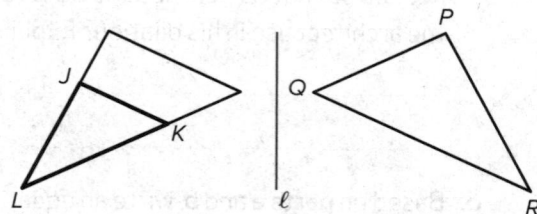

EXAMPLE 1 Determine Whether Figures Are Similar

EXPLORE Malikah is using tiles to make a new patio in her yard. She sees the tiles shown at the right in a catalog, and she wants to know which of the tiles are similar.

a. **PLAN A SOLUTION** Which of the tiles appear to be similar? Why? What do you need to do to prove that the tiles you identified are similar?

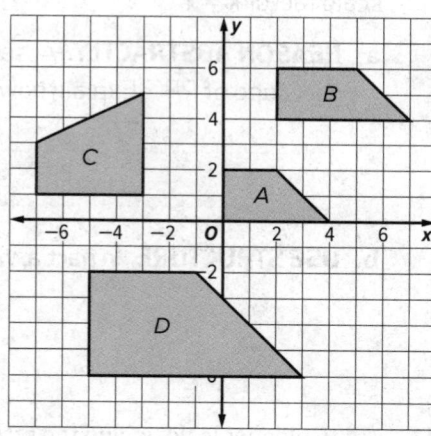

_____

_____

_____

b. **CONSTRUCT ARGUMENTS** Give an argument using specific similarity transformations to prove that the tiles you identified in **part a** are similar. Write an algebraic rule for each transformation you use.

_____

_____

_____

c. **COMMUNICATE PRECISELY** Explain how the angles of figure A and figure D are related. Explain how their side lengths are related.

_____

_____

When you know that two triangles are similar, you can draw certain conclusions about their angles and side lengths. You will investigate this in the next example.

EXAMPLE 2 **Investigate Similar Triangles**

In the figure, △JKL is similar to △PQR. There is a sequence of similarity transformations that maps △JKL onto △PQR. You can first use a dilation to enlarge △JKL and then one or more rigid motions to map the dilated image to △PQR.

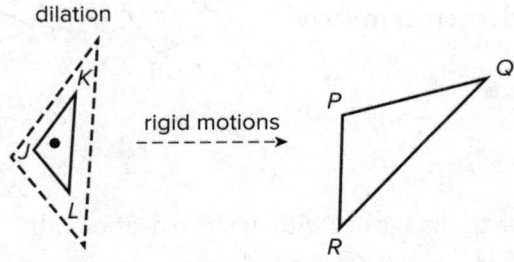

dilation

rigid motions

a. REASON ABSTRACTLY What can you conclude about the angles of △JKL and △PQR? Use properties of dilations and rigid motions to explain why this must be true.

_____

_____

_____

_____

b. USE STRUCTURE Suppose the above dilation has scale factor k. Complete the following.

$PQ = k\,(\boxed{})$ $\qquad QR = k\,(\boxed{})$ $\qquad PR = k\,(\boxed{})$

c. USE STRUCTURE Solve each of the equations in **part b** for k. Use the results to write a set of ratios that must be equal. Then state a generalization based on your findings.

_____

_____

d. EVALUATE REASONABLENESS Assume △JKL ~ △PQR and JL is 2 units long and PR is 5 units long. If JK is 4 units long, is it reasonable to conclude that PQ is about 8 units long? Explain your answer.

_____

_____

_____

Two triangles are similar if and only if corresponding angles are congruent and corresponding side lengths are proportional.
You write △ABC ~ △DEF to show that △ABC is similar to △DEF, and corresponding angles are listed in the same position in the name of each triangle. That is, given the statement △ABC ~ △DEF, you can conclude that ∠A corresponds to ∠D, and ∠A ≅ ∠D. You can also conclude that $\dfrac{DE}{AB} = \dfrac{EF}{BC} = \dfrac{DF}{AC}$.

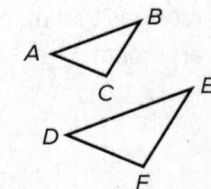

The AA Similarity Postulate states that if two angles of one triangle are congruent to two angles of another triangle, then the triangles are similar.

EXAMPLE 3 **Establish the AA Similarity Postulate**

In the figure, $\angle A \cong \angle D$ and $\angle B \cong \angle E$. You must prove that $\triangle ABC \sim \triangle DEF$.

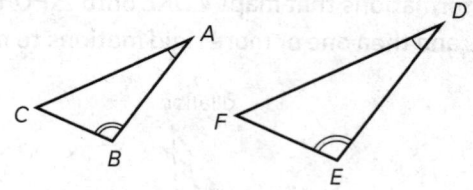

a. **PLAN A SOLUTION** What do you have to do in order to show that $\triangle ABC \sim \triangle DEF$?

_____

_____

b. **CONSTRUCT ARGUMENTS** Let $\triangle A'B'C'$ be the image of $\triangle ABC$ under a dilation with scale factor $\frac{DE}{AB}$, as shown below. Why is $\angle A \cong \angle A'$ and $\angle B \cong \angle B'$?

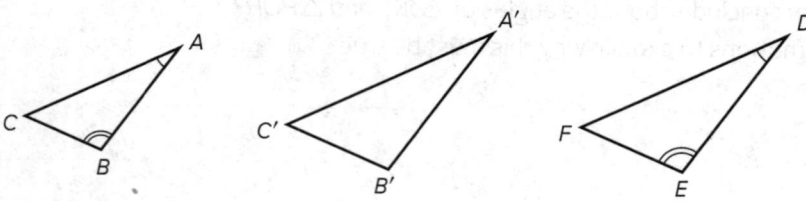

_____

_____

c. **CONSTRUCT ARGUMENTS** Can you conclude that $\angle A' \cong \angle D$ and $\angle B' \cong \angle E$? Explain why or why not.

_____

_____

d. **REASON ABSTRACTLY** Use the scale factor of the dilation to explain how to find the length of $\overline{A'B'}$.

_____

_____

e. **CONSTRUCT ARGUMENTS** Explain why $\triangle A'B'C' \cong \triangle DEF$.

_____

_____

f. **CONSTRUCT ARGUMENTS** Since $\triangle A'B'C' \cong \triangle DEF$, there is a sequence of rigid motions that maps $\triangle A'B'C'$ onto $\triangle DEF$. Explain how you can use this to complete the argument.

_____

_____

In addition to the AA Similarity Postulate, there are two other theorems that you can use to prove two triangles are congruent.

KEY CONCEPT **Similarity Criteria**

Complete the description of each similarity criterion.

Description	Example
AA Similarity Postulate If _____ _____ _____ then the triangles are similar.	If $\angle J \cong \angle M$ and $\angle K \cong \angle N$, then $\triangle JKL \sim \triangle MNP$.
SSS Similarity Theorem If _____ _____ _____ then the triangles are similar.	If $\frac{DE}{AB} = \frac{EF}{BC} = \frac{DF}{AC}$, then $\triangle ABC \sim \triangle DEF$.
SAS Similarity Theorem If _____ _____ _____ _____ then the triangles are similar.	If $\frac{XY}{ST} = \frac{YZ}{TU}$ and $\angle T \cong \angle Y$, then $\triangle STU \sim \triangle XYZ$.

EXAMPLE 4 **Prove Triangles Are Similar**

In the figure, $\overline{QR}$ is parallel to $\overline{ST}$.

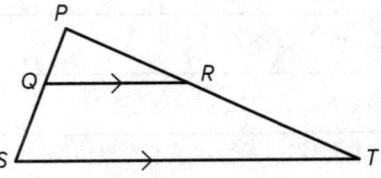

a. **CONSTRUCT ARGUMENTS** Write a paragraph proof to show $\triangle PQR \sim \triangle PST$.

_____

_____

_____

b. **COMMUNICATE PRECISELY** Suppose you know that $PR = 16$ cm, $PT = 40$ cm, and $PS = 20$ cm. Explain how you can find PQ.

_____

_____

_____

COMMUNICATE PRECISELY Determine whether the given figures are similar. If the figures are similar, give a specific sequence of similarity transformations that maps one figure onto the other. If the figures are not similar, explain why not.

1.
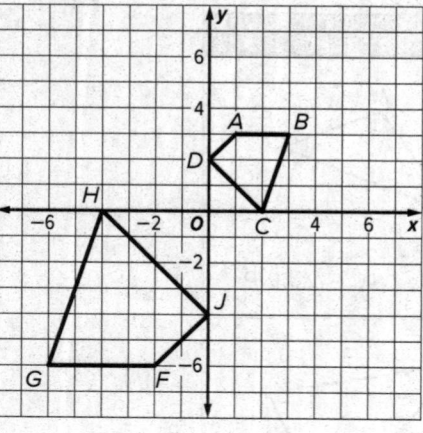

2.
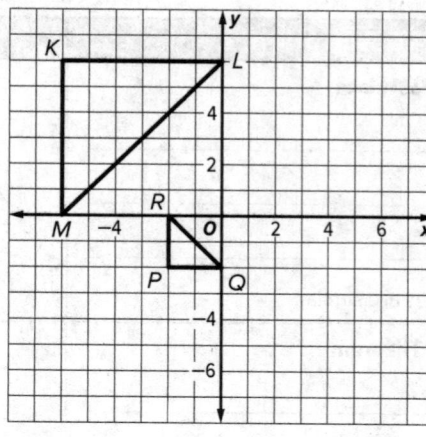

_____
_____
_____

3. **CONSTRUCT ARGUMENTS** Write a two-column proof for the following.

Given: $AD = 3AB$; $AE = 3AC$
Prove: $\triangle ABC \sim \triangle ADE$

Statements	Reasons
1.	1.
2.	2.
3.	3.
4.	4.
5.	5.

4. **CRITIQUE REASONING** Brandon was told that $\triangle PQR \sim \triangle STR$ and he was asked to find the value of x. He wrote the proportion $\frac{x}{12} = \frac{9}{8}$ and he solved the proportion to find that $x = 13.5$. Do you agree with Brandon's reasoning? If not, explain his error and show how to find the correct value of x.

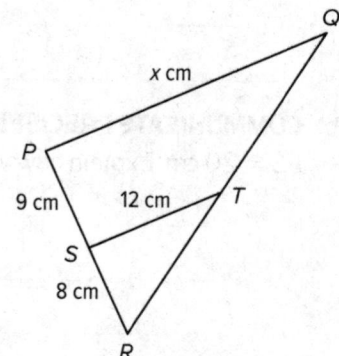

_____
_____
_____

5. Jacob wants to find the height of a building. On a sunny afternoon, he stands near the building and finds that his shadow is 3 feet long, while the building's shadow is 84 feet long. Jacob is 5 feet 6 inches tall.

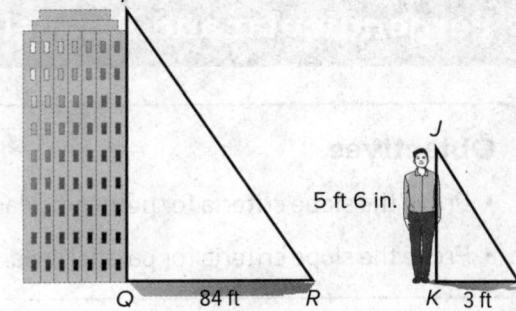

a. **USE A MODEL** Explain why you can conclude that the two triangles in the figure are similar.

_____

_____

_____

_____

_____

b. **USE STRUCTURE** Explain how you can use similar triangles to find the height of the building.

_____

_____

_____

_____

6. **CONSTRUCT ARGUMENTS** Emmaline arranges six wood boards to make two triangular planter boxes, as shown in the figure. She wants to know if she can conclude that ∠A is congruent to ∠D. Explain whether or not she can make this conclusion.

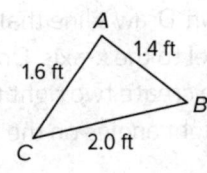
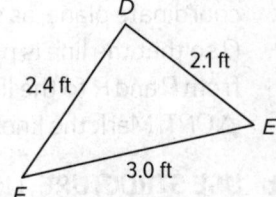

_____

_____

_____

_____

_____

7. **CONSTRUCT ARGUMENTS** In the figure, $KM \perp JL$ and $JK \perp KL$. Is △JKL ~ △JMK? Provide a proof to demonstrate their similarity or give an explanation of why they are not similar.

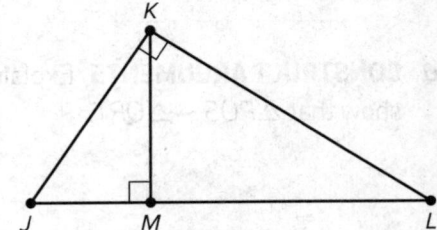

_____

_____

_____

_____

_____

_____

Objectives

• Prove the slope criteria for perpendicular lines.

• Prove the slope criteria for parallel lines.

Mathematical Practices
1, 2, 3, 6, 7

Previously you developed and applied the slope criteria for perpendicular and parallel lines. In this lesson, you will use what you know about similar triangles to prove these criteria.

EXAMPLE 1 Prove a Slope Criterion for Perpendicular Lines

EXPLORE Two nonvertical lines are perpendicular if and only if the product of their slopes is −1. Follow these steps to prove that this conditional statement is true. You will prove the converse of this statement in Exercise 2.

Given: Nonvertical lines $\overleftrightarrow{PQ}$ and $\overleftrightarrow{QR}$ with $\overleftrightarrow{PQ} \perp \overleftrightarrow{QR}$;
slope of $\overleftrightarrow{PQ} = m$, slope of $\overleftrightarrow{QR} = n$.

Prove: $mn = -1$

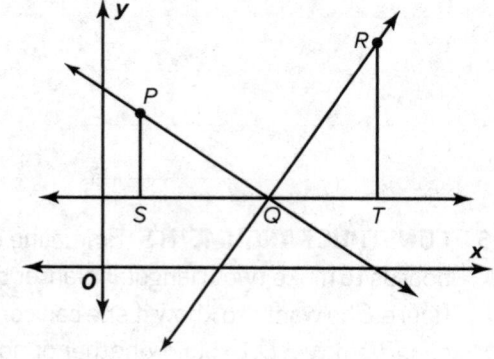

a. **PLAN A SOLUTION** Draw $\overleftrightarrow{PQ}$ and $\overleftrightarrow{QR}$ with $\overleftrightarrow{PQ} \perp \overleftrightarrow{QR}$ on a coordinate plane, as shown. Draw a line that passes through point Q so that the line is parallel to the x-axis. Draw perpendiculars from P and R to the line to create two right triangles, $\triangle PQS$ and $\triangle QRT$. Mark the known right angles on the figure.

b. **USE STRUCTURE** Use side lengths from the right triangles to complete each of the following.

slope of $\overleftrightarrow{PQ} = m = \dfrac{\text{rise}}{\text{run}} = -\dfrac{\boxed{}}{\boxed{}}$ slope of $\overleftrightarrow{QR} = n = \dfrac{\text{rise}}{\text{run}} = +\dfrac{\boxed{}}{\boxed{}}$

c. **REASON ABSTRACTLY** Since $\triangle PQS$ is a right triangle, $\angle QPS$ must be complementary to $\angle PQS$. Also, $\angle SQT$ is a straight angle and $\angle PQR$ is a right angle, so $\angle RQT$ must be complementary to $\angle PQS$. Explain why you can conclude that $\angle QPS \cong \angle RQT$.

_____

d. **CONSTRUCT ARGUMENTS** Explain how you can use the fact that $\angle QPS \cong \angle RQT$ to show that $\triangle PQS \sim \triangle QRT$.

_____

_____

e. **REASON ABSTRACTLY** Use the similar triangles to complete the following.

$\dfrac{\boxed{}}{QT} = \dfrac{\boxed{}}{RT}$; therefore, $\dfrac{RT}{QT} = \dfrac{\boxed{}}{\boxed{}}$.

f. **CALCULATE ACCURATELY** Complete the following to show how you can use your results from **parts b** and **e** to finish the proof.

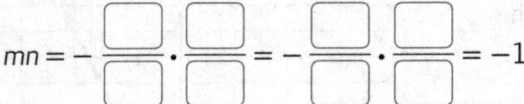

$$mn = - \frac{\square}{\square} \cdot \frac{\square}{\square} = - \frac{\square}{\square} \cdot \frac{\square}{\square} = -1$$

g. **USE STRUCTURE** Look back at the proof. Explain how the proof uses the fact that $\overleftrightarrow{PQ}$ and $\overleftrightarrow{QR}$ are nonvertical lines.

_____

_____

EXAMPLE 2 Prove a Slope Criterion for Parallel Lines

Two nonvertical lines are parallel if and only if they have the same slope. Follow these steps to prove that the converse of this statement is true. You will prove the conditional statement in Exercise 3.

Given: Nonvertical lines $\overleftrightarrow{AB}$ and $\overleftrightarrow{CD}$ with slope of $\overleftrightarrow{AB} = m$, slope of $\overleftrightarrow{CD} = n$, and $m = n$.

Prove: $\overleftrightarrow{AB} \parallel \overleftrightarrow{CD}$

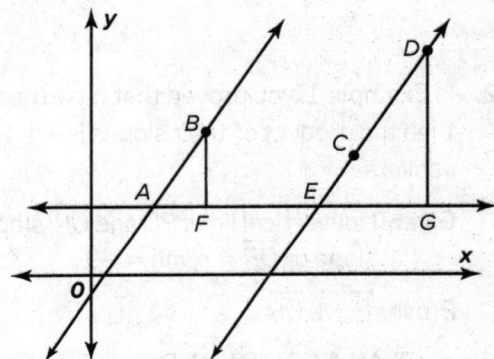

a. **PLAN A SOLUTION** Draw $\overleftrightarrow{AB}$ and $\overleftrightarrow{CD}$ on a coordinate plane, as shown. Draw a line through point A that is parallel to the x-axis. Assume the line intersects $\overleftrightarrow{CD}$ at point E. Draw perpendiculars from B and D to the line to create two right triangles, $\triangle ABF$ and $\triangle EDG$. Mark the known right angles in the figure.

b. **USE STRUCTURE** Use side lengths from the right triangles to complete each of the following.

slope of $\overleftrightarrow{AB} = m = \frac{\text{rise}}{\text{run}} = \frac{\square}{\square}$ slope of $\overleftrightarrow{CD} = n = \frac{\text{rise}}{\text{run}} = \frac{\square}{\square}$

c. **CALCULATE ACCURATELY** Use your results from **part b** to complete the following.

Since $m = n$, $\frac{\square}{\square} = \frac{\square}{\square}$. This proportion can be rewritten as $\frac{EG}{AF} = \frac{\square}{\square}$.

d. **CONSTRUCT ARGUMENTS** Explain how you can use the result of **part c** to show that $\triangle ABF \sim \triangle EDG$.

_____

_____

e. **CONSTRUCT ARGUMENTS** Explain how you can use the fact that $\triangle ABF \sim \triangle EDG$ to draw a conclusion about $\angle BAF$ and $\angle DEG$. Then explain how you can use this conclusion to complete the proof.

_____

_____

_____

1. Bonita examines the graph of the line shown in the figure at the right.

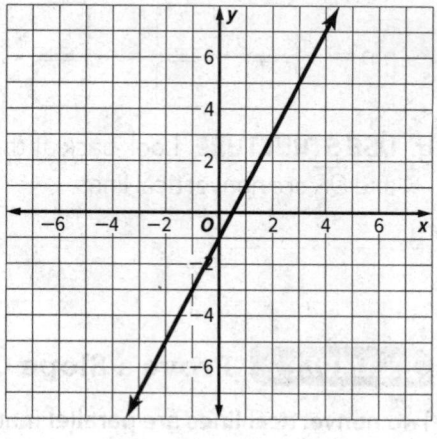

a. REASON ABSTRACTLY What is the equation for the line shown in slope-intercept form?

_____

b. CRITIQUE REASONING Without graphing, Bonita concludes that a line that passes through the points $(4, 4)$ and $(2, -2)$ will be parallel to this line. Is Bonita correct? Explain.

_____

_____

_____

2. In **Example 1**, you proved that if two nonvertical lines are perpendicular, then the product of their slopes is -1. Now you will prove the converse.

Given: Nonvertical lines $\overleftrightarrow{PQ}$ and $\overleftrightarrow{QR}$; slope of $\overleftrightarrow{PQ} = m$, slope of $\overleftrightarrow{QR} = n$, $mn = -1$.

Prove: $\overleftrightarrow{PQ} \perp \overleftrightarrow{QR}$

a. PLAN A SOLUTION Draw a figure based on the given information. Add any auxiliary lines or segments to the drawing that will be necessary to complete the proof.

b. CONSTRUCT ARGUMENTS Use your drawing to write a paragraph proof that $\overrightarrow{PQ} \perp \overleftrightarrow{QR}$.

_____

_____

_____

_____

_____

_____

_____

_____

_____

_____

3. In **Example 2**, you proved that if two nonvertical lines have the same slope, then the lines are parallel. Now you will prove the converse.

> **Given:** Nonvertical lines $\overleftrightarrow{AB}$ and $\overleftrightarrow{CD}$; $\overleftrightarrow{AB} \parallel \overleftrightarrow{CD}$;
>   slope of $\overleftrightarrow{AB} = m$, slope of $\overleftrightarrow{CD} = n$.

Prove: $m = n$

a. **PLAN A SOLUTION** Draw a figure based on the given information. Add to the drawing anything that will be necessary for completing the proof, but that was not given. Refer to **Example 2** for guidance.

b. **CONSTRUCT ARGUMENTS** Use your drawing to write a paragraph proof that $m = n$.

_____

_____

_____

_____

_____

_____

4. REASON QUANTITATIVELY Write the equation in slope-intercept form of each of the lines described below.

a. A line parallel to ℓ and passing through $(-2, 1)$

_____

_____

b. A line perpendicular to ℓ and passing through $(-2, 1)$

_____

_____

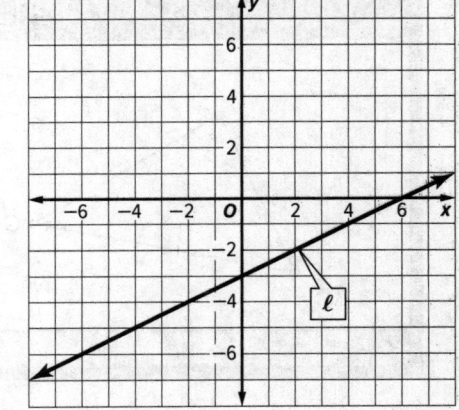

5. Matt wants to write an equation of a line perpendicular to the line represented by the equation $-20x + 5y = 30$ and that crosses at the same y-intercept.

a. **PLAN A SOLUTION** Describe the steps Matt should take to solve the problem.

_____

_____

_____

b. **COMMUNICATE PRECISELY** Write the equation of the perpendicular line in slope-intercept form.

_____

Mathematical Practices
2, 3, 4, 5, 7, 8

Objectives

- Prove the Triangle Proportionality Theorem and its converse.
- Prove the Triangle Midsegment Theorem.
- Use a compass and straightedge to partition a line segment in a given ratio.

EXAMPLE 1 **Investigate Proportionality in Triangles**

EXPLORE Use the Geometer's Sketchpad for this exploration.

a. **USE TOOLS** Draw a triangle. Label the vertices *A*, *B*, and *C*. Then plot a point *D* on $\overline{AB}$, as shown on the left below.

b. **USE TOOLS** Draw a line through point *D* that is parallel to $\overline{BC}$. Construct the intersection of this line and $\overline{AC}$, and label the point of intersection *E*, as shown on the right below.

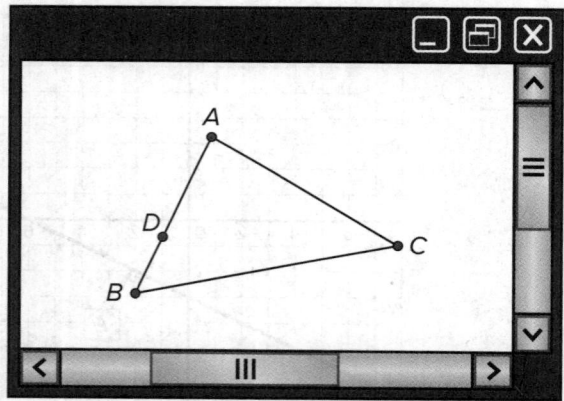
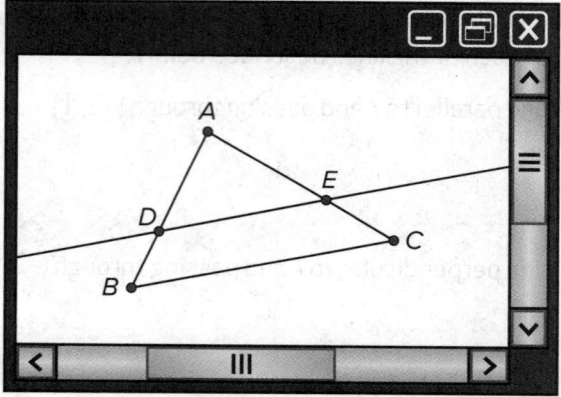

c. **USE TOOLS** Measure $\overline{AD}$, $\overline{DB}$, $\overline{AE}$, and $\overline{EC}$. Then calculate the ratios $\frac{AD}{DB}$ and $\frac{AE}{EC}$.

d. **FIND A PATTERN** Drag point *D* to different locations along $\overline{AB}$. Change the shape of $\triangle ABC$. What do you notice?

_____

e. **MAKE A CONJECTURE** Use your observations to state a conjecture. Be sure to state the conjecture as a conditional statement.

_____

_____

f. **USE STRUCTURE** Predict what will happen if you drag point *D* so that it is the midpoint of $\overline{AB}$.

_____

The relationship that you discovered in the investigation can be stated as a theorem. The converse of the theorem is also true. You will prove the theorem in **Example 2** and prove its converse as an exercise.

KEY CONCEPT **Triangle Proportionality**

Complete the statement of each theorem and complete the examples.

Theorem	Example
Triangle Proportionality Theorem If a line is parallel to one side of a triangle and intersects the other two sides, then _____ _____	If $\overline{DE} \parallel \overline{BC}$, then _____
Converse of the Triangle Proportionality Theorem If a line intersects two sides of a triangle and separates the sides into proportional corresponding segments, then, _____ _____	If $\frac{AD}{DB} = \frac{AE}{EC}$, then _____

EXAMPLE 2 **Prove the Triangle Proportionality Theorem**

Follow these steps to prove the Triangle Proportionality Theorem.

Given: $\overline{DE} \parallel \overline{BC}$

Prove: $\dfrac{AD}{DB} = \dfrac{AE}{AC}$

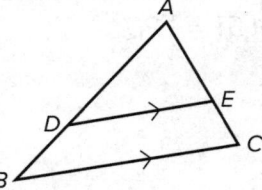

a. **CONSTRUCT ARGUMENTS** Explain how you can show that $\triangle ADE \sim \triangle ABC$.

_____

_____

_____

b. **REASON ABSTRACTLY** Show how to complete the proof. (*Hint:* Use the similar triangles to write a proportion. Then use the Segment Addition Postulate.)

_____

_____

_____

_____

A midsegment of a triangle is a segment with endpoints that are the midpoints of two sides of the triangle. For example, $\overline{ST}$ is a midsegment of $\triangle PQR$.

The Triangle Midsegment Theorem is a special case of the Converse of the Triangle Proportionality Theorem.

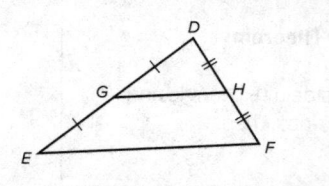

KEY CONCEPT   Triangle Midsegment Theorem

Complete the example of the theorem.

A midsegment of a triangle is parallel to one side of the triangle, and its length is one-half the length of that side.
Example: If G and H are midpoints of $\overline{DE}$ and $\overline{DF}$, then _____

EXAMPLE 3   Prove the Triangle Midsegment Theorem

Follow these steps to prove the Triangle Midsegment Theorem.

Given: S and T are midpoints of $\overline{PR}$ and $\overline{QR}$.
Prove: $\overline{ST} \parallel \overline{PQ}$ and $ST = \frac{1}{2}PQ$

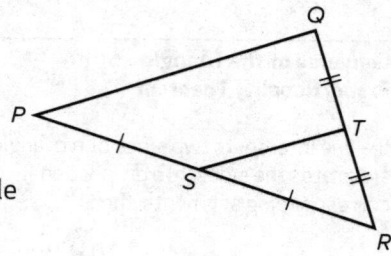

a. CONSTRUCT ARGUMENTS Explain how to use the Converse of the Triangle Proportionality Theorem to show that $\overline{ST} \parallel \overline{PQ}$.

_____

_____

b. CONSTRUCT ARGUMENTS Explain why $\triangle RST \sim \triangle RPQ$.

_____

_____

c. REASON ABSTRACTLY Explain how you can complete the proof by using the similar triangles to show that $ST = \frac{1}{2}PQ$.

_____

_____

The Triangle Proportionality Theorem can be generalized to situations in which three or more parallel lines are cut by two transversals.

KEY CONCEPT   Proportional Parts of Parallel Lines

Complete the example of the theorem.

If three or more parallel lines intersect two transversals, then they cut off the transversals proportionally.
Example: If $\overline{JK} \parallel \overline{LM} \parallel \overline{NP}$, then _____

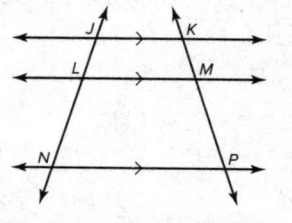

As a corollary of the previous theorem, if three or more parallel lines cut off congruent segments on one transversal, then they cut off congruent segments on every transversal.

EXAMPLE 4 Partition a Line Segment

Locate two points on $\overline{PQ}$ that partition the segment into 3 congruent parts.

a. **USE TOOLS** Use the straightedge to draw a ray, $\overrightarrow{PX}$. Place the point of the compass on P and make an arc that intersects $\overrightarrow{PX}$ at point A.

b. **USE TOOLS** Without changing the compass setting, place the point of the compass on A and make an arc that intersects $\overrightarrow{PX}$ at point B. Without changing the compass setting, place the point of the compass on B and make an arc that intersects $\overrightarrow{PX}$ at point C.

c. **USE TOOLS** Now use the straightedge to draw $\overline{CQ}$. Then construct lines through A and B that are parallel to $\overline{CQ}$. Label the intersection points with $\overline{PQ}$ as D and E. Point E is the required point.

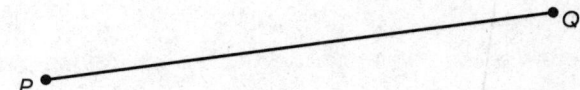

d. **CONSTRUCT ARGUMENTS** Explain why the construction works.

_____

_____

_____

PRACTICE

1. Follow these steps to prove the Converse of the Triangle Proportionality Theorem.

   Given: $\dfrac{AD}{DB} = \dfrac{AE}{EC}$

   Prove: $\overline{DE} \parallel \overline{BC}$

   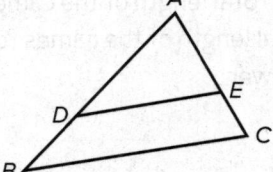

   a. **REASON ABSTRACTLY** Explain how you can use the given proportion to show that $\dfrac{AB}{AD} = \dfrac{AC}{AE}$.

   _____

   _____

   _____

   b. **CONSTRUCT ARGUMENTS** Explain how to use the proportion $\dfrac{AB}{AD} = \dfrac{AC}{AE}$ to complete the proof.

   _____

   _____

   _____

2. EVALUATE REASONABLENESS In the figure, $\overline{DE} \parallel \overline{BC}$, $BD = 12$, $EC = 10$, and $AE = 15$. Eliza found that the length of $\overline{AD}$ is 8. Is Eliza correct? Explain.

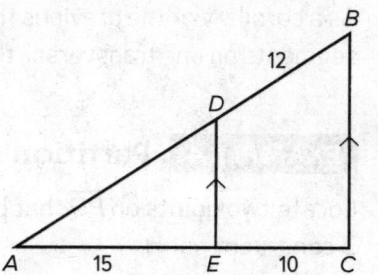

_____

_____

_____

USE TOOLS Use a compass and straightedge to locate a point that partitions the given line segment in the given ratio. Label the required point X.

3. 1:2

4. 3:1

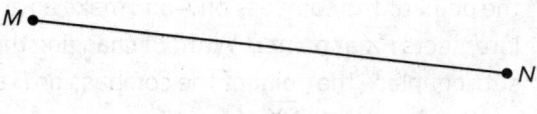

5. USE A MODEL The divider bars between the pieces of colored glass in a stained glass window are called *cames*. In the stained window at the right, the total length of the cames for $\triangle PQR$ is 78 centimeters. What is the total length of the cames for $\triangle JKL$? Give an argument to support your answer.

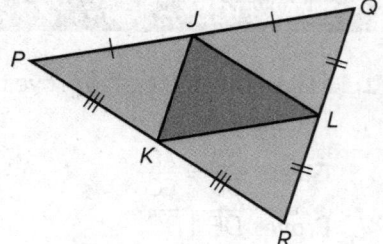

_____

_____

_____

6. USE A MODEL A crew is laying out a shuffleboard court using the plan shown at the right. A member of the crew wants to know the length of $\overline{AB}$, $\overline{BD}$, and $\overline{DF}$ to the nearest tenth of a foot. Explain how the crew member can calculate these lengths. Justify your steps.

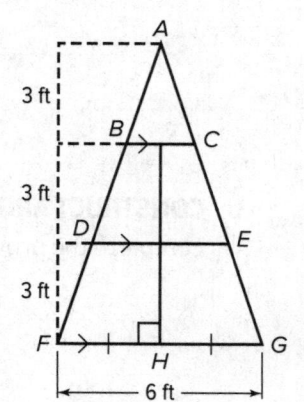

_____

_____

_____

_____

_____

7. In △PQR, the length of the base PQ = 16. A series of midsegments are drawn such that $\overline{ST}$ is the midsegment of △PQR, $\overline{UV}$ is the midsegment of △SRT, and $\overline{WX}$ is the midsegment of △URV.

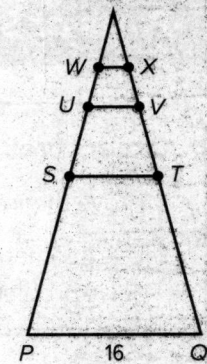

a. **CALCULATE ACCURATELY** What is the length of each midsegment?

   ST = _____   UV = _____   WX = _____

b. **USE STRUCTURE** What would be the measure of midsegment $\overline{YZ}$ of △WXR?

   _____

8. Bert is building the frame of one side of a roof. The roof forms an equilateral triangle with each side 14 feet long. For support, Bert wants to add beams that connect the midpoints of each side of the outer roof frame.

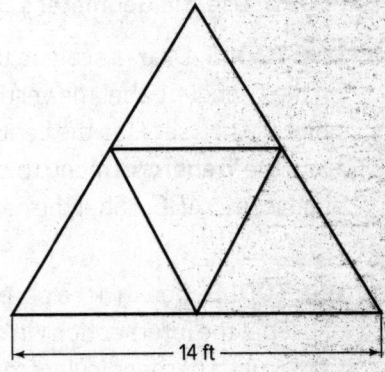

a. **CALCULATE ACCURATELY** How many beams of wood and of what length(s) will Bert need to construct the outer frame of the roof? How many beams of wood and of what length(s) will Bert need to construct the inner frame of the roof? Explain.

   _____
   _____
   _____
   _____

b. **REASON ABSTRACTLY** Write equations for finding the lengths of the outer and inner frames. What is the ratio of the perimeter of the larger outer frame to the perimeter of the inner support frame?

   _____
   _____
   _____
   _____

9. **CONSTRUCT ARGUMENTS** Write a paragraph proof for the following.

   Given: F is the midpoint of $\overline{AB}$, D is the midpoint of $\overline{AF}$, G is the midpoint of $\overline{AC}$, and E is the midpoint of $\overline{AG}$.

   Prove: BC = 4DE

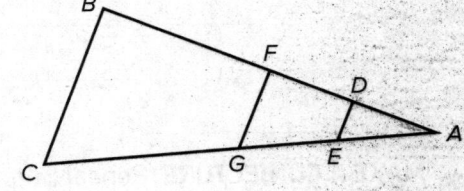

   _____
   _____
   _____

10. **USE TOOLS** Explain how you can use the construction in **Example 4** to locate a point that partitions $\overline{PQ}$ in the ratio 5:1.

   _____
   _____

Objectives

- Prove theorems about parts of similar triangles.

- Use geometric shapes and their measures to describe objects and solve problems related to similarity.

Mathematical Practices
1, 3, 4, 5, 6, 7

EXAMPLE 1 **Investigate Parts of Similar Triangles**

EXPLORE Use the Geometer's Sketchpad for this exploration.

a. **USE TOOLS** Draw a scalene triangle and a point not on the triangle. Label the vertices A, B, and C and the point P. Then set P as the center of dilation. Select **Dilate** from the **Transform** menu to create a triangle that is similar to $\triangle ABC$. Label the new triangle $\triangle DEF$.

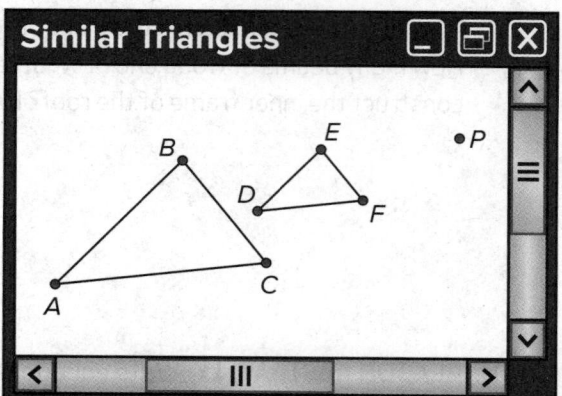

b. **USE TOOLS** Construct a perpendicular from B to $\overline{AC}$ and label the intersection with $\overline{AC}$ as point X. Construct a perpendicular from E to $\overline{DF}$ and label the intersection with $\overline{DF}$ as point Y.

c. **FIND A PATTERN** Calculate the ratios $\frac{AB}{DE}$ and $\frac{BX}{EY}$. Change the shape of $\triangle ABC$ and notice what happens to the ratios.

_____

_____

d. **MAKE A CONJECTURE** Use your observations to make a conjecture about altitudes in similar triangles.

_____

_____

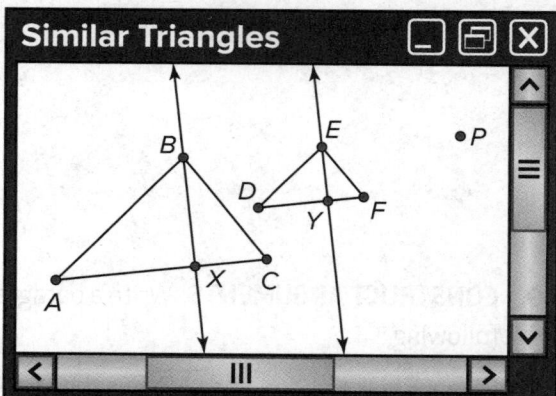

e. **MAKE A CONJECTURE** Repeat Steps a–c, but this time construct angle bisectors from B to $\overline{AC}$ and from E to $\overline{DF}$. Then make a conjecture about angle bisectors in similar triangles.

_____

_____

f. **MAKE A CONJECTURE** Repeat Steps a–c, but this time draw medians from B to $\overline{AC}$ and from E to $\overline{DF}$. Then make a conjecture about medians in similar triangles.

_____

_____

g. CALCULATE ACCURATELY Remember: $\triangle ABC \sim \triangle DEF$. If $AB = 15$, $BX = 12$, and $DE = 5$, what is the length of altitude $\overline{EY}$? Explain your reasoning.

_____

_____

_____

The conjectures you made in **Example 1** are theorems about the special segments in similar triangles. You will prove these theorems in **Example 2**, and **Exercises 1–2**.

KEY CONCEPT **Special Segments of Similar Triangles**

7.8	If two triangles are similar, the lengths of corresponding altitudes are proportional to the lengths of corresponding sides.
7.9	If two triangles are similar, the lengths of corresponding angle bisectors are proportional to the lengths of corresponding sides.
7.10	If two triangles are similar, the lengths of corresponding medians are proportional to the lengths of corresponding sides.

EXAMPLE 2 **Prove a Theorem**

Follow these steps to prove that if two triangles are similar, then the lengths of corresponding angle bisectors are proportional to the lengths of corresponding sides.

Given: $\triangle JKL \sim \triangle PQR$; $\overline{KX}$ bisects $\angle K$; $\overline{QY}$ bisects $\angle Q$.

Prove: $\dfrac{KX}{QY} = \dfrac{JK}{PQ}$

a. PLAN A SOLUTION Describe the main steps you will use in the proof.

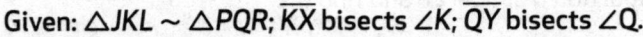

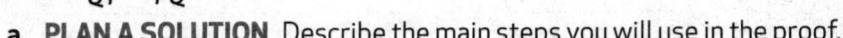

b. CONSTRUCT ARGUMENTS Write the proof as a paragraph proof.

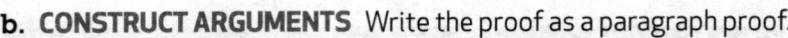

_____

_____

_____

_____

There is another important result about angle bisectors and proportionality.

KEY CONCEPT **Triangle Angle Bisector Theorem**

An angle bisector in a triangle divides the opposite side into two segments that are proportional to the lengths of the other two sides.

If $\overline{AD}$ bisects $\angle A$, then $\dfrac{BD}{DC} = \dfrac{AB}{AC}$.

EXAMPLE 3 Use Proportionality to Solve a Problem

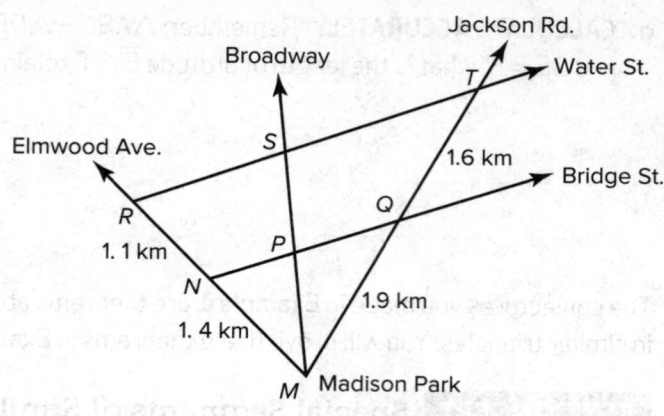

The organizer of a charity walk-a-thon wants to know the total distance of the route. The route starts at Madison Park, then along Elmwood Avenue to Water Street, along Water Street to Jackson Road, and along Jackson Road back to Madison Park. Broadway bisects the angle formed by Elmwood Avenue and Jackson Road. The distance along Water Street from Broadway to Jackson Road is 0.5 km greater than the distance along Water Street from Elmwood Avenue to Broadway.

a. PLAN A SOLUTION Describe the main steps you will use to solve this problem.

b. USE A MODEL Explain how the organizer of the walk-a-thon can write a proportion to determine the total distance of the route. Tell what any variables represent and justify your steps.

c. CALCULATE ACCURATELY Show how to solve the proportion from **part a** and find the total length of the route.

PRACTICE

1. **CONSTRUCT ARGUMENTS** Prove that if two triangles are similar, then the lengths of corresponding altitudes are proportional to the lengths of corresponding sides.

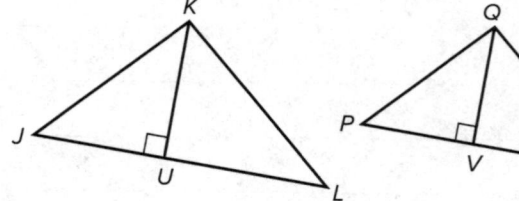

Given: $\triangle JKL \sim \triangle PQR$; $\overline{KU}$ and $\overline{QV}$ are altitudes.

Prove: $\dfrac{KU}{QV} = \dfrac{JK}{PQ}$

2. **CONSTRUCT ARGUMENTS** Prove that if two triangles are similar, then the lengths of corresponding medians are proportional to the lengths of corresponding sides.

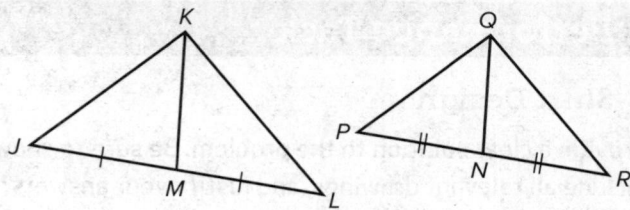

Given: $\triangle JKL \sim \triangle PQR$; $\overline{KM}$ and $\overline{QN}$ are medians.

Prove: $\dfrac{KM}{QN} = \dfrac{JK}{PQ}$

_____

_____

_____

_____

3. **USE A MODEL** An architect is designing the support structure for the roof of a new house, as shown in the figure at the right. According to the plan, $\overline{TY}$ bisects $\angle T$ and $XZ = 3.3$ meters. What are the lengths of $\overline{XY}$ and $\overline{YZ}$? Explain how you know your answer is reasonable.

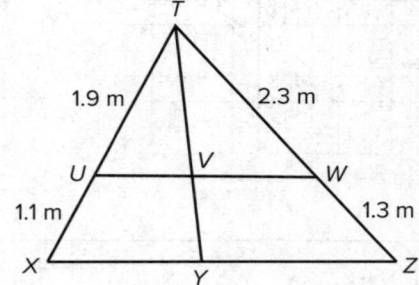

_____

_____

4. **USE A MODEL** Amani uses copper wire to make earrings. She bends a piece of wire 45 mm long to make $\triangle ABC$. Then she adds a piece of wire to make the angle bisector $\overline{BD}$. The lengths of $\overline{AD}$ and $\overline{DC}$ are as shown. Explain how Amani can find the lengths of $\overline{AB}$ and $\overline{BC}$ without measuring.

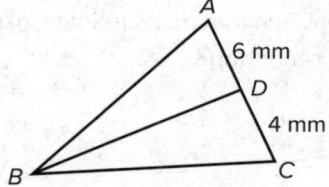

_____

_____

_____

5. **CRITIQUE REASONING** As part of an origami project, Emilio cuts out $\triangle PQR$ and folds it so that $\overline{QR}$ lies on top of $\overline{PR}$. When he unfolds the triangle, the crease $\overline{SR}$ is formed, as shown. Emilio claims that PS must be greater than QS. Do you agree? Explain why or why not.

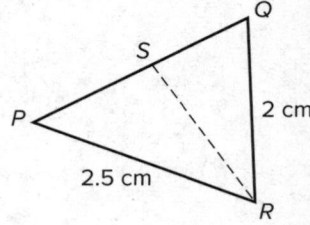

_____

_____

6. **CONSTRUCT ARGUMENTS** Does the angle bisector of a triangle always divide the side opposite it into two proportional and unequal parts? Explain.

_____

_____

_____

T-Shirt Design

Provide a clear solution to the problem. Be sure to show all of your work, include all relevant drawings, and justify your answers.

For the school picnic, the school is selling T-shirts. The logo on the shirt has many similar triangles. After deciding on the T-shirt's initial shape shown in the graph below, the design team creates many other similar triangles.

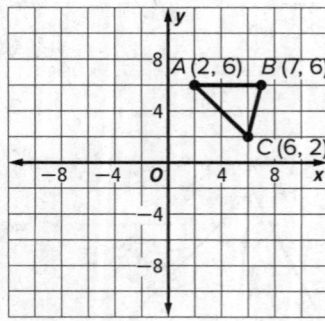

Part A

If the coordinates of the new triangle are $\left(2, \frac{2}{3}\right)$ and $\left(\frac{2}{3}, 2\right)$ and $\left(\frac{7}{3}, 2\right)$ what is the function that maps point (x, y) under this dilation? Is the new image an enlargement or reduction?

Part B

For each coordinate of the new triangle, state which coordinate mapped to it through the function you defined in **Part A**.

Part C

If the image in **Part A** is again scaled by the same factor as the original image, what will the new coordinates be? Use a function to relate the new coordinates to the original coordinates.

Part D

If the original image required 2 mL of ink to print, how much ink would a design that includes the original image, the image in **Part A**, and the image in **Part C** require? Explain your reasoning.

Planting for Privacy

Provide a clear solution to the problem. Be sure to show all of your work, include all relevant drawings, and justify your answers.

Louise has decided to plant a row of trees to block the view of her house from the scenic lookout point *P*. The image shows the valley with her house and the scenic lookout at point *P*. Assume all trees will grow parallel to the front and back of her house.

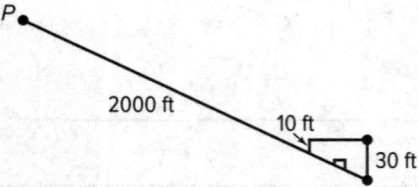

Part A

Sketch the sightline from the lookout point to the front of the house. Sketch the sightline from the lookout point to the back of the house. Which sightline should the trees cover to provide the most privacy?

Part B

If the trees will be 6 feet tall, where can Louise plant the trees so that they are as close to her house as possible while still covering the sightline chosen in **Part A**? Explain.

Part C

Louise decides to plant trees that will be 10 feet tall at a distance of 500 feet from point P. After planting the trees, she decides that she wants to add a new level to her house. How tall can the new level be such that the trees will cover the new sightline from point P?

Part D

Suppose a tall platform was added to the lookout. How would this change the height requirements for the trees? Justify your answer.

1. What is the value of *x* in the following diagram?

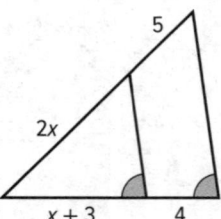

$x = \boxed{}$

2. An angle bisector in a triangle divides the opposite side of the triangle into segments that are 6 and 10 units long. Another side of the triangle is 15 units long. What are the possible lengths of the third side of the triangle?

$\boxed{}$ or $\boxed{}$ units

3. Quadrilateral *ABCD* has vertices *A*(4, 3), *B*(−2, 7), *C*(−3, 0), and *D*(4, −4). After quadrilateral *ABCD* is dilated about the origin, *A′* is located at (6, 4.5). Find the scale factor and complete the following about *A′B′C′D′*.

Scale factor: $\boxed{}$

B′($\boxed{}$, $\boxed{}$)

C′($\boxed{}$, $\boxed{}$)

D′($\boxed{}$, $\boxed{}$)

4. Elena is standing outside near a flagpole on a sunny day. Elena is 54 inches tall, and her shadow is 66 inches long. The shadow of the flagpole is 14 feet 8 inches long. The flagpole is $\boxed{}$ feet $\boxed{}$ inches tall.

5. Consider the following diagram.

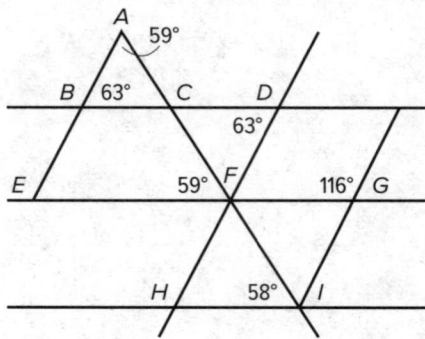

Select the triangles that are similar to △*ABC*.

△*AFE*   △*CDF*

△*FIH*   △*FGI*

6. Consider the following diagram.

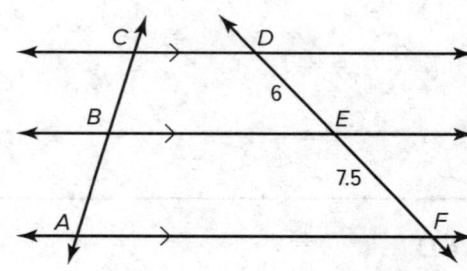

If *AC* = 9, then *AB* = $\boxed{}$.

7. △*JKL* ~ △*MNO*. The perimeter of △*JKL* is 24 and the perimeter of △*MNO* is 60. If *MO* = 15 and *MN* = 20, then *KL* = $\boxed{}$.

8. Point *D* is the midpoint of $\overline{AB}$ and point *E* is the midpoint of $\overline{BC}$. If the perimeter of △*BDE* is 15 units, then the perimeter of △*BAC* is $\boxed{}$ units. If the area of △*BAC* is 56 square units, then the area of △*BDE* is $\boxed{}$ square units.

9. $\triangle X'Y'Z'$ is a dilation of $\triangle XYZ$. In the diagram below, draw lines to show how to find the center of dilation. What are the coordinates of the center of dilation?

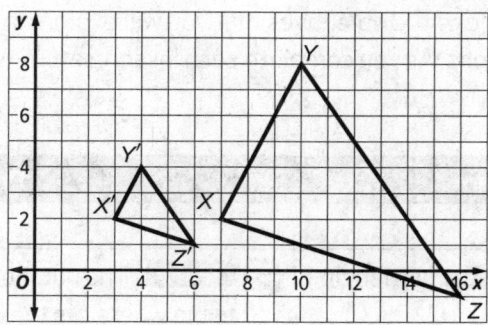

_____

10. Consider the following diagram.

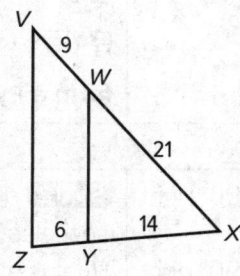

Are $\overline{VZ}$ and $\overline{WY}$ parallel? Explain how you know.

_____
_____
_____
_____

11. $\triangle JKL$ has vertices $J(-9, 4)$, $K(-3, 8)$, and $L(5, -4)$. $\triangle DEF$ has vertices $D(-1, 4)$, $E(2, 2)$, and $F(6, 8)$. Are the triangles similar? Show your work.

_____
_____
_____
_____
_____
_____
_____

CHAPTER FOCUS Learn about some of the objectives that you will explore in this chapter. Answer the preview questions. As you complete each lesson, return to these pages to check your work.

What You Will Learn	Preview Question
The Pythagorean Theorem	
• Prove the Pythagorean Theorem using triangle similarity. • Use the Pythagorean Theorem to solve right triangles in applied problems.	**SPM 3** The numbers a, b, and c form a Pythagorean triple where c is the length of the hypotenuse. Paolo claims that $3a$, $3b$, and $3c$ also form a Pythagorean triple. Do you agree? Justify your answer. **Because a, b, and c form a Pythagorean triple,** **$a^2 + b^2 = c^2$. By the Mult. Prop. of Equality,** **$9a^2 + 9b^2 = 9c^2$ which can be written as** **$(3a)^2 + (3b)^2 = (3c)^2$. So by def. $3a$, $3b$, and $3c$** **form a Pythagorean triple.**
Special Right Triangles	
• Identify and apply side ratios in 45°−45°−90° right triangles. • Identify and apply side ratios in 30°−60°−90° right triangles.	**SMP 2** The rectangle below is composed of an isosceles trapezoid and two 30°−60°−90° triangles. What is the perimeter of the rectangle to the nearest hundredth? 15 8 60° **64.48 units**
Trigonometry	
• Define trigonometric ratios for acute angles in right triangles. • Use trigonometric ratios and the Pythagorean Theorem to solve right triangles. • Use the relationship between the sine and cosine of complementary angles.	**SMP 1** Find the height of the rectangular prism to the nearest hundredth. Justify your answer. 21 ft 35° h **12.05 ft; $\sin 35 = \frac{h}{21}$; 21 $\sin 35 = h$;** **$h \approx 12.05$**

What You Will Learn	Preview Question
Angles of Elevation and Depression	
• Solve problems involving angles of elevation. • Solve problems involving angles of depression.	**SMP 4** A boat is spotted from the top of a lighthouse that is 46 meters tall. The angle of depression from the top of the lighthouse to the boat is 48°. Write and solve an equation to find the distance from the boat to the lighthouse. Round to the nearest meter. **SMP 1** The angle of elevation from a point at the base of a mountain to the peak is 32°. If the straight-line distance from the point to the peak is 21,120 feet, write and solve an equation to find how many feet higher the peak is than the point at the base. Round to the nearest foot.
The Law of Sines and Law of Cosines	
• Derive a trigonometric formula for the area of a triangle. • Prove and apply the Law of Sines. • Prove and apply the Law of Cosines.	**SMP 7** Solve the triangle. Justify your work.

Objectives

• Prove the Pythagorean Theorem using triangle similarity.

• Use the Pythagorean Theorem to solve right triangles in applied problems.

Mathematical Practices
1, 2, 3, 4, 5, 6, 7

EXAMPLE 1 **Investigate Similarity in Right Triangles**

EXPLORE Use the right triangle, $\triangle PQR$, for this investigation.

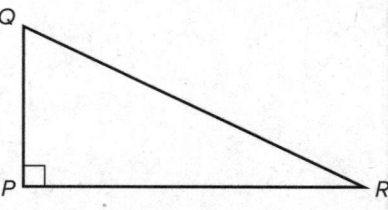

a. **USE STRUCTURE** Draw an altitude from P to the hypotenuse. Label the point of intersection S.

b. **CONSTRUCT ARGUMENTS** Is $\triangle PSR$ similar to $\triangle QPR$? If so, construct an argument to explain why. If not, explain why not.

_____

_____

c. **CONSTRUCT ARGUMENTS** Is $\triangle QSP$ similar to $\triangle QPR$? If so, construct an argument to explain why. If not, explain why not.

_____

_____

d. **CONSTRUCT ARGUMENTS** Is $\triangle QSP$ similar to $\triangle PSR$? If so, construct an argument to explain why. If not, explain why not.

_____

_____

_____

e. **MAKE A CONJECTURE** Use your observations to state a conjecture about the triangles that are formed when you draw an altitude to the hypotenuse of a right triangle.

_____

_____

_____

f. **REASON ABSTRACTLY** Explain how you can use your conjecture to find the value of x.

_____

_____

_____

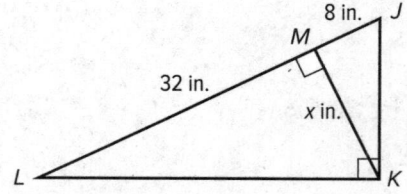

The Pythagorean Theorem states one of the most important relationships in geometry. You can use the similarity relationships that you observed in **Example 1** to prove the Pythagorean Theorem.

KEY CONCEPT The Pythagorean Theorem

Complete the example of the theorem.

The Pythagorean Theorem	Example
In a right triangle, the sum of the squares of the lengths of the legs is equal to the square of the length of the hypotenuse.	 If $\triangle ABC$ is a right triangle with right angle C, then _____

EXAMPLE 2 Prove the Pythagorean Theorem

Follow these steps to prove the Pythagorean Theorem.

Given: $\triangle ABC$ with right angle C.

Prove: $a^2 + b^2 = c^2$

a. **USE STRUCTURE** Draw an altitude from C to the hypotenuse and label the resulting side lengths as shown in the figure. Identify the similar triangles in the figure.

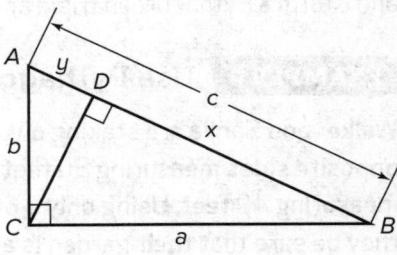

_____

b. **CONSTRUCT ARGUMENTS** Complete the remainder of the proof.

$\dfrac{a}{x} = \dfrac{\square}{a}$ $\dfrac{b}{y} = \dfrac{\square}{b}$ In similar triangles, corresponding side lengths are proportional.

$a^2 = $ _____ $b^2 = $ _____   In a proportion, cross products are equal.

$a^2 + b^2 = $ _____   Addition Property of Equality

$a^2 + b^2 = c($ _____ $)$ Distributive Property

$a^2 + b^2 = c($ _____ $)$ Substitution

$a^2 + b^2 = $ _____   Simplify.

c. **COMMUNICATE PRECISELY** In addition to the Substitution Property of Equality, what postulate or theorem are you using in the next-to-last step of the proof when you substitute c for $x + y$? Explain.

_____

_____

_____

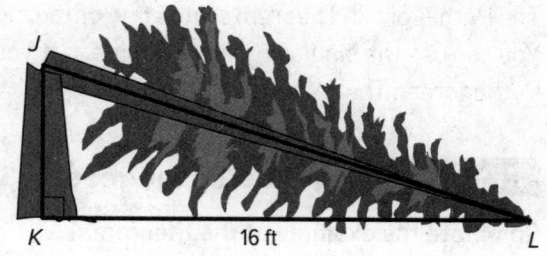

EXAMPLE 3 | Use the Pythagorean Theorem

A redwood tree in a national park is 20 meters tall. After it is struck by lightning, the tree breaks and falls over, as shown in the figure. The top of the tree lands at a point 16 feet from the centerline of the tree. A park ranger wants to know the height of the remaining stump of the tree.

a. USE A MODEL The ranger lets x represent the height of the stump, $\overline{JK}$. Explain how the ranger can write an expression for the length of $\overline{JL}$. Then write an equation that can be used to solve the problem.

_____

_____

b. CALCULATE ACCURATELY Show how to solve the equation from **part a** to find the height of the stump.

_____

_____

_____

KEY CONCEPT

A **Pythagorean triple** is a set of three integers that form the sides of a right triangle. If a, b, and c form a Pythagorean triple and the largest of the three is c, then $a^2 + b^2 = c^2$.

EXAMPLE 4 | Use Pythagorean Triples

Walker and Sanjia are staking out a garden that has one pair of opposite sides measuring 30 feet and the other pair of sides measuring 40 feet. Using only a 60-foot-long tape measure, how can they be sure that their garden is a rectangle?

a. USE A MODEL Draw a model of the garden with diagonal t. Let $p = 30$ and $q = 40$.

b. USE STRUCTURE If the garden is a rectangle, what must be true about p, q, and t? Why?

_____

_____

c. COMMUNICATE PRECISELY Walker measures the diagonal and finds that it is 50 feet long. Is there enough information to determine whether their garden is a rectangle? Explain.

_____

_____

_____

_____

_____

1. There is another way to use similarity to prove the Pythagorean Theorem. Given $\triangle ABC$ with right angle C, let M be the midpoint of $\overline{AB}$. Rotate $\triangle ABC$ $180°$ about point M to create the rectangle $ADBC$. Then draw the perpendicular segment from D to $\overline{AB}$, as shown.

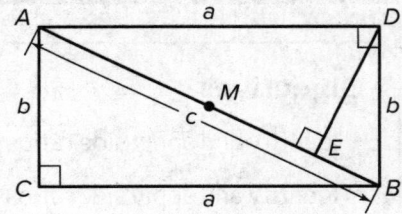
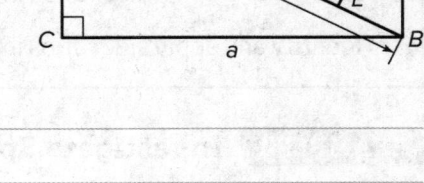

   a. **REASON ABSTRACTLY** Explain why $\triangle DAE \sim \triangle ABC$ and $\triangle BDE \sim \triangle ABC$.

   _____

   _____

   _____

   _____

   _____

   _____

   b. **CONSTRUCT ARGUMENTS** Explain how to complete the proof.

   _____

   _____

   _____

2. **INTERPRET PROBLEMS** An office park has a rectangular lawn with the dimensions shown. Employees often take a shortcut by walking from P to R, rather than from P to S to R. What is the total distance an employee saves in a week by taking the shortcut twice a day for five days? Explain.

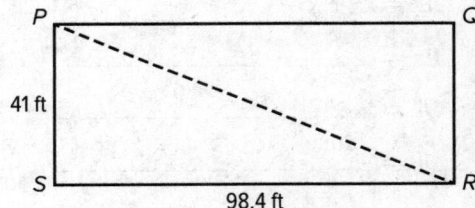

   _____

   _____

   _____

3. **USE A MODEL** Eduardo and Lisa both leave school on their bikes at the same time. Eduardo rides due east at 12 miles per hour for 45 minutes. Lisa rides due south at 16 miles per hour for 30 minutes. Draw a diagram in the space at the right to represent the problem. To the nearest hundredth of a mile, how far apart are they when they stop riding their bikes?

   _____

4. **USE TOOLS** A landscape architect is making a path between two trees, A and B, that are 60 feet apart. He wants the path to be 70 feet long and to contain one right angle, as shown. Write an equation to determine how long the legs of the path should be. Use a calculator to help you solve the equation. Find the lengths to the nearest tenth of a foot.

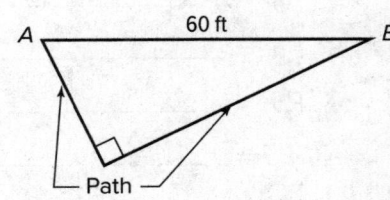

   _____

The Pythagorean Theorem **257**

Objectives

- Identify and apply side ratios in 45°−45°−90° right triangles.

- Identify and apply side ratios in 30°−60°−90° right triangles.

EXAMPLE 1 **Investigate Special Right Triangles**

Some common right triangles have special properties. In this exploration you will examine these right triangles and make generalizations about their side ratios.

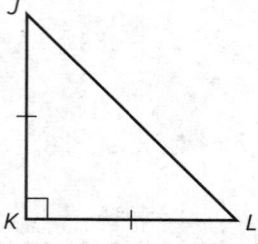

a. USE STRUCTURE What are the measures of the acute angles in $\triangle JKL$? Explain how you know.

_____

_____

_____

_____

b. REASON ABSTRACTLY Suppose $JK = KL = x$. Complete the following to find JL.

$JL^2 = x^2 + x^2$ Pythagorean Theorem

$JL^2 =$ _____   Combine like terms.

$JL =$ _____   Take the square root of each side.

$JL =$ _____   Simplify.

c. USE STRUCTURE $\triangle PQR$ is equilateral and $\overline{PS}$ is an altitude. What are the measures of the acute angles in $\triangle QPS$? Explain how you know.

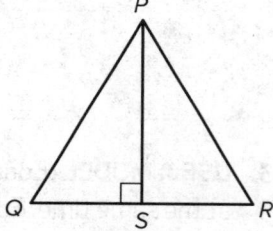

_____

_____

_____

d. USE STRUCTURE Suppose $QS = y$. What is the length of QP in terms of y? Explain.

_____

_____

e. REASON ABSTRACTLY Complete the following to find PS.

$PS^2 + y^2 =$ _____   Pythagorean Theorem

$PS^2 + y^2 =$ _____   Simplify.

$PS^2 =$ _____   Subtract y^2 from each side.

$PS =$ _____   Take the square root of each side.

$PS =$ _____   Simplify.

An isosceles right triangle is known as a 45°–45°–90° right triangle. In the exploration, you also investigated side lengths in a 30°–60°–90° right triangle. The following Key Concept box summarizes relationships among the sides of these special right triangles.

KEY CONCEPT **Special Right Triangles**

Label the missing side lengths in each figure in terms of the given variable.

Theorem	Figure
45°–45°–90° Triangle Theorem In a 45°–45°–90° Triangle, the legs are congruent and the length of the hypotenuse is $\sqrt{2}$ times the length of a leg.	*B* (right angle) 45° *C*, *x*, 45° *A*
30°–60°–90° Triangle Theorem In a 30°–60°–90° Triangle, the length of the hypotenuse is 2 times the length of the shorter leg and the length of the longer leg is $\sqrt{3}$ times the length of the shorter leg.	*F* 30° *D* 60° *y* *E* (right angle)

EXAMPLE 2 **Apply the 45°–45°–90° Triangle Theorem**

A lumber yard sells square pieces of wood for $0.02 per square inch. Adrian wants to buy a square piece of wood and cut it in half along a diagonal to make two tabletops like the one shown. He wants the longest side of each tabletop to be 30 inches, and he wants to know how much the wood will cost.

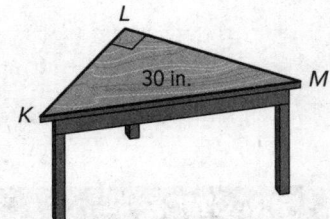

a. **PLAN A SOLUTION** Describe the main steps you will use to solve this problem.

_____

_____

_____

b. **USE STRUCTURE** Explain how to find the lengths of the legs of △KLM.

_____

_____

c. **REASON QUANTITIVELY** Explain how Adrian can find the cost of the wood for the tabletops.

_____

_____

_____

EXAMPLE 3 Apply the 30°—60°—90° Triangle Theorem

Tenisha is in charge of building a ramp for a loading dock. According to the plan, the ramp makes a 30° angle with the ground, as shown. Also, the plan states that $\overline{ST}$ is 4 feet longer than $\overline{RS}$. Tenisha needs to determine the lengths of the three sides of the ramp.

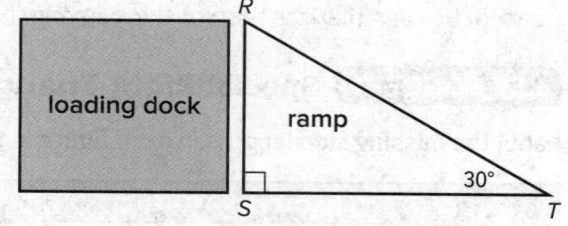

a. **USE A MODEL** Let $RS = x$ ft. Explain how you can write two different expressions for ST. Then use the expressions to write an equation that Tenisha can use to find the value of x.

_____

_____

b. **USE TOOLS** Solve the equation you wrote in **part a**. Use a calculator to find the value of x to the nearest thousandth. Then show how to find the lengths of the sides of the ramp to the nearest thousandth.

_____

_____

c. **EVALUATE REASONABLENESS** Explain how you know the side lengths are reasonable.

_____

_____

_____

PRACTICE

USE STRUCTURE Determine the value of the variable(s) in each figure. Express answers in radical form when necessary.

1.
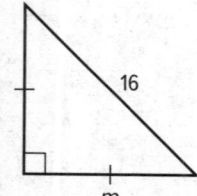

2.
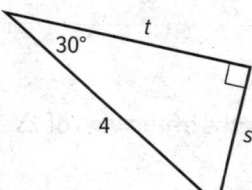

3.
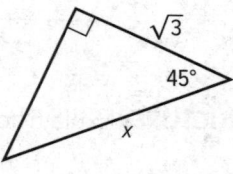

4.
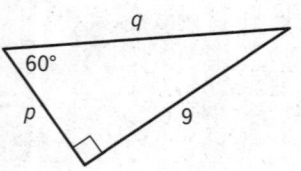

5.
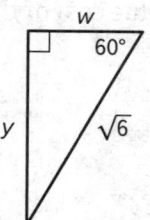

6.
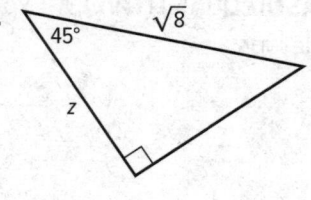

7. USE A MODEL Liling is making a quilt. She starts with two small squares of material and cuts them along the diagonal. Then she arranges the four resulting triangles to make a large square quilt block. She wants the large quilt block to have an area of 36 square inches. What side lengths should she use for the two small squares of material? Explain.

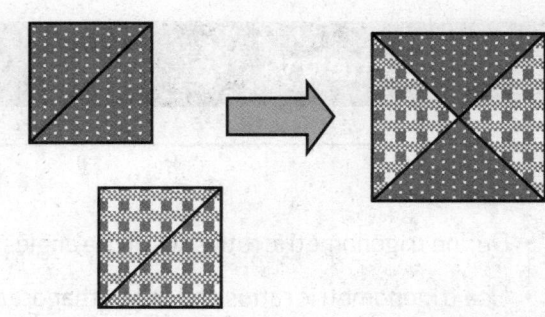

_____

_____

_____

_____

8. CRITIQUE REASONING A student was told that the length of $\overline{QR}$ in the figure is 10 centimeters. He marked the other side lengths as shown. Do you agree with the student's work? If so, explain why. If not, explain how you know the student made an error and then give the correct side lengths.

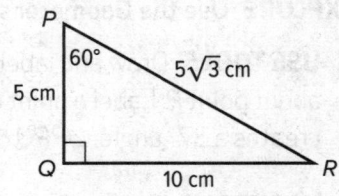

_____

_____

_____

9. DESCRIBE A METHOD The support beams for a bridge have cross sections that are equilateral triangles. An engineer wants to determine the area of a cross section if she knows the length of a side of the triangle. Explain how she can develop a general method that will give her the area in square feet of a cross section that has a side length of x feet.

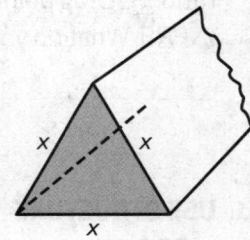

_____

_____

_____

_____

10. USE TOOLS Jeffrey has 40 feet of fencing to make a triangular enclosure for his chickens. He wants the enclosure to be a right isosceles triangle. Explain how Jeffrey can write an equation to find the dimensions of the enclosure, assuming he uses all of the fencing. Then use a calculator to find the dimensions of the sides of the enclosure to the nearest tenth of a foot.

_____

_____

_____

Objectives

- Define trigonometric ratios for acute angles in right triangles.

- Use trigonometric ratios and the Pythagorean Theorem to solve right triangles.

- Use the relationship between the sine and cosine of complementary angles.

Mathematical Practices
1, 2, 3, 4, 5, 6, 7, 8

EXAMPLE 1 **Investigate Ratios in Similar Triangles**

EXPLORE Use the Geometer's Sketchpad for this exploration.

a. **USE TOOLS** Draw and label $\overline{PQ}$. Then rotate $\overline{PQ}$ by 37° about point P. Label a point on the new line point R. This creates a 37° angle, $\angle RPQ$.

b. **USE TOOLS** Draw a perpendicular line from point R to $\overline{PQ}$. Label the intersection point S. You have constructed a right triangle, $\triangle PRS$, with a 37° angle at vertex P.

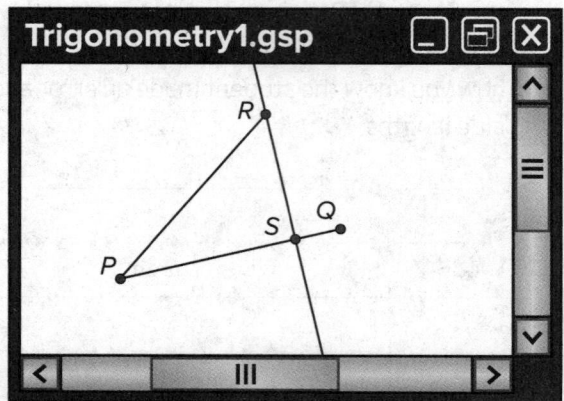

Trigonometry1.gsp

c. **FIND A PATTERN** Measure $\overline{RS}$ and $\overline{PR}$. Then calculate the ratio $\frac{RS}{PR}$. Drag point P to change the size and location of $\triangle PRS$. What do you notice about the ratio $\frac{RS}{PR}$?

_____

_____

d. **USE STRUCTURE** As you drag point P, what is true about all of the triangles named $\triangle PRS$ that you create? Why?

_____

_____

e. **CONSTRUCT ARGUMENTS** The figure shows two right triangles that each have an acute angle that measures 37. Explain why $\frac{BC}{AB} = \frac{EF}{DE}$. What does this tell you about any right triangle with a 37° angle?

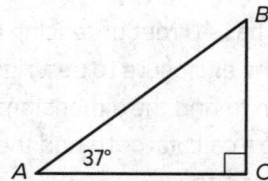
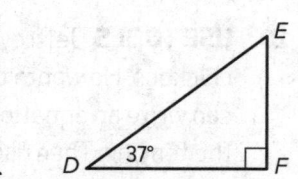

_____

_____

_____

_____

A **trigonometric ratio** is the ratio of the lengths of two sides of a right triangle.

KEY CONCEPT Trigonometric Ratios

Use side lengths in the figure to complete the example of each trigonometric ratio.

Trigonometric Ratio	Example	Figure
In a right triangle, $\triangle ABC$, with an acute angle at vertex A, the **sine** of $\angle A$ (sin A) is the ratio of the length of the leg opposite $\angle A$ to the length of the hypotenuse.	$\sin A =$ _____	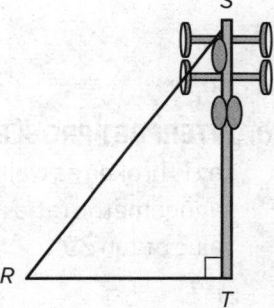
In a right triangle, $\triangle ABC$, with an acute angle at vertex A, the **cosine** of $\angle A$ (cos A) is the ratio of the length of the leg adjacent to $\angle A$ to the length of the hypotenuse.	$\cos A =$ _____	
In a right triangle, $\triangle ABC$, with an acute angle at vertex A, the **tangent** of $\angle A$ (tan A) is the ratio of the length of the leg opposite $\angle A$ to the length of the leg adjacent to $\angle A$.	$\tan A =$ _____	

EXAMPLE 2 Use a Trigonometric Ratio

A cell phone tower is supported by a guy wire as shown. Celia wants to determine the height of the tower. She finds that the guy wire makes an angle of 53° with the ground and it is attached to the ground 65 feet from the base of the tower.

a. USE A MODEL Label the known information in the figure.

b. INTERPRET PROBLEMS Let the height of the tower be x. Which trigonometric ratio should you use to write an equation that you can solve for x? Justify your choice.

_____

_____

_____

c. CALCULATE ACCURATELY Write and solve an equation for the height of the tower. Explain your steps. Round the height of the cell phone tower to the nearest tenth of a foot.

_____

_____

d. CALCULATE ACCURATELY What if Celia had wanted to find the length of the guy wire? What would you have done differently to solve the problem? Explain.

_____

_____

e. INTERPRET PROBLEMS If the tower's height is increased to 121.5 feet and the angle at R remains the same, find the distance from the base of the tower at which the guy wire needs to be attached to the ground. Round the distance to the nearest tenth of a foot.

_____

EXAMPLE 3 Use Complementary Angles

Blake needs to use his calculator to find the value of sin 29°. He finds that his calculator's sine key is broken. He wants to know if there is another way he can use his calculator to find this value.

a. **REASON ABSTRACTLY** In the figure, $m\angle A = 29$. What is $m\angle B$? Why? Mark the angle measure in the figure.

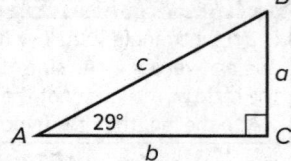

_____

_____

b. **REASON ABSTRACTLY** Using side lengths from the figure, write a ratio for sin 29°. What other trigonometric ratio for an angle in the figure is equal to this? Explain.

_____

_____

c. **USE TOOLS** Explain how Blake can use his calculator to find the value of sin 29°. Then use your calculator to verify that your answer is correct.

_____

_____

d. **INTERPRET PROBLEMS** Unfortunately, Blake discovers that his calculator's tangent key is broken as well. This means that he is only able to use the cosine key to evaluate trigonometric ratios for angles. Explain how Blake can use his calculator to find the value of tan 29°.

_____

_____

_____

If you know the sine, cosine, or tangent of an angle, you can find the measure of the angle.

KEY CONCEPT Inverse Trigonometric Ratios

Complete each example.

Inverse Trigonometric Ratio	Example
If the sine of A is x, then the **inverse sine** of x is the measure of ∠A.	If sin 30° = 0.5, then $\sin^{-1} 0.5 =$ _____.
If the cosine of A is x, then the **inverse cosine** of x is the measure of ∠A.	If cos 60° = 0.5, then $\cos^{-1} 0.5 =$ _____.
If the tangent of A is x, then the **inverse tangent** of x is the measure of ∠A.	If tan 45° = 1, then $\tan^{-1} 1 =$ _____.

When you are given two side lengths of a right triangle, or one side length and the measure of one acute angle, you can use trigonometry to find all of the remaining side lengths and angle measures. This process is known as *solving a right triangle*.

EXAMPLE 4 **Solve a Right Triangle**

Hailey is mounting a new shelf on the wall of her bedroom. The shelf is 18 cm wide. To support the shelf, Hailey will attach a metal brace that is 21 cm long to the edge of the shelf and to the wall, as shown. Hailey wants to know how far below the shelf she should attach the brace to the wall. She also wants to know the angle the brace will make with the wall and the shelf.

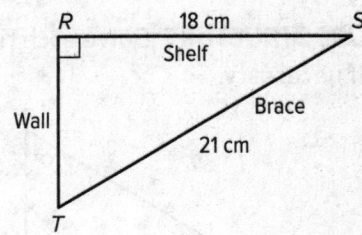

a. **PLAN A SOLUTION** Describe the steps you can use to find the two missing angle measures and the length of $\overline{RT}$.

_____

_____

_____

b. **USE TOOLS** Explain how you can use your calculator to find $m\angle S$ to the nearest degree.

_____

_____

c. **COMMUNICATE PRECISELY** How can Hailey find the $m\angle T$? Describe two different methods.

_____

_____

_____

d. **REASON QUANTITATIVELY** How far below the shelf should Hailey attach the brace to the wall? Explain your steps. Round to the nearest tenth of a centimeter.

_____

_____

_____

e. **REASON QUANTITATIVELY** Explain how you can solve **part d** using trigonometric ratios.

_____

_____

f. **DESCRIBE A METHOD** Compare your answers in **part d** and **part e**. Why might using the trigonometric method be better in some cases? Why might the Pythagorean Theorem method be better in some cases?

_____

_____

_____

USE STRUCTURE Solve each right triangle. Round to the nearest tenth, if necessary.

1.

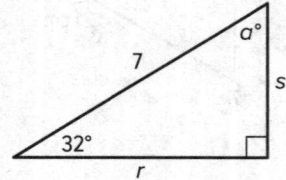

2.

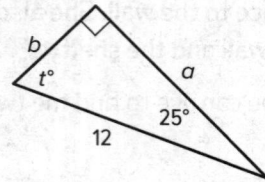

3.

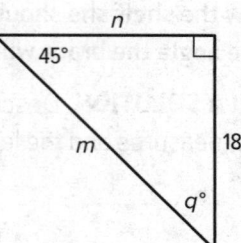

4. **CONSTRUCT ARGUMENTS** Lawrence wants to know the value of cos 40°, but he does not have a calculator. He uses software to draw two right triangles that have an acute angle that measures 40, as shown. He plans to use the software to measure the length of the side adjacent to the 40° angle and the length of the hypotenuse and then calculate the ratio of these lengths. Does it matter which triangle Lawrence uses when he measures these lengths and finds the cosine ratio? Explain.

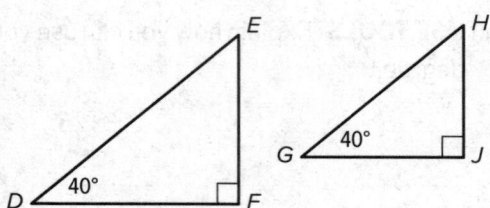

_____

_____

_____

5. a. **CRITIQUE REASONING** Kate was asked to find RQ to the nearest tenth. Her answer was 14.9 in. Without doing any calculations, how can you tell that she made an error? Explain how to find the correct value for RQ.

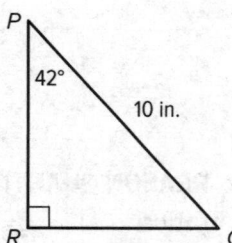

_____

_____

b. **CRITIQUE REASONING** Kate believes that she has enough information to find PR and $m\angle PQR$. Is she correct? If yes, explain and find PR and $\angle PQR$.

_____

_____

6. **REASON ABSTRACTLY** Explain how you can use only the table at the right to find the value of cos 20°.

_____

_____

_____

$m\angle A$	$\sin A$
65°	0.9063
70°	0.9397
75°	0.9659
80°	0.9848
85°	0.9962

7. INTERPRET PROBLEMS A right triangle with legs of length a and $2a$ and an angle θ is shown. If a is a positive real number, does θ depend on the value of a? If not, find the measure of θ. Explain your reasoning.

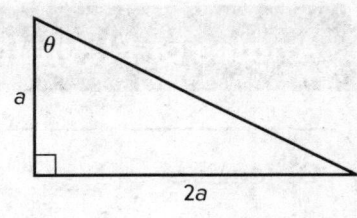

_____

_____

_____

8. REASON QUANTITATIVELY In the right triangle shown, $\sin \alpha = 0.6428$ and $\cos \alpha = 0.7660$. Find $\sin \beta$ and $\cos \beta$ and explain your reasoning.

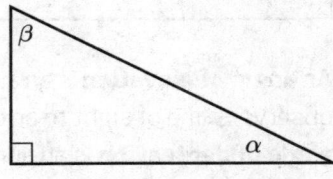

_____

_____

_____

9. An engineer is designing the support structure for the ceiling of an indoor stadium. The structure consists of several similar right triangles. All of the triangles contain a horizontal beam, a vertical beam, and a 28° angle, as shown.

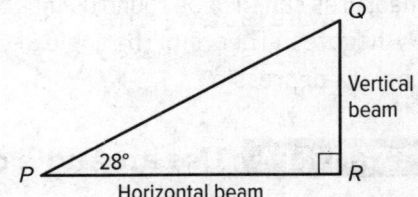

a. DESCRIBE A METHOD The engineer wants a simple formula that she can use to find the length of the vertical beam v when she knows the length of the horizontal beam h. Explain how she can derive such a formula.

_____

_____

b. CALCULATE ACCURATELY In one of the triangles in the structure, the horizontal beam is 2 meters longer than the vertical beam. Explain how the engineer can find the lengths of the two beams.

_____

_____

c. EVALUATE REASONABLENESS How can you check that your answer to **part b** is reasonable?

_____

_____

_____

d. CALCULATE ACCURATELY When sketching the plans for the support structure, the engineer realizes that she needs the length of the third side PQ as well as the measure of $\angle Q$. Find these measurements.

_____

_____

Angles of Elevation and Depression

Objectives

- Solve problems involving angles of elevation.
- Solve problems involving angles of depression.

An **angle of elevation** is an angle formed by a horizontal line and an observer's line of sight to an object above the horizontal line. An **angle of depression** is an angle formed by a horizontal line and an observer's line of sight to an object below the horizontal line.

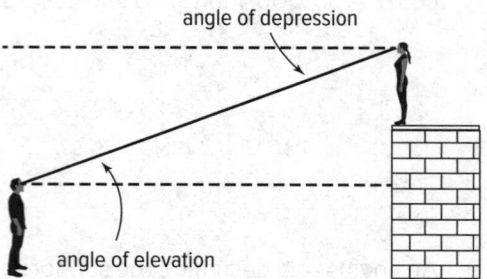

The angle of elevation or depression is part of a description of a real world problem that is represented by a right triangle. Missing measures can then be found using a trigonometric ratio or the Pythagorean Theorem. The angle of elevation is congruent to the angle of depression.

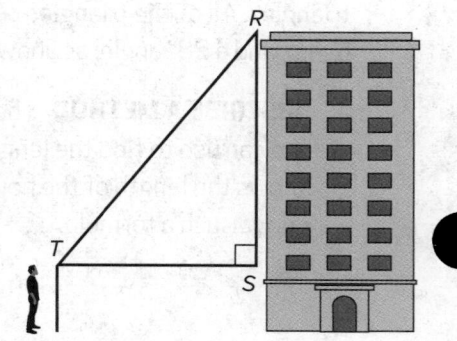

EXAMPLE 1 Use an Angle of Elevation

EXPLORE Tomas is 5 feet 9 inches tall. He stands 40 feet from the base of an apartment building and looks up to the top of the building using an angle of elevation of 80°. Tomas wants to know the height of the building to the nearest foot.

a. INTERPRET PROBLEMS Construct and label a right triangle with the given information. (Note that the figure is not drawn to scale.)

b. PLAN A SOLUTION Explain the steps you can use to solve this problem.

_____

_____

c. USE A MODEL Explain how you can find the length of $\overline{RS}$ to the nearest hundredth. Find the height of the building to the nearest foot.

_____

_____

_____

_____

d. REASON QUANTITATIVELY Suppose that the surface that Tomas is standing on is level and he take a few steps back so that he is farther from the base of the building. Will the angle of elevation increase or decrease? Why does the distance from the building not affect the calculation of the height?

_____

_____

Some applications use two angles of elevation or depression relative to the same line. When this is the case, you may need two triangles to analyze.

EXAMPLE 2 Use Angles of Depression

Towers *A* and *B* are forest lookout towers with the same height. *A* and *B* are one mile apart. Rangers in each tower spot a fire at point *F* between the towers. For the ranger in Tower A, the angle of depression to the fire is 2°. For the ranger in Tower B, the angle of depression to the fire is 6°. The rangers want to know the distance, to the nearest foot, from the base of each tower to the fire.

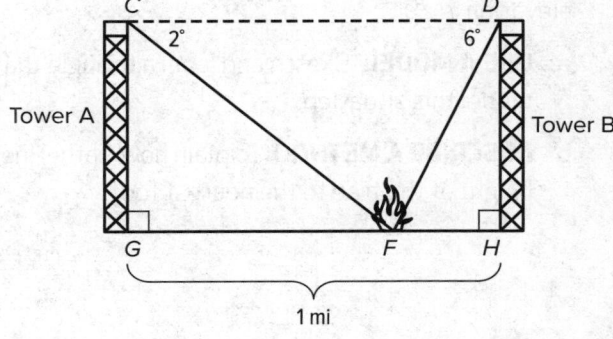

a. **USE STRUCTURE** Explain how to find $m\angle GCF$ and $m\angle HDF$. Mark these angle measures on the figure.

_____

_____

_____

_____

b. **USE A MODEL** Mark the distance in feet from the base of Tower A to the fire *x*. Mark the height in feet of each of the towers *y*. Use a trigonometric ratio in △*GCF* to write an equation that relates *x* and *y*. Then use a trigonometric ratio in △*HDF* to write an equation that relates *x* and *y*. Solve each of the equations for *y*. Explain how you can use the two equations you to find the value of *x*.

_____

_____

_____

_____

_____

_____

c. **REASON QUANTITATIVELY** What is the distance from the base of each tower to the fire?

_____

_____

d. **EVALUATE REASONABLENESS** How do you know the distances you found are reasonable?

_____

_____

_____

1. Catherine is 5 feet 3 inches tall. She stands 40 feet from the
   base of a tree and looks up to the top of the tree. The angle of
   elevation is 29°.

   a. **USE A MODEL** Sketch and label a triangle that can be used to
      model this situation.

   b. **DESCRIBE A METHOD** Explain how Catherine can find the
      height of the tree to the nearest foot.

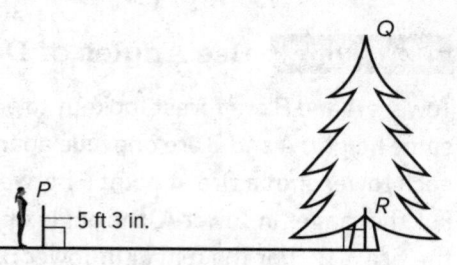

_____

_____

2. **CALCULATE ACCURATELY** Latrell is 6 feet tall. He stands
   on the beach and looks at the top of a cliff that he knows is
   95 feet tall. The angle of elevation to the top of the cliff is
   33°. How far is Latrell standing from the base of the cliff?
   Round your answer to the nearest foot.

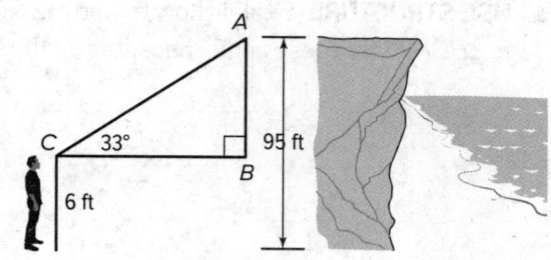

_____

3. Lucia is 5.5 feet tall. She is standing on the roof of a building that
   is 80 feet tall. She spots a fountain at ground level that she knows
   to be 122 feet away from the base of the building.

   a. **INTERPRET PROBLEMS** Describe how the sketch to the right
      models this situation.

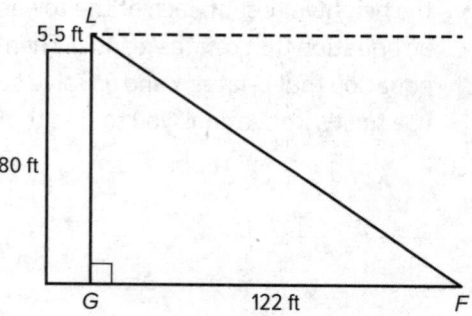

   b. **CALCULATE ACCURATELY** Explain how to use the figure to find the measure of
      the angle of depression formed by Lucia's horizontal line of sight and *LF*. Write the
      measure of the angle of depression to the nearest tenth.

_____

_____

4. A geologist wants to determine the height of a rock formation.
   He stands at a distance of *d* meters from the formation and
   sight the top of the formation at an angle of *x*°, as shown. The
   geologist's height is 1.8 m.

   a. **DESCRIBE A METHOD** Write a general formula that the
      geologist can use to find the height *h* of the rock formation
      if he knows the values of *d* and *x*.

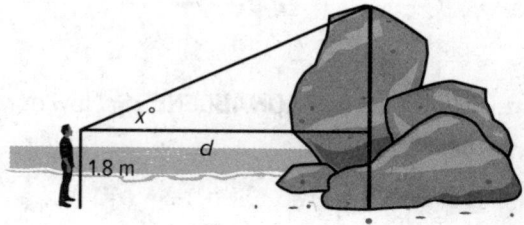

_____

b. CALCULATE ACCURATELY Show how to use your formula to find the height of the rock formation if the geologist stands 23.5 meters from it and the angle of elevation is 72°.

5. Janelle is piloting a helicopter at an altitude of 450 meters. She spots a lake on the ground. The angle of depression to the nearest edge of the lake is 59° and the angle of depression to the farthest edge of the lake is 42°.

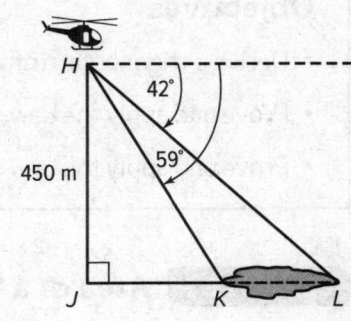

a. PLAN A SOLUTION Describe the steps you can use to find the width of the lake, *KL*.

b. CALCULATE ACCURATELY Find the width of the lake to the nearest meter.

6. **CRITIQUE REASONING** Yoshio is watching a hot air balloon as it rises vertically. He is standing 120 feet from the location where the balloon began its ascent. When he starts watching, the angle of elevation to the hot air balloon is 35°. Later, the angle of elevation is 60°. Yoshio claims that he can find the distance the hot air balloon rose during the time he was watching by calculating 120tan(60° − 35°). Do you agree? If so, use this expression to find the distance. If not, write and evaluate a correct expression to find the distance.

7. Ann and Brett are both 6 feet tall. They stand 100 feet apart from each other, face the same direction, and sight the top of an office building, as shown in the figure. The angle of elevation for Ann is 31° and 52° for Brett.

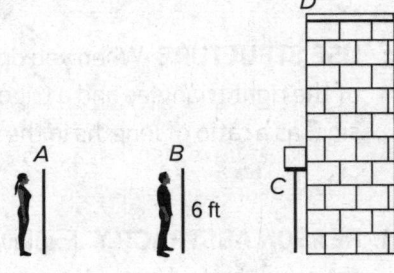

a. USE A MODEL On the picture to the right, sketch and label two triangles that can be used to model this situation.

b. CALCULATE ACCURATELY Find the height of the office building to the nearest foot.

8. **USE A MODEL** An amusement park has three observation towers that lie along a straight line, as shown in the figure. Towers P and R are 20 meters tall and are 200 meters apart. From the top of Tower P to the top of Tower Q, the angle of elevation is 20°. From the top of Tower R to the top of Tower Q, the angle of elevation is 30°. Find the height of Tower Q to the nearest meter.

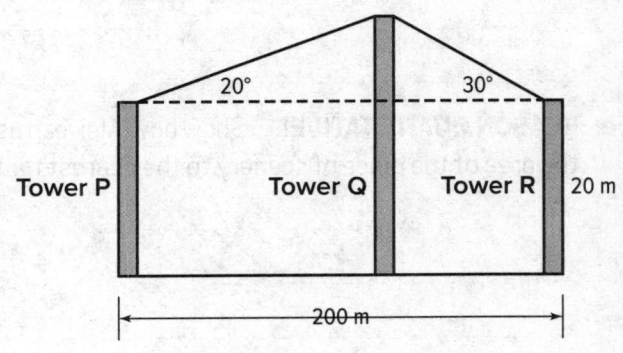

Objectives

- Derive a trigonometric formula for the area of a triangle.
- Prove and apply the Law of Sines.
- Prove and apply the Law of Cosines.

Mathematical Practices
1, 2, 3, 4, 5, 6, 7, 8

EXAMPLE 1 Area of a Triangle

EXPLORE Alex is helping to build the set for a play. One piece of scenery is a large triangle that will be constructed out of wood and be painted to represent a mountain. The plan gives only the information shown in the figure. Alex would like to know if this is enough information to determine the area of the piece of scenery so that he can buy the right amount of paint.

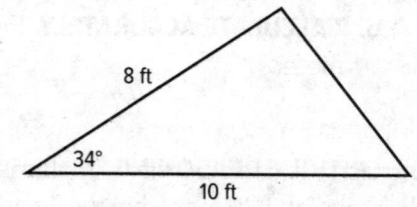

a. **USE STRUCTURE** Alex is planning to build several triangular pieces of scenery, so he draws $\triangle ABC$ and tries to derive a general formula for the area of the triangle. To begin, draw an altitude from point A to $\overline{CB}$. Label the length of the altitude h.

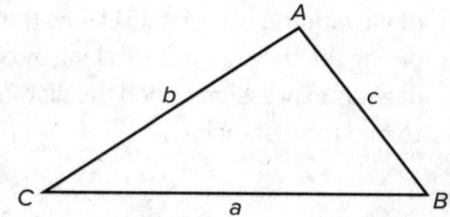

b. **COMMUNICATE PRECISELY** What formula can you use to find the area A of $\triangle ABC$? Write the formula in terms of the variables that are used in the figure.

_____

_____

c. **USE STRUCTURE** When you drew the altitude you created two right triangles. Use one of the right triangles and a trigonometric ratio to write an equation that expresses $\sin C$ as a ratio of lengths in the triangle.

_____

d. **REASON ABSTRACTLY** Explain how you can solve the equation for h and substitute the result in the formula you wrote in **part b**.

_____

_____

_____

e. **REASON QUANTITATIVELY** Show how Alex can use the formula you developed to find the area of the piece of scenery to the nearest tenth of a square foot.

_____

_____

_____

The Key Concept box summarizes the formula you developed in the previous exploration.

KEY CONCEPT Area of a Triangle

Complete the formula.

If △ABC has lengths *a*, *b*, and *c*, representing the lengths of the sides opposite the angles with measures *A*, *B*, and *C*, then the area

of △ABC is given by the formula A = _____.

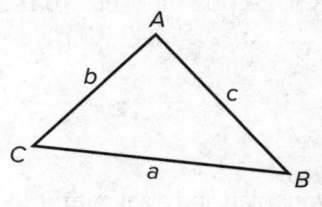

You can use the formula for the area of a triangle to prove the following relationship.

KEY CONCEPT Law of Sines

Complete the statement of the Law of Sines.

If △ABC has lengths *a*, *b*, and *c*, representing the lengths of the sides opposite the angles with measures *A*, *B*, and *C*, then

$$\frac{\sin A}{\Box} = \frac{\sin B}{\Box} = \frac{\sin C}{\Box}.$$

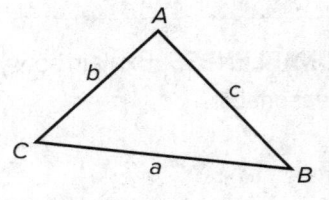

EXAMPLE 2 Prove the Law of Sines

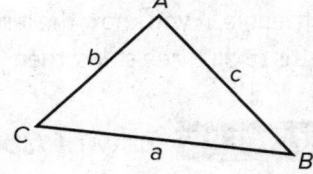

Follow these steps to prove the Law of Sines.

Given: △ABC with sides of length *a*, *b*, and *c*

Prove: $\dfrac{\sin A}{a} = \dfrac{\sin B}{b} = \dfrac{\sin C}{c}$

a. REASON ABSTRACTLY Write three different expressions for the area of △ABC. One expression should involve sin *A*, one should involve sin *B*, and one should involve sin *C*.

_____

b. CONSTRUCT ARGUMENTS Explain how you can set two different expressions for the area equal to each other to derive one of the proportions in the Law of Sines.

_____

_____

_____

c. CONSTRUCT ARGUMENTS Explain how to complete the proof.

_____

_____

_____

EXAMPLE 3 Apply the Law of Sines

Two surveyors stand 150 meters apart and measure the angle to a hill *B* in the distance, as shown in the figure. Each surveyor wants to calculate his or her distance to the hill.

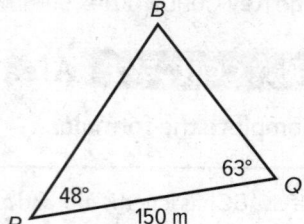

a. PLAN A SOLUTION Describe the steps you can use to solve this problem.

_____

_____

b. USE TOOLS Use your plan and a calculator to find the distance of the surveyor at point *P* to the hill to the nearest tenth of a meter. Show your work.

_____

_____

c. USE TOOLS Find the distance of the surveyor at point *Q* to the hill to the nearest tenth of a meter. Show your work.

_____

d. EVALUATE REASONABLENESS Explain how you know that the distances you found in **parts b** and **c** are reasonable.

_____

_____

Given a triangle, if you know the lengths of two sides and the measure of the included angle or the lengths of all three sides, then you can use the Law of Cosines to solve the triangle.

KEY CONCEPT Law of Cosines

Complete the statement of the Law of Cosines.

If $\triangle ABC$ has lengths a, b, and c, representing the lengths of the sides opposite the angles with measures A, B, and C, then

$a^2 = b^2 + c^2 - 2bc \cos A$

$b^2 = $ _____

$c^2 = $ _____

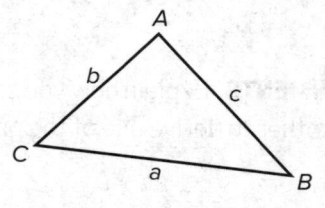

EXAMPLE 4 Prove the Law of Cosines

Follow these steps to prove the Law of Cosines.
Given: $\triangle ABC$ with sides of length a, b, and c
Prove: $c^2 = a^2 + b^2 - 2bc \cos C$

a. USE STRUCTURE On the figure, draw a perpendicular from point *A* to $\overline{CB}$. Label its length h. Label the lengths of the segments formed on $\overline{CB}$ as x and $a - x$.

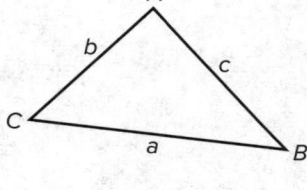

b. CONSTRUCT ARGUMENTS Complete the two-column proof of the theorem.

Statements	Reasons
1. Draw a perpendicular from A to $\overline{CB}$.	1.
2. $c^2 = (a - x)^2 + h^2$	2.
3. $c^2 = a^2 - 2ax + x^2 + h^2$	3.
4. $b^2 = x^2 + h^2$	4.
5. $c^2 = a^2 - 2ax + b^2$	5.
6. $\cos C = \dfrac{\square}{\square}$	6.
7. $b \cos C = $ _____	7.
8. $c^2 = $ _____	8. Substitution
9. $c^2 = $ _____	9. Commutative Property

EXAMPLE 5 **Apply the Law of Cosines**

Naomi is designing a park for a triangular piece of land that is formed by three streets. She knows one of the angles formed by the streets and the lengths of two sides of the park. She wants to determine the other two angle measures and the length of the third side of the park.

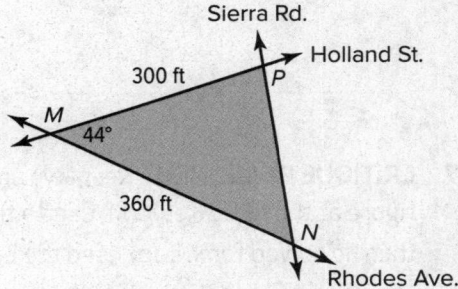

a. REASON QUANTITATIVELY Show the steps Naomi can use to find NP to the nearest tenth of a foot.

_____

_____

_____

_____

b. INTERPRET PROBLEMS Next, Naomi wants to find $m\angle P$. Should she use the Law of Cosines, the Law of Sines, or some other theorem? Explain.

_____

_____

_____

c. REASON QUANTITATIVELY Show the steps for finding $m\angle P$ and $m\angle N$ to the nearest degree.

_____

_____

_____

USE STRUCTURE Determine the value of the variable in each triangle to the nearest tenth.

1.

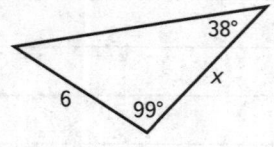

2.

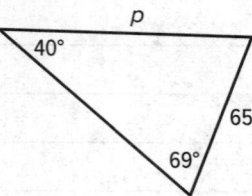

3.

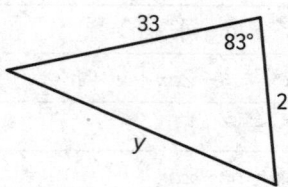

4.

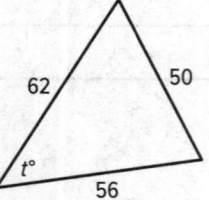

5.

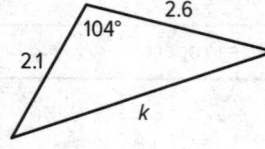

6.

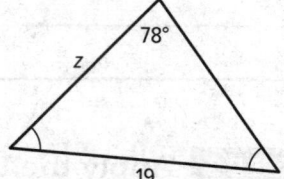

7. **CRITIQUE REASONING** Keshawn and Lacy were asked to find the value of x in the figure at the right. Keshawn used a trigonometric ratio to write $\sin 35° = \frac{x}{10}$, and then he solved for x. Lacy used the Law of Sines to write $\frac{\sin 35°}{x} = \frac{\sin 90°}{10}$, and then she solved for x. Is either student correct? Explain.

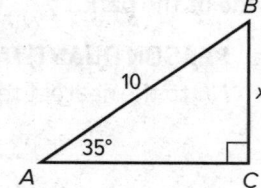

8. To find the distance EF across a pond, a surveyor stands at point D and measures the distance to points E and F and the angle formed by $\overline{DE}$ and $\overline{DF}$. The surveyor's measurements are shown in the figure.

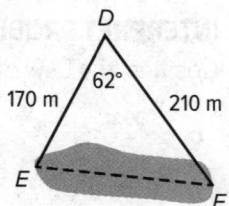

a. **USE A MODEL** What is the distance across the lake to the nearest tenth of a meter?

b. **CALCULATE ACCURATELY** The surveyor spots a rowboat starting across the pond from point F. To travel from point F to point E, what angle should the rowboat make with DF?

9. Mitchell drives due east at 55 miles per hour for 2 hours. Gabriela starts at the same point as Mitchell, but she drives northeast at 60 miles per hour for 3 hours. They want to know how far apart they are when they stop driving.

a. **INTERPRET PROBLEMS** Make a sketch that represents the problem in the space provided.

b. **CALCULATE ACCURATELY** To the nearest mile, how far apart are Mitchell and Gabriela when they stop driving?

10. **USE A MODEL** A 21-foot tree leans at an angle of 82°, as shown. The owner of the tree plans to attach a cable to the top of the tree to support it. The cable will be attached 8 feet from the base of the tree. To the nearest tenth of a foot, what will be the length of the cable? To the nearest degree, what angle will the cable make with the ground?

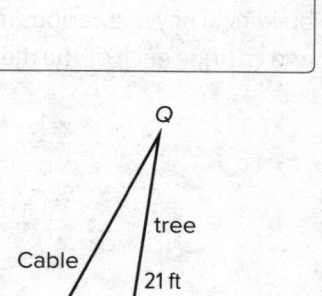

11. Jennifer is buying triangular tiles to make a mosaic. According to the catalog, each tile has the side lengths shown in the figure. Jennifer wants to know the angle measures in the tiles.

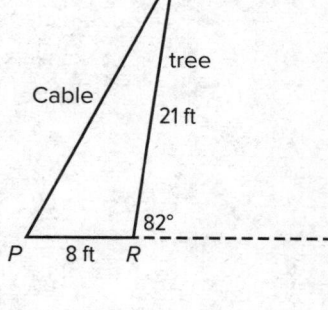

a. **CRITIQUE REASONING** Jennifer says the only way to find the angle measures is to use the Law of Cosines three times. Do you agree? Explain.

b. **CALCULATE ACCURATELY** What are the angle measures in the tiles? Round to the nearest degree.

12. **USE STRUCTURE** The vertices of △FGH are F(1, 2), G(0, −2), and H(3, −2). Which angle has the greatest measure? Explain your solution process.

13. **INTERPRET PROBLEMS** Two shoreline towers 100 feet apart measure the angle to an incoming ship. Find the distance d that the ship is from Tower A to the nearest tenth of a foot.

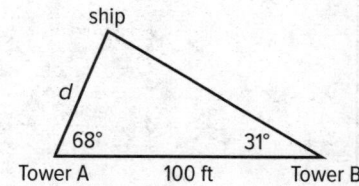

Various Methods for Height

Provide a clear solution to the problem. Be sure to show all of your work, include all relevant drawings, and justify your answers.

For a school project, classmates are asked to measure the height of a school building. They use various methods to find the height. Imagine you are a teacher and critique each of the methods.

Part A

One student measures the shadow of the school, his own shadow, and his height. He comes up with the following proportion.

$$\frac{\text{Shadow of school}}{\text{Shadow of self}} = \frac{\text{Height of school}}{\text{Height of self}}$$

Part B

Another student measures the angle of elevation at spot A, then (after moving 30 feet closer) the angle of elevation at spot B. She uses the following logic.

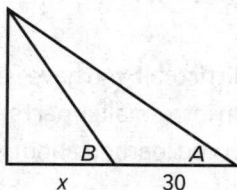

$\tan A = \dfrac{\text{opposite side}}{(30 + x)}$; $\tan B = \dfrac{\text{opposite side}}{x}$

$30 \tan A + x \tan A = \text{opposite side}$; $x \tan B = \text{opposite side}$

$30 \tan A + x \tan A = x \tan B$

$30 \tan A = x \tan B - x \tan A$

$30 \tan A = x(\tan B - \tan A)$

So, the height of the building is $\dfrac{30 \tan A}{(\tan B - \tan A)}$.

Part C

A student walks 30 feet from the foot of the building, finds the angle of elevation A to the top of the building and writes the following equation:

$\tan A = \dfrac{x}{30}$

So $x = \dfrac{\tan A}{30}$.

Taking Sides

Provide a clear solution to the problem. Be sure to show all of your work, include all relevant drawings, and justify your answers.

Sometimes finding a missing measurement can seem difficult if you have very little given information. However, if you divide a problem into smaller parts, you may find it is easier than it seemed. Use the information you learned about right triangles to find the missing value.

Part A
Solve for x. Give the exact value as well as an approximate value.

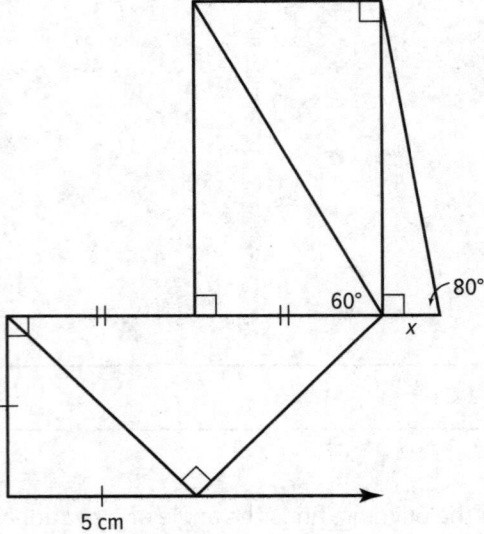

Part B

Explain the steps you used to calculate *x*. Justify each step.

Part C

Are there any similar triangles in the figure? Explain.

Part D

In the triangle with the 80° angle, would you expect the ratio of the hypotenuse to *x* to be greater or less than 2:1? Explain your reasoning.

1. Find the exact value of x in the following diagram.

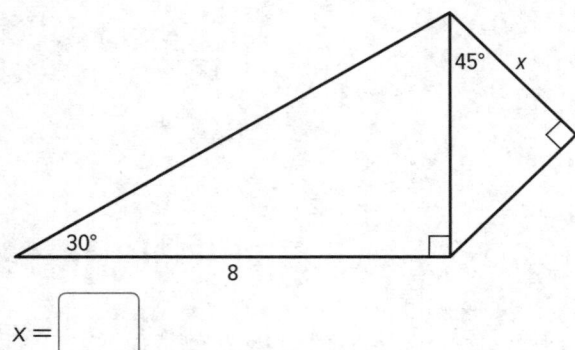

$x = \boxed{}$

2. In $\triangle ABC$, $m\angle A = 82$, $AB = 6$, and $BC = 11$. Find the following, rounding answers to the nearest tenth.

$m\angle B = \boxed{}$ $m\angle C = \boxed{}$

$AC = \boxed{}$

3. Find exact values for x, y, and z in the following diagram.

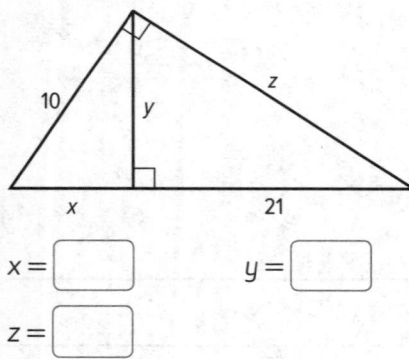

$x = \boxed{}$ $y = \boxed{}$

$z = \boxed{}$

4. The logo on the door of a video game company is an equilateral triangle with a height of 10 cm. The exact perimeter of the logo is $\boxed{}$ cm.

5. Martha is looking out a window that is 190 feet above the ground, and she sees her house. The angle of depression is 27°. Rounded to the nearest tenth, Martha's house is $\boxed{}$ feet from the building.

6. The diagram represents a roof truss designed for a park pavilion.

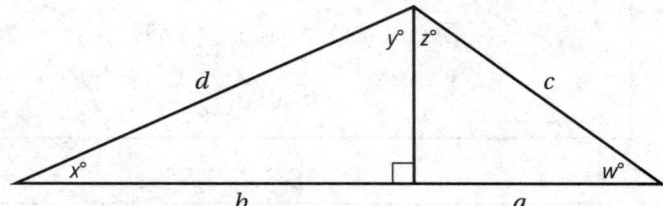

Select the expressions that are equivalent to d.

$\dfrac{\sin(y)}{b}$	$\dfrac{c\sin(w)}{\sin(x)}$
$\dfrac{b}{\cos(x)}$	$(a + b)\sin(w)$
$\sqrt{(a + b)^2 - c^2}$	
$(a + b)^2 + c^2 - 2c(a + b)\cos(w)$	

7. In $\triangle DEF$, $DE = 8$, $EF = 11$, and $DF = 17$. Find the following, rounding answers to the nearest tenth.

$m\angle D = \boxed{}$ $m\angle E = \boxed{}$

$m\angle F = \boxed{}$

8. Complete the missing information for each vector in the table below. Round your answers to the nearest tenth.

Component Form	Magnitude	Direction	
		Degrees	Map Direction
	18	52°	E of N
$\langle -5, 8 \rangle$			N of W
	13	18°	W of S
$\langle 7, -20 \rangle$			E of S

9. Cheng is standing 210 feet from the base of a cliff. A TV antenna is at the top of the cliff. The angle of elevation to the top of the cliff is 26°, and the angle of elevation to the top of the TV antenna is 37°.

    a. Rounded to the nearest tenth, how tall is the cliff? Show your work.

    _____

    b. Rounded to the nearest tenth, how tall is the TV antenna? Show your work.

    _____

    _____

10. Lenecia is rowing a boat across a lake. She sets out rowing at a constant rate of 5 miles per hour at a bearing of 32° west of north. There is a 4 mile per hour wind blowing due north. What are Lenecia's actual speed and direction? Round to the nearest tenth. Show your work.

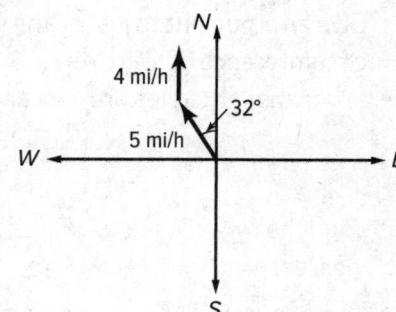

    _____

    _____

    _____

    _____

11. A group of surveyors is taking measurements on a triangular piece of land. The diagram at right shows the measurements they have taken.

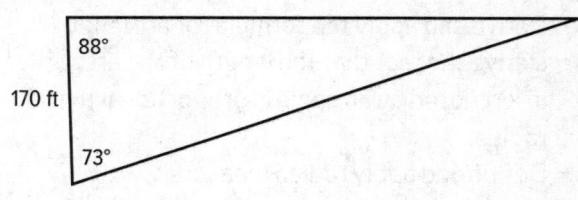

    a. What is the perimeter of the land? Round your answer to the nearest tenth. Show your work.

    _____

    b. What is the area of the land? Round your answer to the nearest tenth. Show your work.

    _____

    _____

12. As a plane is approaching an air base, the angle of elevation to the plane is measured from two points at the base 100 meters apart. The diagram shows the angles that were recorded. Show all work when answering the questions.

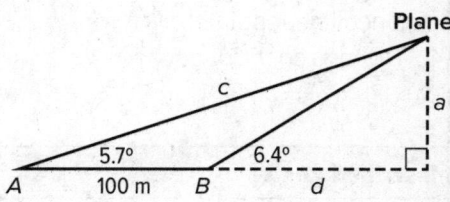

    a. What is the distance c from Point A to the plane? Round your answer to the nearest tenth.

    _____

    b. What is the altitude a of the plane when these measurements are taken? Round your answer to the nearest tenth.

    _____

    c. What is the horizontal distance d from Point B to the point directly under the plane? Round your answer to the nearest tenth.

    _____

9 Circles

CHAPTER FOCUS Learn about some of the objectives that you will explore in this chapter. Answer the preview questions. As you complete each lesson, return to these pages to check your work.

What You Will Learn	Preview Question
Circles and Circumference	
• Give an argument to justify the formula for the circumference of a circle. • Prove that all circles are similar.	SMP 2 The centers of four congruent circles with radii of 1.5 feet lie on the diameter of a larger circle. What is the circumference of the largest circle in terms of pi? **12π feet**
Angles and Arcs	
• Derive and apply the formula for arc length. • Derive the fact that the length of the arc intercepted by an angle is proportional to the radius. • Define and apply radian measure.	SMP 4 A sprinkler has a circular range of 8 feet. If the sprinkler is set so that it only sprays a sector of the circular region that is one-sixth of the entire circle, what is the length of the intercepted arc? Round your answer to the nearest hundredth. **8.38 feet**
Arcs and Chords	
• Identify and describe relationships among chords, arcs, and radii. • Construct the circle that passes through three noncollinear points.	SMP 6 The radius of a circle is perpendicular to a chord 16 units long. If the radius of the circle is 12 units, what is the perpendicular distance from the center of the circle to the chord? Round your answer to the nearest hundredth. **8.94 units**
Inscribed Angles	
• Identify and describe relationships involving inscribed angles. • Prove properties of angles for a quadrilateral inscribed in a circle.	SMP 1 *DEFG* is inscribed in circle *C*. What is the measure of ∠*G*? **87**

Copyright © McGraw-Hill Education

What You Will Learn	Preview Question

Tangents

• Identify and describe relationships among tangents and radii. • Identify and describe relationships among circumscribed angles and central angles. • Construct a tangent line from a point outside a circle to the circle.	**SMP 3** Explain why $\overline{AB}$ is not tangent to circle O at point B. 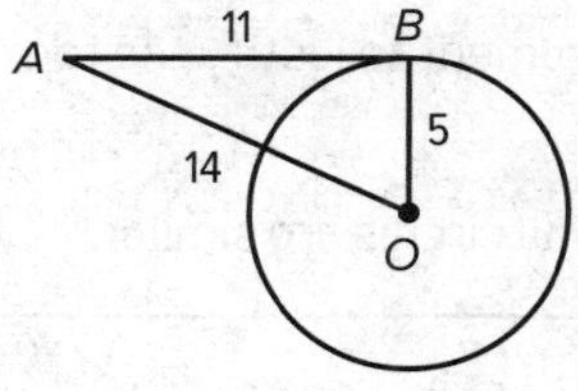 _____ _____ _____ _____

Inscribed and Circumscribed Circles

• Construct an equilateral triangle, a square, and a regular hexagon inscribed in a circle. • Construct the inscribed and circumscribed circles of a triangle.	**SMP 3** An equilateral triangle is inscribed in a circle with diameter of 8 inches. What is the length of each side of the triangle? Explain. _____ _____ _____ _____ _____

Equations of Circles

• Derive the equation of a circle given the center and radius. • Complete the square to find the center and radius of a circle given by an equation.	**SMP 7** What are the center and radius of a circle whose equation is shown below? $$x^2 + y^2 - 4x + 6y = 8$$

Equations of Parabolas

• Derive the equation of a parabola given a focus and directrix. • Use coordinates to prove geometric theorems algebraically.	**SMP 5** What is the equation of a parabola with focus $(1, -3)$ and directrix $y = 2$? Describe its graph.

Objectives

- Give an argument to justify the formula for the circumference of a circle.
- Prove that all circles are similar.

A **circle** is the set of all points in a plane that are equidistant from a given point called the **center** of the circle.

A **radius** of a circle is a segment with one endpoint at the center of the circle and one endpoint on the circle. A **diameter** of a circle is a segment that passes through the center with endpoints on the circle.

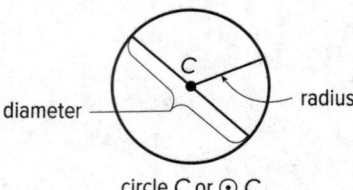

circle C or $\odot C$

EXAMPLE 1 Investigate the Circumference of a Circle

EXPLORE Use Geometer's Sketchpad for this exploration. Before beginning, change the precision to hundred thousandths by selecting Preferences in the Edit menu.

a. **USE TOOLS** Draw a circle with center labeled C. Place a point on the circle and label it.

b. **USE TOOLS** Construct a line through points C and A. Then construct the intersection of this line with the circle, and label it as point B, as shown on the right below.

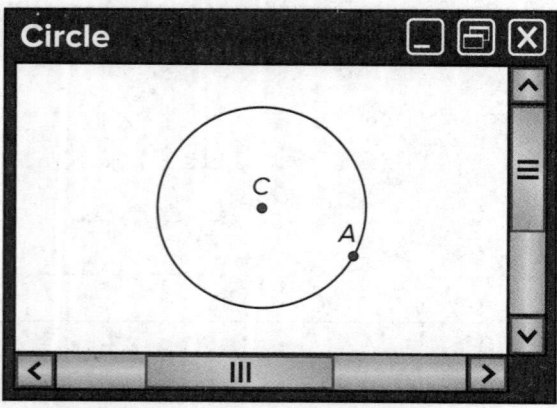

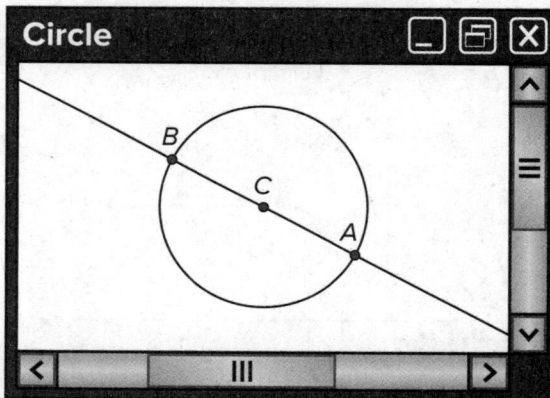

c. **USE TOOLS** Measure the circumference of the circle. Draw $\overline{AB}$ and find the length. Then calculate the ratio of the circumference to AB

d. **USE TOOLS** Drag point A to change the size of the circle. Describe what you notice about the ratio of the circumference to the diameter.

_____

_____

e. **CONSTRUCT ARGUMENTS** Based on your results, describe how you can find the circumference of a circle if the diameter is known. Describe how you can find the diameter of a circle if the circumference is known.

_____

_____

KEY CONCEPT | Circumference of a Circle

Complete the statement and use symbols to the complete the formulas.

If a circle has diameter *d* or radius *r*, the circumference *C* equals _____ _____.	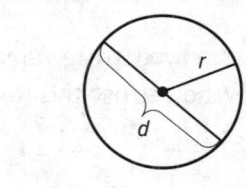 $C = $ _____ or $C = $ _____

A *circumscribed circle*, also called a *circumcircle*, of a polygon is a circle that passes through all vertices of the polygon. The polygon that is circumscribed by the circle is called an *inscribed polygon*.

EXAMPLE 2 | Justify the Formula for the Circumference of a Circle

Complete the following argument to justify the formula for the circumference of a circle.

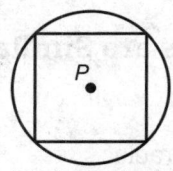 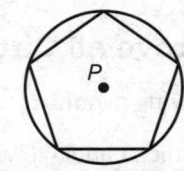 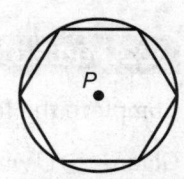

inscribed polygon with 4 sides   inscribed polygon with 5 sides   inscribed polygon with 6 sides

a. **PLAN A SOLUTION** Consider a sequence of regular polygons that are inscribed in the circle, as shown. Describe what happens to the perimeters of the polygons as the number of sides increases.

_____

b. **CONSTRUCT ARGUMENTS** Let the radius of the circle be *r*. The next step of the argument is to write a general formula for the perimeter of an inscribed polygon in terms of *r* and the number sides of the polygon, *n*. To do so, first let $\overline{QR}$ be one side of the inscribed polygon, as shown, and let *M* be the midpoint of $\overline{QR}$. Explain why $\triangle PMQ \cong \triangle PMR$.

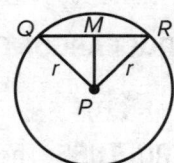

_____

_____

_____

c. **CONSTRUCT ARGUMENTS** What can you conclude about ∠MPQ and ∠MPR? Why?

_____

d. **REASON ABSTRACTLY** Explain how to use your result from **part c** and the fact that the inscribed polygon can be partitioned into *n* triangles congruent to △QPR to write an expression for *m*∠MPQ. The expression should only depend on the variable *n*.

_____

_____

e. **CONSTRUCT ARGUMENTS** Explain why △PMQ must be a right triangle.

_____

_____

Copyright © McGraw-Hill Education

Circles and Circumference   **287**

f. REASON ABSTRACTLY The figure shows $\triangle PMQ$. Use the figure to help you complete the following.

$\sin P = \dfrac{\square}{\square}$, so $x =$ _____

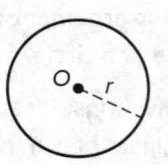

g. REASON ABSTRACTLY In the last expression you wrote, replace P with the expression you wrote for $m\angle MPQ$ in **part d**. Then show how to use this to write an expression for the perimeter of the inscribed polygon.

_____

h. USE TOOLS Enter the expression $n\sin\dfrac{180}{n}$ in your calculator as function Y_1. Use the table feature to explore what happens to the value of the expression as n grows larger and larger. Explain how this completes the argument.

_____

_____

_____

EXAMPLE 3 **Prove All Circles are Similar**

Complete the following proof.

Given: $\odot O$ with radius r and $\odot P$ with radius s
Prove: $\odot O \sim \odot P$

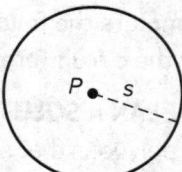

a. PLAN A SOLUTION What do you need to do to show that the circles are similar?

_____

b. CONSTRUCT ARGUMENTS Describe a transformation that maps point O to point P.

_____

c. USE STRUCTURE The result of the transformation you identified in **part b** is shown in the figure. The image of $\odot O$ is $\odot O'$. What must be true about the radius of $\odot O'$? Why?

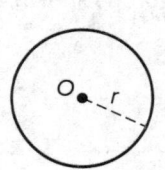
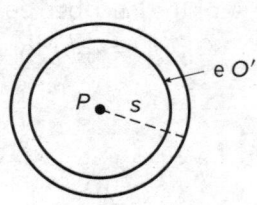

_____

_____

d. CONSTRUCT ARGUMENTS What transformation maps $\odot O'$ to $\odot P$? Explain why this completes the proof.

_____

_____

REASON ABSTRACTLY Determine whether each statement is *always*, *sometimes*, or *never* true. Explain.

1. If points G and H lie on $\odot C$, then $CG = CH$.

_____

2. If points P, Q, and R lie on $\odot C$, then points P, Q, and R are coplanar.

_____

3. If points A and B lie on $\odot C$, then the line segment with endpoints A and B is a diameter.

_____

4. If points X and Y lie on $\odot C$, then $\triangle XYC$ is a scalene triangle.

_____

5. CONSTRUCT ARGUMENTS Describe a specific sequence of transformations you can use to prove that $\odot D$ is similar to $\odot E$.

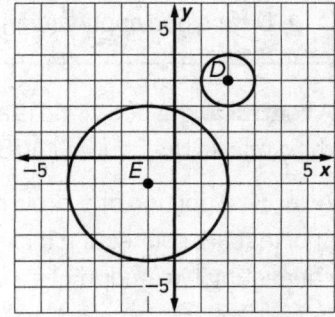

_____

_____

_____

6. You can give an argument to justify the formula for the circumference of a circle using a sequence of regular circumscribed polygons, as shown in the figure.

 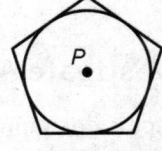 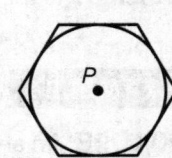

circumscribed polygon with 4 sides   circumscribed polygon with 5 sides   circumscribed polygon with 6 sides

a. PLAN A SOLUTION Describe what happens to the perimeters of the polygons as the number of sides increases.

_____

b. REASON ABSTRACTLY As in **Example 2**, you can show that $m\angle MPQ = \frac{180}{n}$. Use the tangent ratio to write an expression for x that only involves the variables r and n. Then write an expression for the perimeter of the circumscribed polygon.

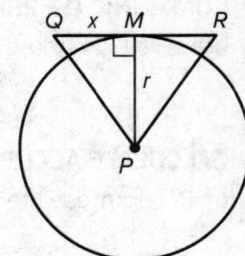

_____

_____

c. USE TOOLS Explain how to use a calculator to complete the argument.

_____

_____

d. REASON ABSTRACTLY Example 2 and this exercise provide estimates for the circumference of a circle. Discuss whether the estimates are greater than or less than the actual circumference. Which estimates seem to be closer to the actual value?

_____

_____

_____

_____

Objectives

- Derive and apply the formula for arc length.
- Derive the fact that the length of the arc intercepted by an angle is proportional to the radius.
- Define and apply radian measure.

Mathematical Practices

1, 2, 3, 4, 6, 7, 8

A **central angle** of a circle is an angle in which the vertex is the center of the circle. In the figure, $\angle ACB$ is a central angle.

An **arc** is a portion of a circle defined by two endpoints. A **minor arc** is the shortest arc connecting two endpoints. A **major arc** is the longest arc connecting two endpoints.

Arc length is the distance between endpoints along an arc.

minor arc $\overset{\frown}{AB}$

major arc $\overset{\frown}{ADB}$

EXAMPLE 1   Investigate Arc Length

EXPLORE An architect is designing the seating area for a theater. The seating area is formed by a region that lies between two circles, as shown in the figure. The architect is planning to place a brass rail in front of the first row of seats. She wants to know the length of the rail.

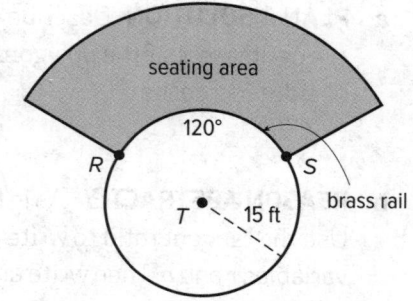

seating area

120°

brass rail

15 ft

a. COMMUNICATE PRECISELY In the figure, the arc that represents the brass rail is marked 120°. Explain what this means.

_____

b. CALCULATE ACCURATELY Explain how to find the circumference of ⊙ T. Express the circumference in terms of π.

_____

c. REASON ABSTRACTLY Explain how you can use your answer to **part b** and proportional reasoning to find the length of the brass rail. Express the length in terms of π and to the nearest tenth of a foot.

_____

_____

d. DESCRIBE A METHOD The architect is considering changing the radius of ⊙ T or changing the measure of $\overset{\frown}{RS}$. Describe a general method she can use to find the length of $\overset{\frown}{RS}$.

_____

_____

_____

e. USE A MODEL The seating area is 15 feet deep, so the distance from the center of the circle to the edge of the seating area is 30 feet. If the architect wants to surround the entire seating area with a brass rail, what length of rail does she need? Explain your reasoning.

_____

_____

_____

The Key Concept box summarizes the relationship you discovered in the previous exploration.

KEY CONCEPT **Arc Length**

Complete the proportion and the equation.

The ratio of the length of an arc ℓ to the circumference of the circle is equal to the ratio of the degree measure of the arc to 360.

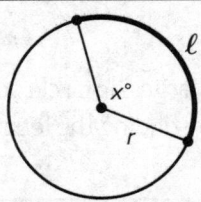

Proportion: $\dfrac{\ell}{2\pi r} = \dfrac{\boxed{}}{\boxed{}}$ Equation: $\ell = $ _____

EXAMPLE 2 **Apply Arc Length**

A model train is set up on a circular track with a diameter of 8 feet. As the train travels from the station to the coal mine, it moves 84° around the track. The train moves at a constant rate of 6 inches per second. Follow these steps to determine how long it takes the train to travel from the station to the coal mine.

a. INTERPRET PROBLEMS Sketch and label a figure that represents this situation in the space at the right.

b. CALCULATE ACCURATELY Explain how to find the length of the track from the station to the coal mine to the nearest hundredth of a foot.

_____

_____

c. CALCULATE ACCURATELY Explain how to find the time it takes the train to travel from the station to the coal mine to the nearest second.

_____

_____

d. CRITIQUE REASONING A student said that if the radius of the track were doubled, then the amount of time it takes the train to travel from the station to the coal mine would also be doubled. Do you agree? Explain.

_____

_____

e. CRITIQUE REASONING A student said that if the angle were doubled, then the amount of time it takes the train to travel from the station to the coal mine would also be doubled. Do you agree? Explain.

_____

_____

_____

EXAMPLE 3 **Investigate Proportionality**

Follow these steps to investigate how the length of the arc intercepted by a central angle of a circle is related to the radius.

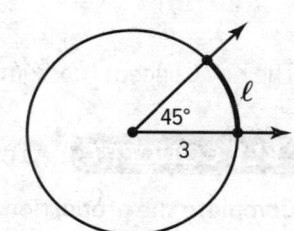

a. CALCULATE ACCURATELY The figure shows an arc intercepted by a central angle of 45°. Show how to find the length of the arc in terms of π.

_____

b. FIND A PATTERN Enlarge the circle so the measure of the central angle remains 45. Complete the table by finding the length of the arc (in terms of π) for each of the given radii.

Radius of circle, r	3	5	11	15	r
Length of arc, ℓ					

c. COMMUNICATE PRECISELY Look for a pattern. When the measure of the central angle is 45, what type of relationship do you notice between the arc length and the radius? Explain.

_____

_____

d. FIND A PATTERN Repeat the above process and complete the following table, but this time consider arcs with a central angle of 60°.

Radius of circle, r	2	4	5	10	r
Length of arc, ℓ					

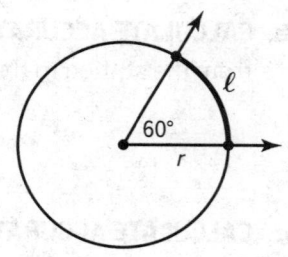

e. COMMUNICATE PRECISELY Look for a pattern in your table. When the measure of the central angle is 60, what type of relationship do you notice between the arc length and the radius? Explain.

_____

_____

f. DESCRIBE A METHOD Now generalize your findings. Suppose the measure of the central angle is x. What type of relationship will exist between the arc length and the radius? How can you find the length of the arc if you know the radius?

_____

_____

You can use your findings from **Example 3** to define the radian measure of an angle. Much like a foot and an inch are two different units for measuring the length of a line, a radian and a degree are two different units for measuring the size of an angle. An angle that measures 1 radian will mark off an arc on a circle that is equal in length to the radius of the circle. Because degrees and radians are two units of measurement, there is a conversion factor between them.

KEY CONCEPT **Radian Measure**

Complete the formula.

The length of the arc intercepted by an angle is proportional to the radius. The constant of proportionality for this relationship is the **radian measure** of the angle.

Formula: The radian measure of an angle of $x°$ = _____

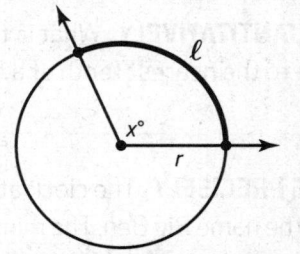

EXAMPLE 4 **Apply Degrees and Radians**

A carpenter is ordering metal brackets from a specialty company. The brackets are available in a variety of angles, as shown. The carpenter wants to know which of the brackets has the greatest angle measure and which of the brackets, if any, are a right angle.

Bracket	Angle
A	70°
B	$\frac{3\pi}{4}$ radians
C	1 radian
D	$\frac{\pi}{2}$ radians

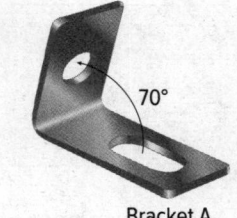

Bracket A

a. **CALCULATE ACCURATELY** Explain how to express the measure of Bracket A in radians.

_____

_____

_____

b. **REASON QUANTITATIVELY** Without doing any further calculations, can the carpenter determine which of the brackets has the greatest angle measure? Explain.

_____

_____

_____

c. **DESCRIBE A METHOD** Suppose you know the radian measure of an angle. Explain how you can convert the measure to degrees.

_____

d. **CALCULATE ACCURATELY** Are any of the brackets a right angle? Explain.

_____

e. **CRITIQUE REASONING** The carpenter orders a bracket that can be adjusted to any angle measure from 0 radians to π radians. The carpenter claims that this bracket can be adjusted to form a straight angle. Do you agree? Explain.

_____

1. The figure shows a circular flower bed. Part of the fence surrounding the flower bed, $\overset{\frown}{AB}$, is damaged and needs to be replaced. The fencing costs $8.75 per linear foot.

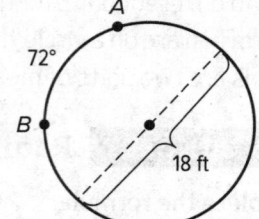

   a. **USE A MODEL** What will it cost to replace the damaged portion of the fence?

   _____

   b. **REASON QUANTITATIVELY** What is the length of the undamaged portion of the fence to the nearest tenth of a foot?

   _____

2. **COMMUNICATE PRECISELY** The clock at the Palace of Westminster in London is best known by the name Big Ben. The minute hand of the clock is 14 feet long. Explain how you can determine how far the tip of the minute hand moves between 3 PM and 3:10 PM. Round to the nearest tenth.

   _____

   _____

   _____

   _____

3. At an amusement park, go-karts travel around a circular track with a radius of 90 feet. As the go-karts travel from point P to point Q, they cover a distance of 165 feet.

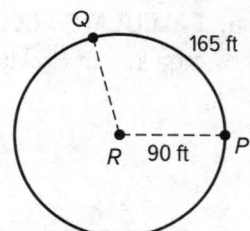

   a. **USE A MODEL** What is the measure of $\angle QRP$ to the nearest degree?

   _____

   b. **DESCRIBE A METHOD** Suppose you know the distance in feet, d, that a go-kart travels. How can you find the measure x (in degrees) of the arc that describes the go-kart's path? Include a written and algebraic description.

   _____

4. **CRITIQUE REASONING** In $\odot C$, $\overset{\frown}{AB}$ has a central angle of 25°. In $\odot D$, $\overset{\frown}{EF}$ has a central angle of 25°. A student claimed that the arc length of $\overset{\frown}{AB}$ must be equal to the arc length of $\overset{\frown}{EF}$. Do you agree? If so, explain why. If not, explain whether there is ever a situation in which you can conclude that the arc lengths are equal.

   _____

   _____

CALCULATE ACCURATELY Convert each degree measure to the equivalent measure in radians.

5. 135°

6. 18°

7. 65°

8. 120°

_____   _____   _____   _____

CALCULATE ACCURATELY Convert each radian measure to the equivalent measure in degrees.

9. $\frac{\pi}{6}$ radians   **10.** $\frac{5\pi}{6}$ radians   **11.** $\frac{\pi}{12}$ radians   **12.** 3 radians

_____   _____   _____   _____

REASON ABSTRACTLY Determine whether each statement is *always*, *sometimes*, or *never* true. Explain.

13. If a central angle measures 2 radians, then the length of the arc it intercepts is twice the length of the circle's radius.

_____

_____

14. The radian measure of an obtuse angle is less than $\frac{\pi}{2}$.

_____

15. In a circle, the length of the arc intercepted by a central angle is equal to the radius of the circle.

_____

16. An angle's measure is 1 in radians and 45 in degrees.

_____

17. **USE A MODEL** As a pendulum swings from point *A* to point *B*, it sweeps out an angle of $\frac{5\pi}{18}$ radians. What is the distance the weight at the end of the pendulum travels as it swings from *A* to *B* and back to *A*? Round to the nearest tenth.

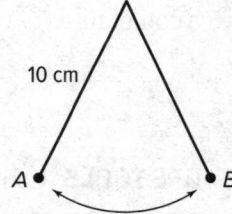

_____

18. **CRTIQUE REASONING** An archery target consists of three concentric circles, as shown. Latanya says that $\overset{\frown}{ST}$ must be 3 times as long as $\overset{\frown}{QR}$. Melanie says that it is not possible to determine this relationship without knowing the measure of $\angle SPT$. Who is correct? Explain.

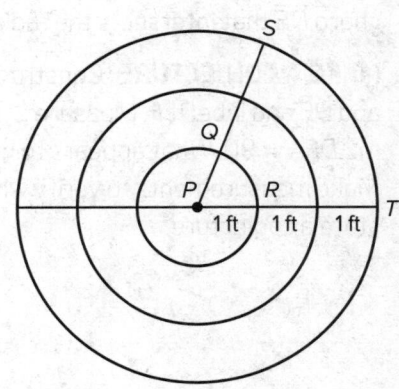

_____

_____

_____

_____

19. **REASON ABSTRACTLY** A bicycle wheel has a diameter of 27.5 inches. On a trip through town, the wheel rotated 2143 times around plus an additional 62°. How far did the bicycle travel?

_____

_____

Objectives

- Identify and describe relationships among chords, arcs, and radii.
- Construct the circle that passes through three noncollinear points.

Mathematical Practices
1, 2, 3, 4, 5, 6

A **chord** is a line segment with endpoints on a circle. In the figure, $\overline{JK}$ is a chord. Note that any chord that is not a diameter divides the circle into a minor arc and a major arc. The minor arc is the smaller of the two, and the major arc is the larger of the two.

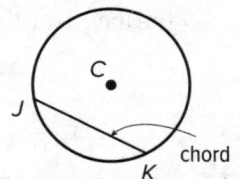

EXAMPLE 1 **Investigate Relationships Among Chords, Arcs, and Radii**

EXPLORE Use The Geometer's Sketchpad for this exploration.

a. **USE TOOLS** Construct a circle C. Then plot four points on the circle and use the points to draw chords $\overline{AB}$ and $\overline{DE}$, as shown.

b. **MAKE A CONJECTURE** Measure the lengths of $\overline{AB}$ and $\overline{DE}$ and measure the minor arcs $\overset{\frown}{AB}$ and $\overset{\frown}{DE}$. Drag points so that the chords are congruent. What do you notice? State a conjecture.

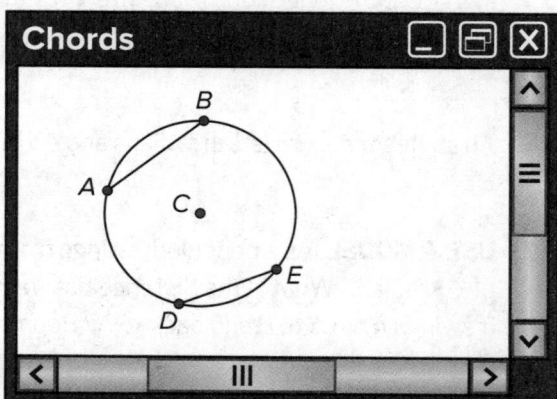

c. **USE TOOLS** Construct a new circle G. Plot a point A on the circle and construct the radius $\overline{AG}$. Then construct a chord $\overline{DE}$ that intersects the radius, as shown.

d. **MAKE A CONJECTURE** Construct the intersection of $\overline{AG}$ and $\overline{DE}$ and label it F. Measure $\angle DFA$. Drag points so that $m\angle DFA = 90$. What appears to happen to $\overline{DE}$ and $\overset{\frown}{DE}$? Make measurements to verify what you observe. Then state a conjecture.

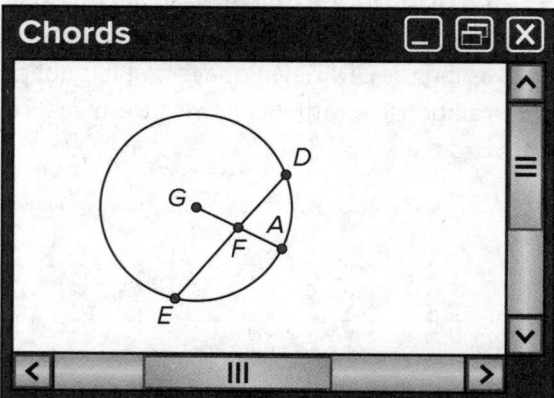

e. **MAKE A CONJECTURE** Construct a new circle and chord. Construct the midpoint of the chord and then construct a line perpendicular to the chord through the midpoint to form the perpendicular bisector. What do you notice about the perpendicular bisector? State a conjecture.

The Key Concept box summarizes the relationships you identified in the previous exploration.

Complete the statement of each theorem and the example for each theorem.

In a circle, two minor arcs are congruent if and only if

_____

Example: $\overarc{AB} \cong \overarc{CD}$ if and only if _____

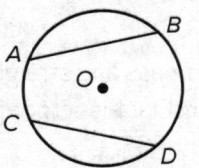

If a diameter (or radius) of a circle is perpendicular to a chord, then

_____

Example: If $\overline{AB} \perp \overline{CD}$, then _____

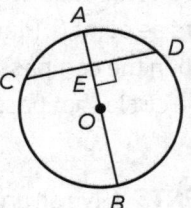

The perpendicular bisector of a chord is

_____

Example: If $\overline{AB}$ is the perpendicular bisector of $\overline{CD}$, then _____

_____

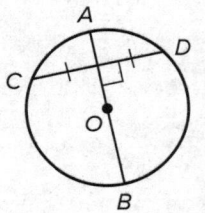

EXAMPLE 2 Apply Relationships Among Chords, Arcs, and Radii

An amusement park is installing a new pirate ship ride in which the ship begins at point M and swings back and forth between points P and Q. The manager of the park knows that $CM = 60$ feet and $PQ = 100$ feet. He would like to know $m \overarc{PM}$.

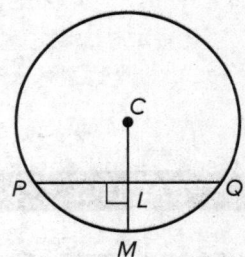

a. **PLAN A SOLUTION** Describe the main steps you can use to find $m \overarc{PM}$.

_____

_____

b. **USE A MODEL** Mark the figure as needed and use your plan to find $m \overarc{PM}$ to the nearest degree. Show your work.

_____

_____

_____

c. **REASON ABSTRACTLY** If we are interested in $m \overarc{MQ}$, is it necessary to repeat all of the calculations in **part b**? Why or why not?

_____

_____

EXAMPLE 3 **Construct a Circle Through Three Points**

Follow these steps to construct a circle that passes
through three noncollinear points. Work directly on the
three points shown at the right.

a. USE TOOLS Use a straightedge to draw $\overline{RS}$ and $\overline{ST}$.

b. USE TOOLS Use a compass and straightedge to
construct the perpendicular bisector of $\overline{RS}$ and the
perpendicular bisector of $\overline{ST}$.

R •   • T

c. USE TOOLS Label the intersection of the
perpendicular bisectors as point C. Place the point
of the compass at C, open the compass to the
distance CR, and draw a circle that passes through
points R, S, and T.

•
S

d. CONSTRUCT ARGUMENTS Give an argument to
explain why this construction works.

_____

_____

_____

e. REASON ABSTRACTLY Is it always possible to construct a circle through three points?
Explain your answer.

_____

_____

_____

PRACTICE

REASON ABSTRACTLY Find the indicated length or angle measure. Round to the
nearest tenth, if necessary

1. $BC = 26$, $EF = 22$. Find OD.   **2.** $\overline{PR} \cong \overline{ST}$, $m\widehat{ST} = 80$. Find $m\angle PQR$.   **3.** $LM = 50$, $JK = 38$. Find $m\angle JNG$.

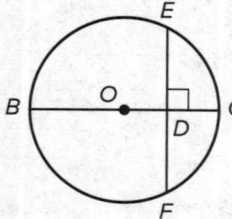

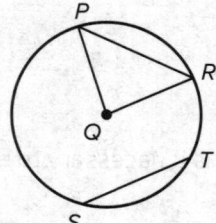

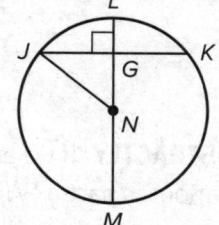

_____   _____

4. **CRITIQUE REASONING** Miyuki saw the figure at the right in her textbook. She concluded that ∠ACD ≅ ∠BCD. Do you agree? Explain.

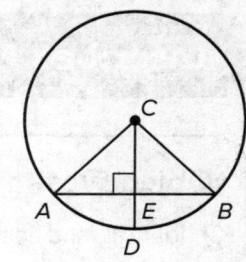

_____

_____

_____

USE TOOLS Construct a circle that passes through the three given points.

5.

6.

N•

•L

K•

J•

•P

M•

7. **COMMUNICATE PRECISELY** Explain what would happen if you tried to do the construction with three collinear points.

_____

_____

8. **CONSTRUCT ARGUMENTS** Write a paragraph proof for the following.

Given: $\overline{OA}$ is a radius and $\overline{OA}$ bisects $\overline{BC}$.

Prove: $\overline{OA}$ is perpendicular to $\overline{BC}$.

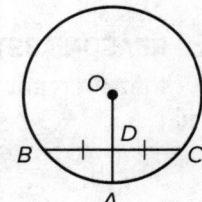

_____

_____

_____

9. **USE A MODEL** A company manufactures bicycle stands in the shape of a rectangle. A wheel with a 26-inch diameter fits perfectly into the stand shown in the figure. What is the length, PQ, of the stand rounded to the nearest tenth? Explain your work.

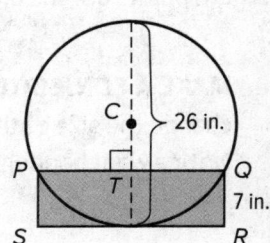

_____

_____

Objectives

- Identify and describe relationships involving inscribed angles.
- Prove properties of angles for a quadrilateral inscribed in a circle.

An **inscribed angle** of a circle has its vertex on the circle and has sides that are chords of the circle. An **intercepted arc** has endpoints on the sides of an inscribed angle and lies in the interior of the inscribed angle.

In the figure, $\angle JKL$ is an inscribed angle and $\overset{\frown}{JL}$ is the arc intercepted by $\angle JKL$.

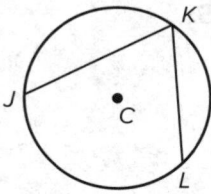

EXAMPLE 1 Investigate Inscribed Angles

EXPLORE Use The Geometer's Sketchpad for this exploration.

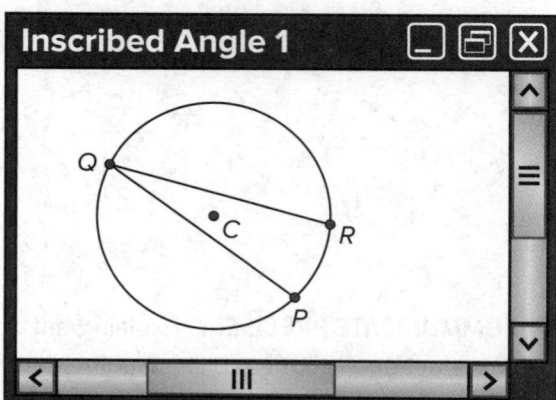

Inscribed Angle 1

a. **USE TOOLS** Construct circle C. Then plot three points on the circle and label them P, Q, and R. Use the points to construct $\angle PQR$, as shown.

b. **MAKE A CONJECTURE** Measure $\angle PQR$ and the arc angle of the intercepted arc $\overset{\frown}{PR}$. Drag points to change the size of the circle and the location of the angle. Look for a relationship in the angle measure and arc measure. Then state a conjecture.

_____

_____

c. **REASON ABSTRACTLY** What can you conclude about two inscribed angles that intercept the same arc? Why?

_____

_____

d. **USE TOOLS** Construct a new circle C. Construct the diameter of the circle, $\overline{AB}$. Plot a point D on the circle and construct an angle whose vertex is point D and whose endpoints are the endpoints of the diameter, as shown.

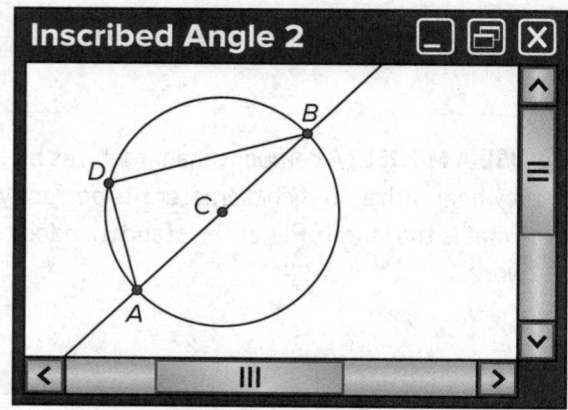

Inscribed Angle 2

e. **MAKE A CONJECTURE** Measure $\angle ADB$. Drag points to change the size of the circle and the location of $\angle ADB$. Notice what happens to $m\angle ADB$. State a conjecture.

_____

_____

KEY CONCEPT Inscribed Angle Theorem

Complete the statement of the theorem and the example.

> If an angle is inscribed in a circle, then the measure of the angle equals
>
> _____
>
> Example: $m\angle EFG =$ _____

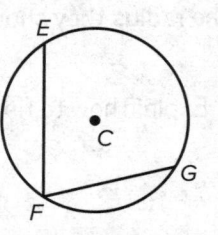

You can apply the relationships among inscribed angles and intercepted arcs to determine arc and angle measures.

EXAMPLE 2 Determine Arc and Angle Measures

Consider ∠HPJ in the figure at the right.

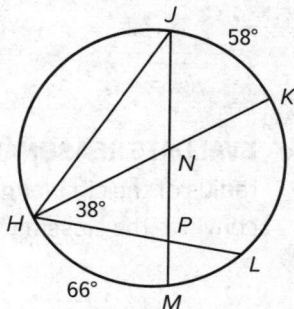

a. PLAN A SOLUTION Describe the steps you can use to determine $m\angle HPJ$. Name any theorems, postulates, or other relationships you will use.

_____

_____

_____

_____

b. CALCULATE ACCURATELY Determine $m\angle HPJ$. _____

c. CRITIQUE REASONING A student used his result from **part b** to conclude that $m\,\widehat{HJ} = 160$. Do you agree? If so, why. If not, explain the student's error.

_____

KEY CONCEPT Corollaries of the Inscribed Angle Theorem

Complete the statement of each corollary. Then use the figure to give an example. These corollaries were explored in Example 1 parts c–e.

> If two inscribed angles of a circle intercept the same arc or congruent arcs, then
>
> _____
>
> Example: _____

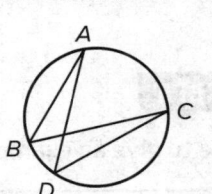

> If an inscribed angle of a circle intercepts a diameter, then
>
> _____
>
> Example: _____

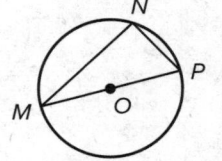

EXAMPLE 3 Apply Arc and Angle Measures

A landscaping crew is installing a circular garden with three paths that form a triangle. The figure shows the plan for the garden and the paths. The leader of the crew wants to determine the radius they should use when they install the circular garden.

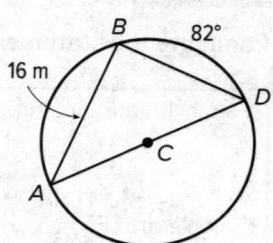

a. **REASON ABSTRACTLY** Explain how to find the measures of the angles in △ABD.

_____

_____

_____

b. **USE A MODEL** Find the radius of the circular garden. Explain your steps.

_____

_____

c. **EVALUATE REASONABLENESS** A classmate solved the problem and found that the radius of the circular garden was 7.1 m. Without doing any calculations, how could you convince the classmate that this answer is not reasonable?

_____

_____

_____

d. **REASON QUANTITATIVELY** Given that a path will go along the outside of the garden (along the circle), as well as along the three sides of the triangle, find the length of path that will need to be installed.

_____

_____

_____

Recall that a polygon is inscribed in a circle if all of its vertices lie on the circle. The Key Concept box describes a relationship among the angles of a quadrilateral that is inscribed in a circle.

KEY CONCEPT

Use the figure to give an example of the theorem.

If a quadrilateral is inscribed in a circle, then its opposite angles are supplementary.

Example: _____

_____

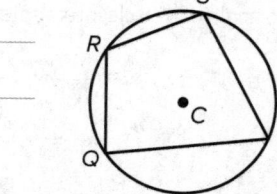

Similar to the way that lengths of collinear segments that have only a single point in common can be added or the measures of adjacent angles can be added, the measures of arcs that have only a single point in common can be added. This is stated formally in the Key Concept box below.

KEY CONCEPT **Arc Addition Postulate**

In any circle, if $\overarc{AB}$ and $\overarc{BC}$ only overlap at point B, then $m\,\overarc{AB} + m\,\overarc{BC} = m\,\overarc{AC}$.

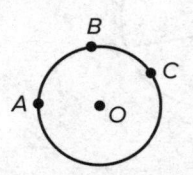

EXAMPLE 4 **Prove Theorems About Inscribed Quadrilaterals**

Follow these steps to prove that the opposite angles of a quadrilateral inscribed in a circle are supplementary.

Given: Quadrilateral $QRST$ is inscribed in $\odot C$.

Prove: $\angle Q$ is supplementary to $\angle S$.

a. **CONSTRUCT ARGUMENTS** Complete the two-column proof of the theorem.

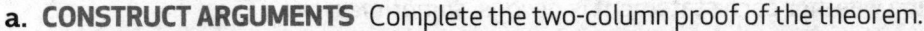

Statements	Reasons
1. Quadrilateral $QRST$ is inscribed in $\odot C$.	**1.**
2. $m\,\overarc{RST} + m\,\overarc{RQT} = 360$	**2.**
3. $m\angle Q = \frac{1}{2}m\,\overarc{RST}; m\angle S = \frac{1}{2}m\,\overarc{RQT}$	**3.**
4. $2m\angle Q = m\,\overarc{RST}; 2m\angle S = m\,\overarc{RQT}$	**4.**
5.	**5.** Substitution Property of Equality
6.	**6.**
7.	**7.**
8.	**8.**

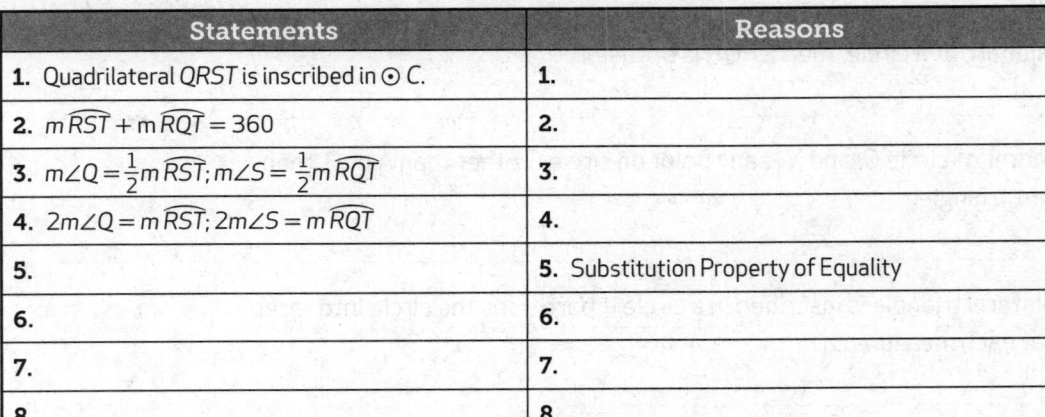

b. **COMMUNICATE PRECISELY** Explain why you do not need to write a separate proof to show that $\angle R$ is supplementary to $\angle T$.

_____

c. **COMMUNICATE PRECISELY** Explain why this theorem shows that the angles in a quadrilateral inscribed in a circle add to 360. Explain why this does not show that the angles in *all* quadrilaterals add to 360.

_____
_____
_____
_____

REASON ABSTRACTLY Find the indicated measure.

1. m $\overset{\frown}{CE}$

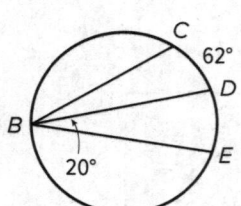

2. m $\overset{\frown}{JM}$

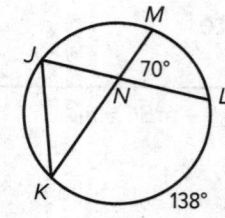

3. m∠QPR

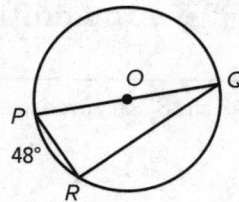

4. m $\overset{\frown}{UW}$

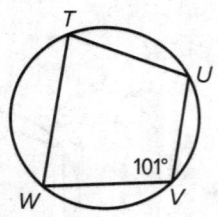

5. m∠AFE

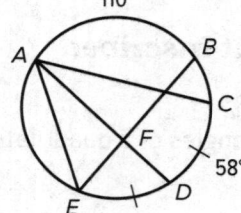

6. m∠KJH

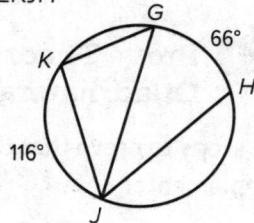

REASON ABSTRACTLY Determine whether each statement is *always*, *sometimes*, or *never* true. Explain.

7. If $\overset{\frown}{PQR}$ is a major arc of a circle, then ∠PQR is obtuse.

8. If $\overline{AB}$ is a diameter of circle O, and X is any point on circle O other than A or B, then △AXB is a right triangle.

9. When an equilateral triangle is inscribed in a circle it partitions the circle into three minor arcs that each measure 120°.

10. Use The Geometer's Sketchpad for this exercise.

  a. **USE TOOLS** Construct a circle, and plot four points J, K, L, and M on the circle. Connect the points so that you have an inscribed quadrilateral. Drag the points around the circle until you form a parallelogram.

  b. **MAKE A CONJECTURE** Make a conjecture based on what you notice. Confirm your conjecture by measuring the sides and angles of the quadrilateral.

  c. **USE TOOLS** Can you construct a counterexample to your conjecture in **part b**? Explain.

d. CONSTRUCT ARGUMENTS Use your sketch to help you write a paragraph proof to justify your conjecture.

_____

_____

_____

_____

_____

11. CONSTRUCT ARGUMENTS Write a paragraph proof to prove that if *PQRS* is an inscribed quadrilateral and $\angle Q \cong \angle S$, then $\overline{PR}$ is a diameter of the circle.

_____

_____

_____

_____

12. USE A MODEL Kendrick is designing a logo for a company that makes equipment for windsurfing, as shown. Kendrick knows that the diameter of the circle is 12 centimeters and that $\widehat{BE} \cong \widehat{DE}$. He also knows that $m\widehat{BD} = 136$. What is the length of $\widehat{AB}$ to the nearest tenth of a centimeter?

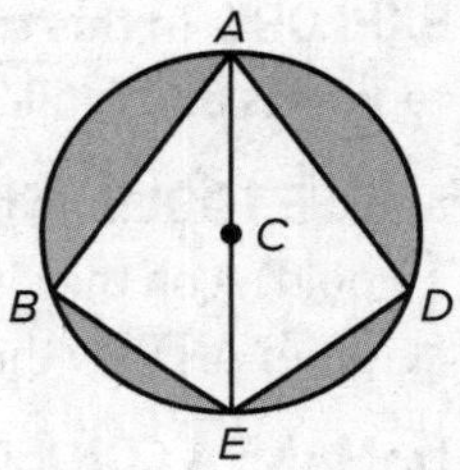

_____

13. USE A MODEL The figure shows the steering wheel in Sarita's car. The diameter of the steering wheel is 16 inches. Sarita turns the wheel so that point *J* rotates clockwise to point *K*. How many inches around the circle does point *J* travel?

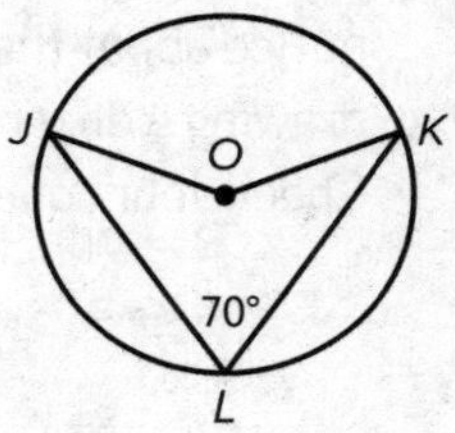

_____

14. CRITIQUE REASONING Ari said that if a quadrilateral has two 50° angles and a 100° angle, then the quadrilateral cannot be inscribed in a circle. Do you agree? Explain why or why not.

_____

_____

_____

_____

15. DESCRIBE A METHOD Alyssa makes earrings by bending wire into various shapes. She often bends the wire to form a circle with an inscribed quadrilateral, as shown. She would like to know how she can find $m\widehat{ADC}$ if she knows $m\angle ADC$. Explain how to write a formula for $m\widehat{ADC}$ given that $m\angle ADC = x$.

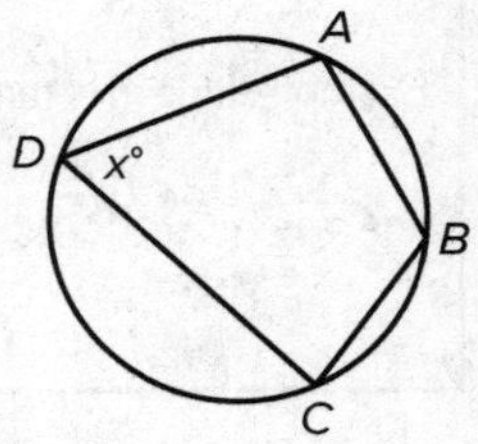

_____

_____

_____

_____

Objectives

- Identify and describe relationships among tangents and radii.
- Identify and describe relationships among circumscribed angles and central angles.
- Construct a tangent line from a point outside a circle to the circle.

Mathematical Practices
1, 2, 3, 4, 5

A **tangent** is a line in the same plane as a circle that intersects the circle in exactly one point. This point is called the **point of tangency.**
In the figure, line ℓ is tangent to circle C. Point P is the point of tangency.

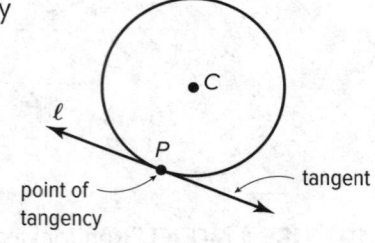

point of tangency

tangent

EXAMPLE 1   Investigate Tangents

EXPLORE In this exploration, you will investigate the relationship between tangents and radii.

a. **USE TOOLS** In the space at the right, use a compass to draw a circle, $\odot C$. Plot a point A on the circle. Then use a straightedge to a draw a line tangent to $\odot C$ at point A. Draw the tangent as accurately as possible.

b. **MAKE A CONJECTURE** Use the straightedge to draw radius $\overline{CA}$. What do you notice about the angle formed by $\overline{CA}$ and the tangent? Repeat **parts a** and **b** by drawing a different circle and different tangent on a separate sheet of paper. Check your conclusion with your classmates.

_____

_____

The Key Concept box summarizes the relationship you identified in the exploration.

KEY CONCEPT

Complete the statement of the theorem and the example.

In a plane, a line is tangent to a circle if and only if

it is _____ to a _____ drawn to the point of tangency.

Example: Line m is tangent to $\odot O$ if and only if

line m is _____ to $\overline{OP}$.

Because the view from the top of a tower or other point of elevation can be modeled with a line tangent to Earth, the fact that tangent lines are perpendicular to the radius of the circle is very useful for finding how far a person can see from that particular elevation.

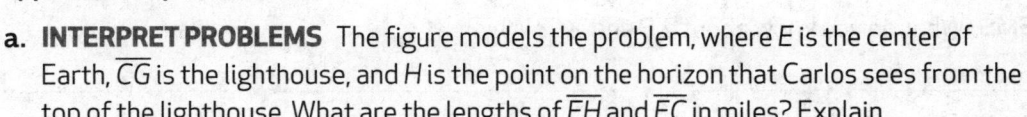

EXAMPLE 2 Apply Relationships Among Tangents and Radii

The Cape Hatteras Lighthouse in North Carolina is 200 feet tall, making it the tallest lighthouse in North America. Carlos stands at the top of the lighthouse and looks at the horizon. He wants to know the distance, to the nearest tenth of a mile, from his eyes to the farthest point he can see on the horizon. The average radius of Earth is approximately 3959 miles.

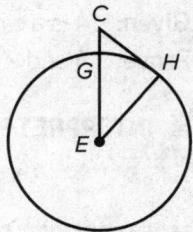

a. **INTERPRET PROBLEMS** The figure models the problem, where E is the center of Earth, $\overline{CG}$ is the lighthouse, and H is the point on the horizon that Carlos sees from the top of the lighthouse. What are the lengths of $\overline{EH}$ and $\overline{EC}$ in miles? Explain.

_____

_____

b. **USE A MODEL** Explain how to use the model to find the distance CH to the nearest tenth of a mile.

_____

_____

_____

c. **REASON ABSTRACTLY** Explain why Carlos would be able to see farther if he were standing on a taller building.

_____

_____

A **circumscribed angle** is an angle formed by two tangents to a circle. In the figure, $\overrightarrow{AB}$ and $\overrightarrow{AC}$ are tangents to circle O, so $\angle BAC$ is a circumscribed angle.

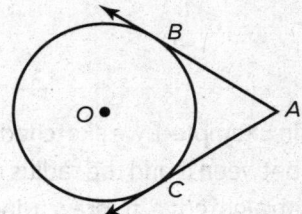

The Key Concept box states two useful theorems about circumscribed angles.

KEY CONCEPT

Use the figures to write an example of each theorem.

A circumscribed angle is supplementary to its corresponding central angle. Example: _____ You will prove this theorem in **Example 3**.	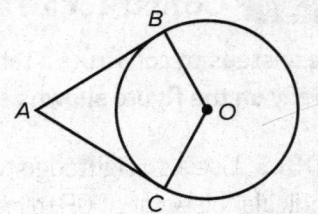
The measure of a circumscribed angle is one-half the difference of the measures of the intercepted arcs. Example: _____ You will prove this theorem in **Exercise 7**.	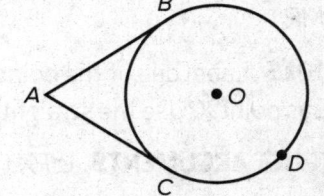

EXAMPLE 3 Prove That a Circumscribed Angle Is Supplementary to a Central Angle

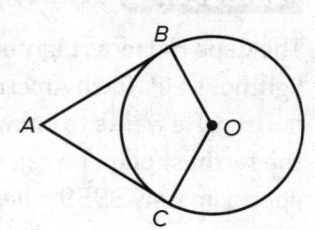

Follow these steps to write a paragraph proof.

Given: ∠A is a circumscribed angle.
Prove: ∠A and ∠O are supplementary.

a. INTERPRET PROBLEMS What do you know about ∠B and ∠C? Why?

_____

b. INTERPRET PROBLEMS What do you know about the four angles in a quadrilateral?

_____

c. INTERPRET PROBLEMS Once we account for the total of ∠B and ∠C, how much is left for ∠A and ∠O? Explain.

_____

_____

_____

d. CONSTRUCT ARGUMENTS Write a paragraph proof.

_____

_____

_____

_____

_____

In **Example 1** we sketched a tangent line so that we could investigate the relationship between it and the radius of the circle. You likely experienced how difficult it is, even with a straightedge, to draw a line that is perfectly tangent to a circle. However, using a straightedge and a compass will allow us to draw a much more accurate diagram.

EXAMPLE 4 Construct a Tangent to a Circle

Follow these steps to construct a tangent to a circle from a point outside the circle. Work directly on the figure shown at the right.

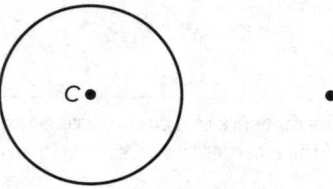

a. USE TOOLS Use a straightedge to draw $\overline{CP}$. Then construct the perpendicular bisector of $\overline{CP}$ to locate the midpoint M of $\overline{CP}$.

b. USE TOOLS Use the compass to draw the circle with center M and radius MP.

c. USE TOOLS Label one of the points of intersection of this circle with circle C as point X. Use the straightedge to draw $\overleftrightarrow{PX}$.

d. CONSTRUCT ARGUMENTS Is $\overleftrightarrow{PX}$ tangent to ⊙C? Justify your reasoning.

_____

_____

REASON ABSTRACTLY Find the indicated measure. Assume that segments that appear to be tangent are tangent.

1. $m\widehat{DE} = 121$. Find $m\angle B$.

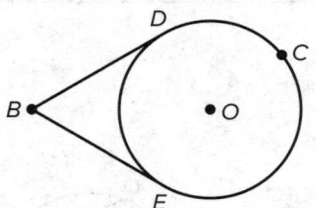

2. Find $m\widehat{PTR}$.

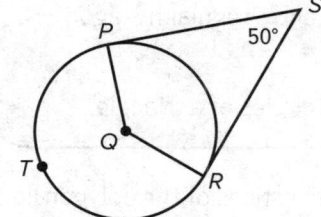

3. $AB = \sqrt{3}$ and $BC = 1$. Find $m\widehat{BD}$.

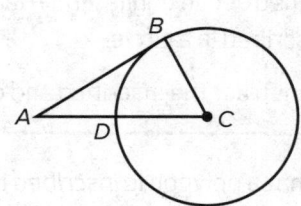

_____   _____   _____

USE TOOLS Construct a tangent to circle O from point P.

4.

5.

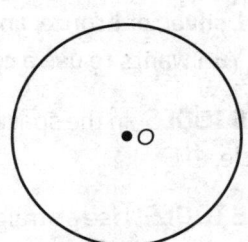

6. USE A MODEL A machine in a factory is operated by two wheels and a belt that surrounds the wheels, as shown. Wheel A has a radius of 10 inches, and wheel B has a radius of 6 inches. The length of the belt between the wheels, $\overline{CD}$, is 24 inches. Assume that $\overline{CD}$ is tangent to both wheels. What is the distance between the centers of the wheels to the nearest tenth of an inch? (*Hint:* Draw a perpendicular line from B to $\overline{AC}$.)

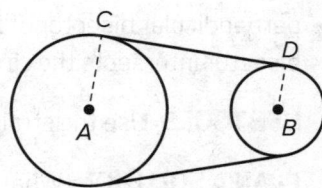

_____

7. CONSTRUCT ARGUMENTS Write a paragraph proof to show that in circle O, $m\angle P = \frac{1}{2}(m\widehat{QSR} - m\widehat{QR})$.

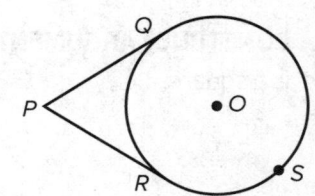

_____

_____

_____

_____

8. USE A MODEL If an airplane is flying at 20,000 feet above Earth, how far can the pilot see until the horizon cuts off his view?

_____

_____

_____

Inscribed and Circumscribed Circles

Objectives

- Construct an equilateral triangle, a square, and a regular hexagon inscribed in a circle.

- Construct the inscribed and circumscribed circles of a triangle.

Mathematical Practices
1, 3, 5, 6, 7

Recall that a polygon is *inscribed* in a circle if the vertices of the polygon lie on the circle. In the figure, the pentagon *PQRST* is inscribed in circle *C*. Because circle *C* passes through each vertex, you can also say that circle *C circumscribes* pentagon *PQRST*.

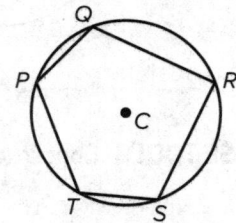

EXAMPLE 1 Construct a Square Inscribed in a Circle

EXPLORE Yuri is designing the medals for an athletic event. The medals will be circles of gold, silver, or bronze, and the design will be based on a square that is inscribed in the circle. Yuri wants to use a compass and straightedge to draw a prototype for the medals.

a. USE TOOLS In the space at the right, use a compass to draw a circle, $\odot C$.

b. USE TOOLS Use a straightedge to draw a diameter, $\overline{JK}$.

c. USE TOOLS Use the compass and straightedge to construct the perpendicular bisector of $\overline{JK}$. Label the points where the perpendicular bisector intersects the circle as points *L* and *M*.

d. USE TOOLS Use the straightedge to draw $\overline{JL}, \overline{LK}, \overline{KM},$ and $\overline{MJ}$.

e. PLAN A SOLUTION What do you need to show in order to prove that the quadrilateral *JLKM* is a square?

_____

_____

f. CONSTRUCT ARGUMENTS Write a paragraph proof to prove that *JLKM* is a square.

_____

_____

_____

_____

_____

A square is not the only regular polygon that can be inscribed in a circle using a straight edge and a compass. We can also inscribe a regular hexagon.

EXAMPLE 2 **Construct a Regular Hexagon Inscribed in a Circle**

Follow these steps to construct a regular hexagon inscribed in a circle.

a. **USE TOOLS** In the space at the right, use a compass to draw a circle, ⊙ C.

b. **USE TOOLS** Using the same compass setting, place the point of the compass on the circle and make an arc that intersects the circle. Label the point of intersection P.

c. **USE TOOLS** Using the same compass setting, place the point of the compass on point P and make an arc that intersects the circle. Label the point of intersection Q.

d. **USE TOOLS** Continue the process to locate points R, S, T, and U. Use a straightedge to draw a hexagon.

e. **PLAN A SOLUTION** What do you need to show in order to prove that the hexagon PQRSTU that you constructed is a regular hexagon?

_____

f. **CONSTRUCT ARGUMENTS** Write a paragraph proof to prove that polygon PQRSTU is a regular hexagon.

_____

_____

_____

_____

_____

g. **USE STRUCTURE** Explain how you can use the method used in the above construction to construct an equilateral triangle inscribed in a circle. Then show the construction in the space provided at the right.

_____

_____

_____

h. **USE STRUCTURE** Using a similar idea, how can you start with the construction of an inscribed square from **Example 1** and construct a regular inscribed octagon?

_____

_____

_____

_____

EXAMPLE 3 Construct the Inscribed Circle of a Triangle

Follow these steps to construct a circle inscribed in △*DEF*.

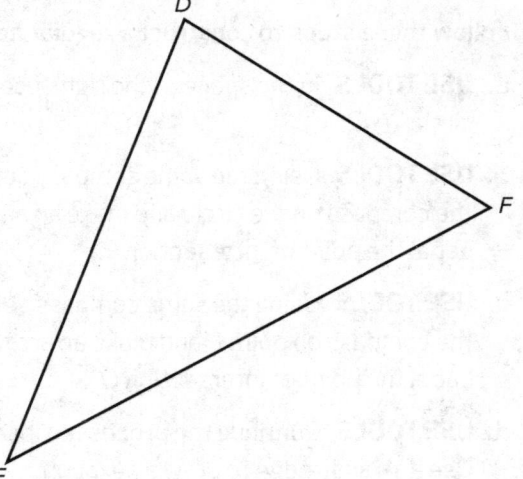

a. PLAN A SOLUTION What must be true about the center of the circle that you will construct? What does this tell you about how you should locate the center of the circle?

_____

_____

_____

_____

b. USE TOOLS Describe the method you will use for this construction. Then perform the construction directly on △*DEF*.

_____

_____

c. COMMUNICATE PRECISELY Describe why it is not necessary to construct the third angle bisector.

_____

_____

EXAMPLE 4 Construct the Circumscribed Circle of a Triangle

Follow these steps to construct a circle circumscribed about △*JKL*.

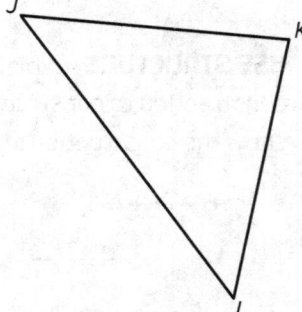

a. PLAN A SOLUTION What must be true about the center of the circle that you will construct? What does this tell you about how you should locate the center of the circle?

_____

_____

_____

_____

_____

b. USE TOOLS Describe the method you will use for this construction. Then perform the construction directly on △*JKL*.

_____

_____

USE TOOLS Use a compass and straightedge to perform each construction in the space provided.

1. Construct a regular hexagon inscribed in circle O.

2. Construct a square inscribed in circle P.

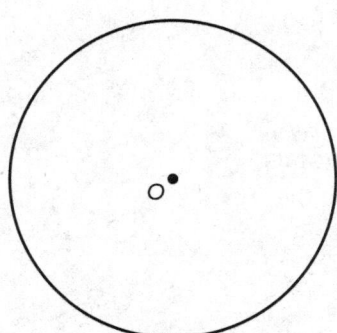

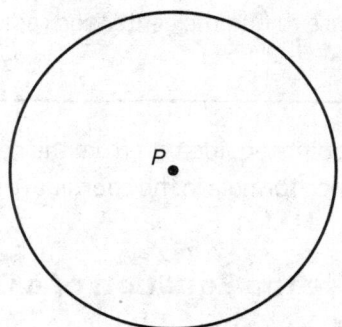

USE TOOLS Use a compass and straightedge to perform each construction in the space provided.

3. Construct a circle inscribed in $\triangle XYZ$.

4. Construct a circle circumscribed about $\triangle FGH$.

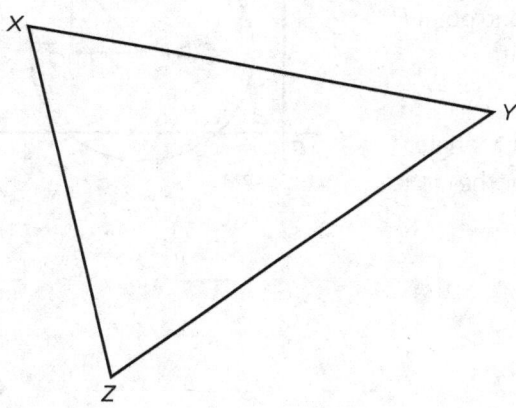

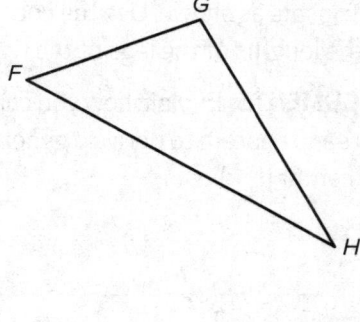

5. CRITIQUE REASONING Rafael constructs the circle inscribed in $\triangle ABC$ by constructing the angle bisectors of $\angle A$ and $\angle B$. He claims that he will get a different inscribed circle if he constructs the bisectors of $\angle B$ and $\angle C$. Do you agree? Explain.

_____

_____

6. COMMUNICATE PRECISELY Describe the steps you can use to construct a regular octagon inscribed in a circle.

_____

_____

_____

Objectives

Mathematical Practices
1, 2, 3, 4, 5, 7, 8

- Derive the equation of a circle given the center and radius.
- Complete the square to find the center and radius of a circle given by an equation.

A circle is the set of all points equidistant from the center of the circle. Using this fact, together with the distance formula in the coordinate plane, we are able to derive the equation of a circle.

EXAMPLE 1 Derive the Equation of a Circle

EXPLORE A radio station broadcasts from a studio located at $(-2, -3)$ on a state map where each unit represents 10 miles. The station's signal reaches 50 miles in all directions. The station's manager wants to write an equation that represents the boundary of the region that the signal can reach.

a. **REASON ABSTRACTLY** Consider a circle whose center C is located at (h, k) and whose radius is r, as shown in the figure. Let (x, y) be the coordinates of a point P on the circle. You can use the center and point P to create a right triangle as shown. Use the coordinates to write expressions for the lengths of the legs of the right triangle.

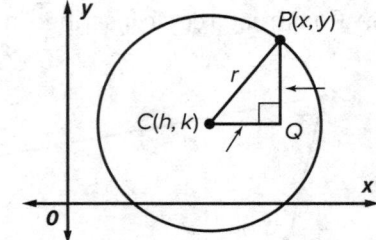

b. **CONSTRUCT ARGUMENTS** Explain how you can use the right triangle and the Pythagorean Theorem to derive a general equation for the circle with radius r and center (h, k).

_____

_____

_____

_____

c. **USE A MODEL** On the coordinate plane at the right, graph the location of the radio station's studio and the boundary of the region that the station's signal can reach.

d. **USE A MODEL** Explain how to write an equation that represents the boundary of the region that the stations' signal can reach.

_____

_____

_____

_____

_____

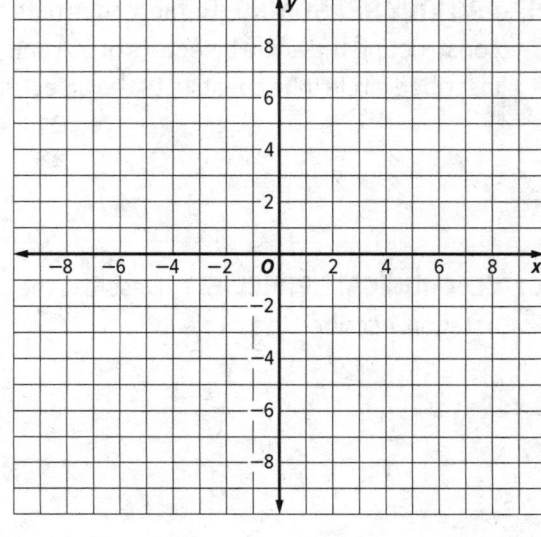

The Key Concept box summarizes the relationship you discovered in the previous exploration.

KEY CONCEPT Equation of a Circle in Standard Form

Complete the statement of the theorem.

| The standard form of the equation of a circle with center at (h, k) and radius r is _____ | 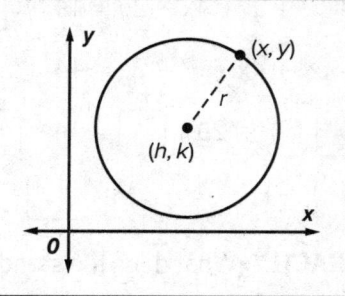 |

EXAMPLE 2 Use Coordinates to Prove a Theorem

Follow these steps to prove or disprove that the point $(0, 3)$ lies on the circle centered at $(3, -1)$ that passes through the point $(-1, 2)$.

a. REASON ABSTRACTLY Explain how you can find the radius of the circle.

_____

_____

b. REASON ABSTRACTLY Explain how you can find the equation of the circle.

_____

_____

c. CONSTRUCT ARGUMENTS Write a paragraph proof to prove or disprove that the point $(0, 3)$ lies on the circle.

_____

_____

_____

_____

d. EVALUATE REASONABLENESS Graph the circle on the coordinate plane at the right. Then explain how you can use your graph to check that your answer to **part c** is reasonable.

_____

_____

_____

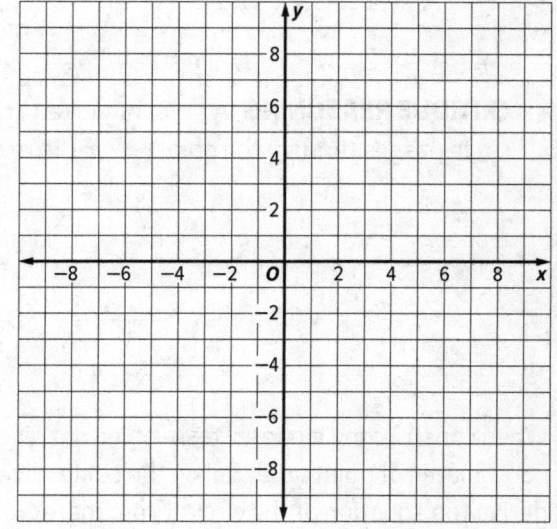

EXAMPLE 3 Complete the Square to Graph an Equation

Follow these steps to graph the circle whose equation is $x^2 + y^2 + 4x - 2y = 44$.

a. USE STRUCTURE Write the equation in standard form by completing the square.

$$x^2 + y^2 + 4x - 2y = 44 \qquad \text{Original equation}$$

$$x^2 + 4x + y^2 - 2y = 44 \qquad \text{Group terms with the same variable.}$$

$$x^2 + 4x + \boxed{} + y^2 - 2y + \boxed{} = 44 + \boxed{} + \boxed{} \qquad \text{Complete the square.}$$

$$(\underline{\hspace{1.5cm}})^2 + (\underline{\hspace{1.5cm}})^2 = \boxed{} \qquad \text{Factor and simplify.}$$

b. REASON ABSTRACTLY What does the standard form of the equation tell you about the circle?

_____

_____

_____

c. USE TOOLS Explain how you can use a compass to graph the circle. Draw the graph on the coordinate plane at the right.

_____

_____

_____

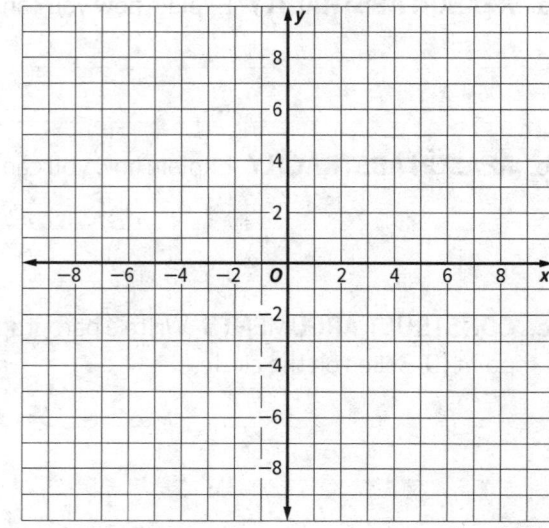

d. EVALUATE REASONABLENESS Explain how you can use the graph to check that you found the standard form of the equation correctly.

_____

_____

_____

_____

e. CRITIQUE REASONING A student looked at the graph of the circle and stated that the circle passes through the point $(-7, 6)$. Do you agree? Explain why or why not.

_____

_____

_____

_____

When constructing circles, we are often not given the center or the radius, but instead the coordinates of points that fall on the circle. Given three points on the circle, we are able to derive the equation of the circle. From there we can find the radius and the center.

EXAMPLE 4 **Solve a Real-World Problem**

An urban planner is designing a new circular road for a housing development. The road will pass through the points $P(-1, 2)$, $Q(5, 2)$, and $R(7, -2)$. The city planner wants to prove or disprove that the new road will intersect an existing road that lies along the line $y = 3$.

a. PLAN A SOLUTION Describe the steps you can use to solve the problem.

_____

_____

_____

_____

b. USE A MODEL Plot the given points on the coordinate plane at the right. Draw the perpendicular bisectors of the two segments. Use a point and a slope to find the equation of these two lines.

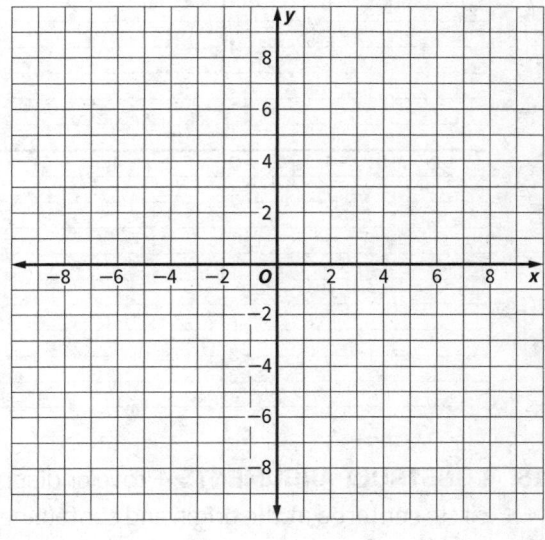

c. USE A MODEL Solve for the intersection of these two lines to find the center of the circle. Find the radius of the circle algebraically, and then write the equation that models the new road. Draw the graph of the new road on the coordinate plane.

d. CONSTRUCT ARGUMENTS Write an argument to prove or disprove that the new road will intersect the existing road that lies along the line $y = 3$.

_____

_____

_____

_____

PRACTICE

REASON ABSTRACTLY Write the equation of each circle.

1. The circle with center $(4, -3)$ and radius 3

2. The circle with center $(0, -5)$ and radius $\sqrt{2}$

3. The circle with center $(3, 1)$ that contains $(3, -1)$

4. The circle with center $(-8, 0)$ that contains the origin

5. The circle passing through $(0, 2)$, $(-2, 0)$, and $(0, -2)$

6. The circle passing through $(-2, 2)$, $(-3, 3)$, and $(-4, 2)$

REASON ABSTRACTLY Find the center and radius of a circle with each of the following equations.

7. $x^2 + y^2 - 8x + 2y = -2$ **8.** $x^2 + y^2 - 6x + 10y = -30$ **9.** $x^2 + y^2 + 12x + 2y = 63$

_____   _____   _____

10. $x^2 + y^2 - 8y = -5$ **11.** $x^2 + y^2 - 14x - 4y = 0$ **12.** $x^2 + y^2 + 22x = -118$

_____   _____   _____

REASON ABSTRACTLY Graph each circle.

13. $x^2 + y^2 - 4x - 6y = 23$

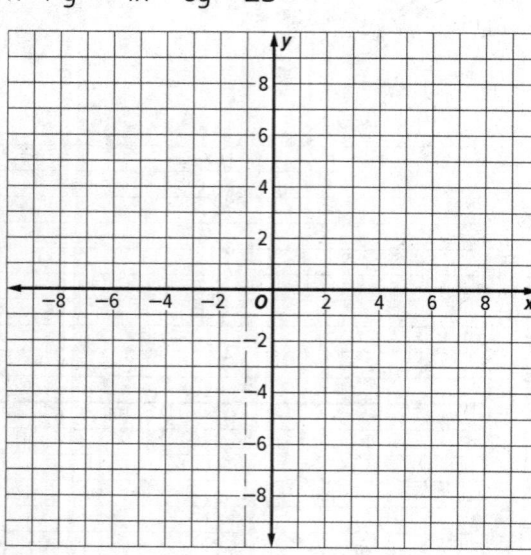

14. $x^2 + y^2 + 2x + 2y = 47$

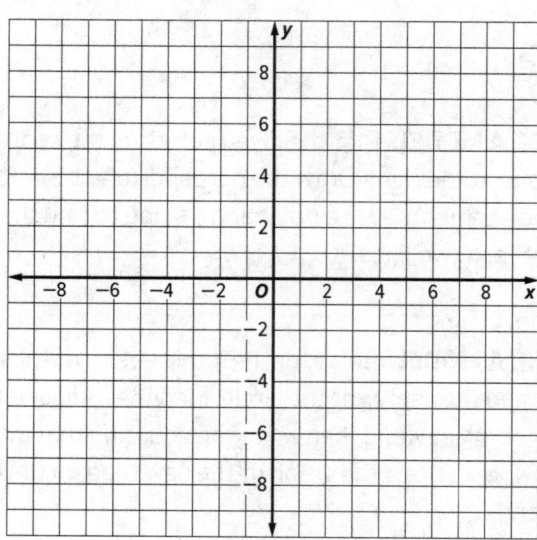

15. CONSTRUCT ARGUMENTS Prove or disprove that the point $(5, \sqrt{39})$ lies on the circle centered at the origin and containing the point $(0, -8)$.

_____

_____

_____

_____

_____

_____

16. CONSTRUCT ARGUMENTS Prove or disprove that the origin lies on the circle whose diameter has endpoints $(-1, 2)$ and $(5, 2)$.

_____

_____

_____

_____

17. **REASON ABSTRACTLY** A circle has center at $(2, 3)$. The point $(2, 1)$ lies on the circle. Find three other points with integer coordinates that lie on the circle. Explain how to do this without finding the equation of the circle.

_____

_____

_____

_____

18. **CRITIQUE REASONING** Adam said that the equation $x^2 + y^2 + 4x - 10y = k$ is the equation of a circle for any value of k since it is always possible to complete the square to find the center and the radius. Do you agree? Explain.

_____

_____

_____

_____

19. **USE A MODEL** A supermarket chain has a warehouse at point W and a store at point A. The president of the supermarket chain wants to open a new store that is the same distance from the warehouse as store A.

    a. An employee suggests opening the new store at $B(1, -5)$. Explain whether this is a possible location for the store.

    _____

    _____

    _____

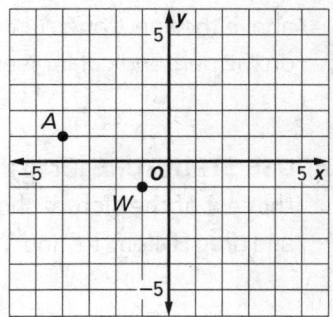

    b. Give the coordinates of a possible location for the new store and justify your choice.

    _____

    _____

20. **CONSTRUCT ARGUMENTS** A circle has center $(6, 2)$ and the y-axis is tangent to the circle. Write the equation of the circle. Then give an argument to prove that your equation is correct.

_____

_____

_____

_____

21. **DESCRIBE A METHOD** Kalina is working with several circles that all have equations in the form $x^2 + y^2 + ax + by = c$. She would like a quick way to find the center of the circle based on the coefficients of the equation. Describe a method she can use to find the center.

_____

Objectives

- Derive the equation of a parabola given a focus and directrix.
- Use coordinates to prove geometric theorems algebraically.

You have seen that the graph of a quadratic function is a parabola. The equation $y = (x - h)^2 + k$ is a parabola with vertex (h, k) and an axis of symmetry $x = h$. A **parabola** is the **locus**, or set of all points, in a plane that are equidistant from a fixed point, called the **focus**, and a fixed line, called the **directrix**. Recall that the distance from a point to a line is the length of the perpendicular segment from the point to the line. You can use this idea and the Distance Formula to derive the equation of a parabola.

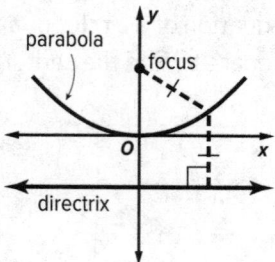

EXAMPLE 1   Investigate the Definition of a Parabola

EXPLORE Use The Geometer's Sketchpad to investigate the shape of a parabola.

a. **USE STRUCTURE** Construct a horizontal line and a point B on the line. Draw a point A above the line. Construct the perpendicular bisector of $\overline{AB}$. What is true of every point on the perpendicular bisector of $\overline{AB}$?

_____

b. **USE STRUCTURE** Drag the point B along the line. If you have done this correctly, then the rest of the picture should move with B. Select **Display | Trace Perpendicular Line** and drag B along its line. What do you observe?

_____

_____

c. **USE STRUCTURE** Select **Construct | Locus**. Drag point B across its line. What shape emerges as B moves?

_____

d. **USE STRUCTURE** Move point A; the locus of points should adjust automatically. How does the curve change when A moves closer to the line? When A moves away from the line?

_____

_____

e. **USE STRUCTURE** Pick a point on the parabola and measure its distance to A. Measure the distance from the point to the line. What do you notice? Pick another point. Does this seem to be always true?

_____

_____

In the previous lesson, we were able to derive the equation of a circle using the center, the radius, and the definition of the circle. We can do the same thing with a parabola. Given the equation of the directrix and the coordinates of the focus, we can use the distance formula and the definition of a parabola to derive its equation.

EXAMPLE 2 Derive the Equation of a Parabola

EXPLORE A parabola has directrix $y = -p$ and focus $F(0, p)$. **Follow these steps to derive the equation of the parabola.**

a. USE STRUCTURE Let $P(x, y)$ be a point on the parabola, as shown in the figure. Draw a perpendicular segment from P to the directrix and let point D be the point of intersection of this segment with the directrix. Explain how to write an expression for the length of $\overline{PD}$.

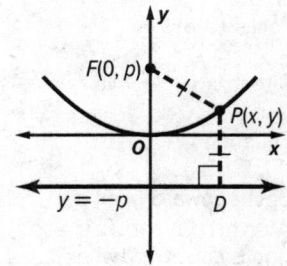

_____

_____

_____

b. REASON ABSTRACTLY Explain how to write an expression for the length of $\overline{FP}$.

_____

_____

_____

c. CALCULATE ACCURACTELY By the definition of a parabola, $PD = FP$. Use the expressions you wrote in **parts a** and **b** to solve for y.

_____	$PD = FP$
_____	Square both sides of the equation.
_____	Square the expressions in parentheses.
_____	Subtract y^2 and p^2 from both sides.
_____	Add $2yp$ to both sides.
_____	Divide both sides by $4p$.

d. REASON ABSTRACTLY Explain how to use your result from **part c** to find the equation of the parabola with focus $(0, 1)$ and directrix $y = -1$.

_____

_____

e. REASON ABSTRACTLY What is the equation of a parabola with focus $(0, -1)$ and directrix $y = 1$? How is the graph of this parabola different from the graph of the parabola in **part d**?

_____

_____

_____

Your work in the previous exploration assumes that the vertex of the parabola is at the origin and that the directrix is a horizontal line. The Key Concept box summarizes your findings.

KEY CONCEPT **Equation of a Parabola with Focus on y-axis**

Complete the following.

Characteristics	Graph
Vertex: $(0, 0)$ Focus: $(0, p)$ Directrix: $y = -p$ Equation: _____ Opens upward when _____ Opens downward when _____	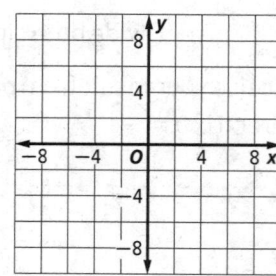

EXAMPLE 3 **Derive an Equation to Graph a Parabola**

Follow these steps to graph the parabola with focus $(0, -2)$ and directrix $y = 2$.

a. REASON ABSTRACTLY Explain how to write the equation of the parabola.

_____

_____

b. CALCULATE ACCURATELY Use the table to list some ordered pairs that lie on the parabola. Then use the table to help you graph the parabola on the coordinate plane.

x	y

The directrix of a parabola may be a horizontal line or a vertical line. The steps for deriving the equation of a parabola with a vertical directrix are similar to the steps you used in **Example 2**.

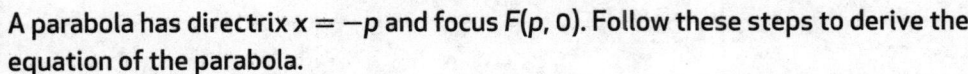

EXAMPLE 4 Derive the Equation of a Parabola with a Vertical Directrix

A parabola has directrix $x = -p$ and focus $F(p, 0)$. Follow these steps to derive the equation of the parabola.

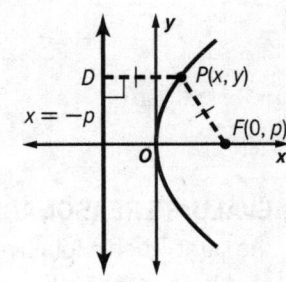

a. **USE STRUCTURE** Let $P(x, y)$ be a point on the parabola, as shown in the figure. Draw a perpendicular segment from P to the directrix and let point D be the point of intersection of this segment with the directrix. Write an expression for the length of $\overline{PD}$.

_____

b. **REASON ABSTRACTLY** Explain how to write an expression for the length of $\overline{FP}$.

_____

c. **CALCULATE ACCURATELY** By the definition of a parabola, $PD = FP$. Use the expressions you wrote in **parts a** and **b** to solve for x.

_____	$PD = FP$
_____	Square both sides of the equation.
_____	Square the expressions in parentheses.
_____	Subtract x^2 and p^2 from both sides.
_____	Add $2xp$ to both sides.
_____	Divide both sides by $4p$.

KEY CONCEPT Equation of a Parabola with Focus on x-axis

Complete the following.

Characteristics	Graph
Vertex: $(0, 0)$ Focus: $(p, 0)$ Directrix: $x = -p$ Equation: _____ Opens right when _____ Opens left when _____	$x = -p$ $F(p, 0)$

EXAMPLE 5 Use Coordinates to Prove a Theorem

Follow these steps to prove or disprove that the point $(-6, \sqrt{3})$ lies on the parabola with focus $\left(-\frac{1}{8}, 0\right)$ and directrix $x = \frac{1}{8}$.

a. **REASON ABSTRACTLY** Does the parabola open upward, downward, left, or right? Explain how you know.

_____

_____

b. CONSTRUCT ARGUMENTS Write a paragraph proof to prove or disprove that the point $(-6, \sqrt{3})$ lies on the parabola.

_____

_____

_____

_____

c. EVALUATE REASONABLENESS Graph the parabola on the coordinate plane at the right. Then explain how you can use your graph to check that your answer to **part b** is reasonable.

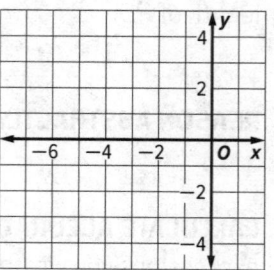

_____

_____

_____

PRACTICE

REASON ABSTRACTLY Write the equation of each parabola.

1. focus: $(0, 5)$

directrix: $y = -5$

_____

2. focus: $(1, 0)$

directrix: $x = -1$

_____

3. focus: $(-3, 0)$

directrix: $x = 3$

_____

4. focus: $\left(0, \frac{1}{2}\right)$

directrix: $y = -\frac{1}{2}$

_____

5. focus: $\left(0, -\frac{1}{16}\right)$

directrix: $y = \frac{1}{16}$

_____

6. focus: $\left(\frac{1}{20}, 0\right)$

directrix: $x = -\frac{1}{20}$

_____

REASON ABSTRACTLY Derive the equation of each parabola to help you graph the parabola.

7. focus: $\left(0, -\frac{1}{8}\right)$

directrix: $y = \frac{1}{8}$

equation: _____

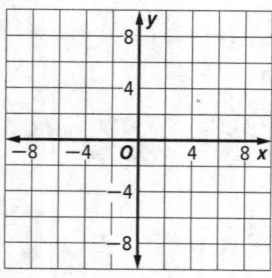

8. focus: $(2, 0)$

directrix: $x = -2$

equation: _____

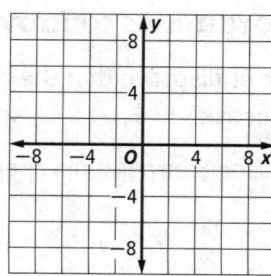

9. **CONSTRUCT ARGUMENTS** Prove or disprove that the point $(\sqrt{3}, -4)$ lies on the parabola with focus $(0, -3)$ and directrix $y = 3$.

10. **CONSTRUCT ARGUMENTS** Prove or disprove that the parabola with focus $(-16, 0)$ and directrix $x = 16$ intersects the line $y = 20$.

11. An engineer is using a coordinate plane to design a tunnel in the shape of a parabola. The line shown in the figure represents the top of the wall that will contain the tunnel. This line will be the directrix of the tunnel, and the focus will be point F. The base of the tunnel (ground level) is represented by the line $y = -10$.

    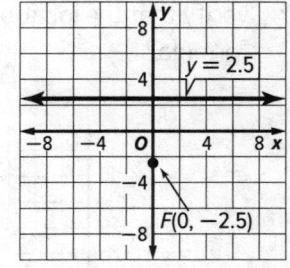

    a. **USE A MODEL** Write the equation of the parabola. Then graph the parabola on the coordinate plane.

    b. **CONSTRUCT ARGUMENTS** Each unit of the coordinate plane represents one foot. Prove or disprove that the width of the tunnel at a height 5 feet above the ground is exactly 14 feet.

12. **REASON ABSTRACTLY** The equation of parabola A is $y = \frac{1}{24}x^2$. Parabola B has the same vertex as parabola A and opens in the same direction as parabola A. However, the focus and directrix for parabola B are twice as far apart as they are for parabola A. Write the equation for parabola B. Explain your steps.

13. **CONSTRUCT ARGUMENTS** Prove or disprove that if a parabola opens up or down, then the length of the segment with endpoints on the parabola that passes through the focus and is parallel to the directrix is $|4p|$.

Where Is the Snack Bar?

Provide a clear solution to the problem. Be sure to show all of your work, include all relevant drawings, and justify your answers.

The manager of an amusement park wants to place a new snack bar equidistant from three rides. The locations of the rides are shown.

Part A

Construct the location of the snack bar on the graph provided. Find the equation of the circle containing the three locations. How could you use the equation of the circle to verify that the locations are equidistant from the snack bar, instead of using the distance formula?

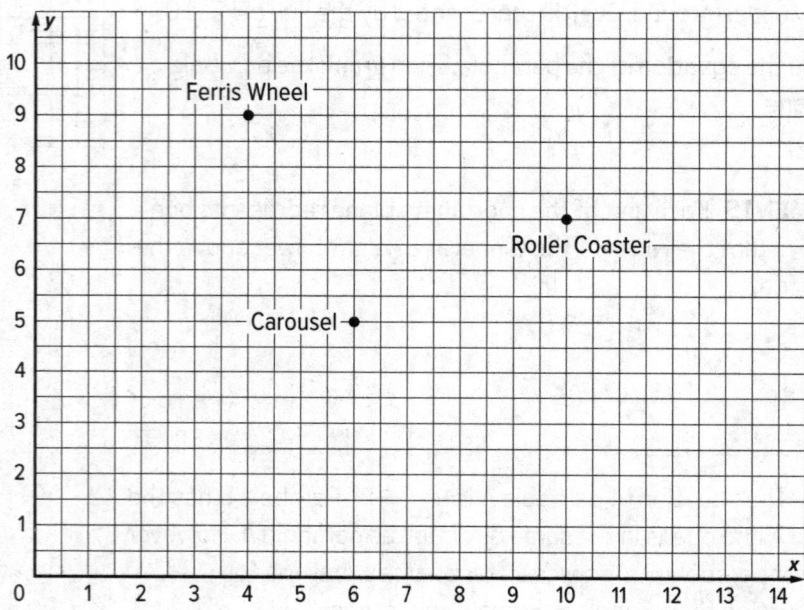

Part B

The manager decides he would like the snack bar to also be equidistant from the fun house which is located at the point (10, 8). Use the graph to place the fun house and new snack bar, if possible. Justify your answer.

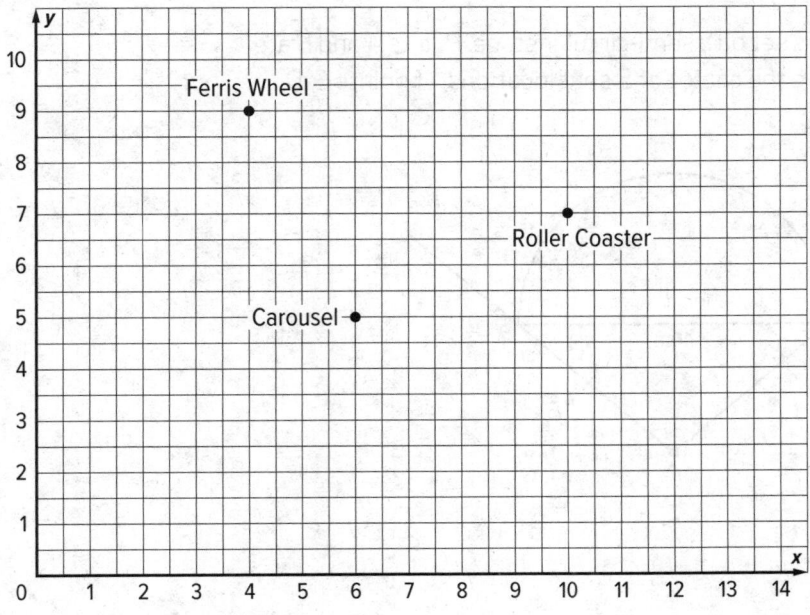

Part C

The manager changes his mind and decides to only use the three original locations, the Ferris wheel, the roller coaster, and the carousel, and now wants to add a fourth location—the new water ride, located at the point (8, 11). Use the graph to place the water ride and new snack bar, if possible. Justify your answer.

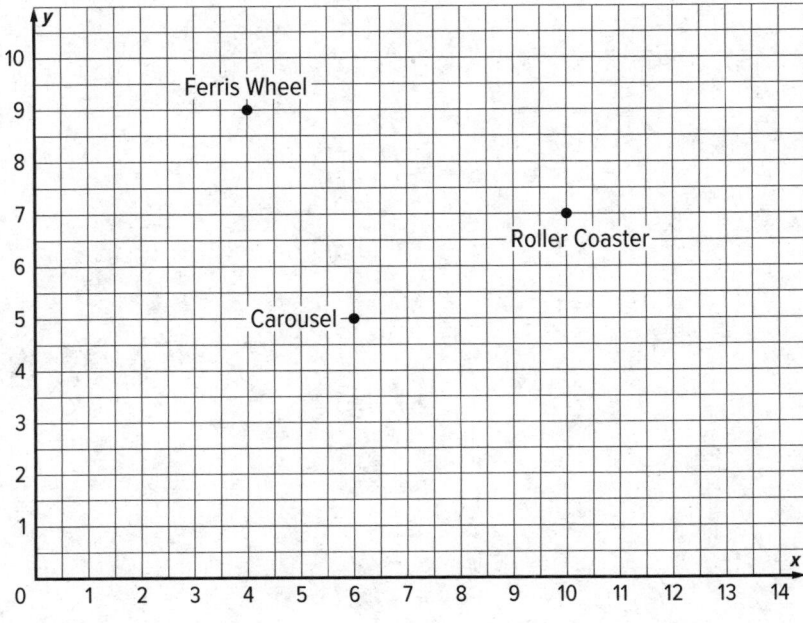

The Best View

Provide a clear solution to the problem. Be sure to show all of your work, include all relevant drawings, and justify your answers.

You are attending a play that is set on a semi-circular stage. Points A and B are the ends of the stage. You have the choice of 5 seat locations, at points C, D, E, F, or G.

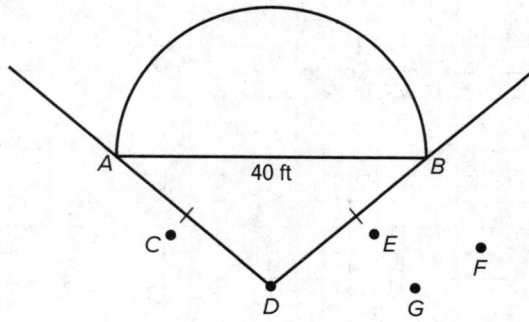

Part A

If you sit in seat D, your viewing area is formed by $\angle ADB$. What is the measure of $\angle ADB$? How can you determine whether seats C, D, and E have the same viewing area? Do they? Justify your answer.

Part B

Leo says seat *G* lies on the line that is tangent to the stage at point *B*. Describe two ways you can use a construction to determine whether Leo is correct? Is he correct?

Part C

How much area in square feet is covered by the viewing area of seat *D*? Justify your answer.

Part D

If seat *C* lies on the circle formed by extending the semicircle of the stage and is 15 feet from point *A*, find the viewing area of seat *C*. Justify your reasoning and round to the nearest tenth.

1. Consider the diagram of circle *E*.

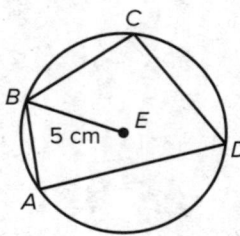

If $\overset{\frown}{BCD} = 15$ centimeters, complete the following. Round to the nearest tenth.

$m\overset{\frown}{BCD} = \boxed{}$ $m\angle BAD \approx \boxed{}$

$\overset{\frown}{BAD} = \boxed{}$ cm

2. A circle has a circumference of 15.5 inches. Rounded to the nearest tenth, what is its diameter?

$\boxed{}$ in.

3. Complete the following.

A circle is a set of $\boxed{}$ that are

$\boxed{}$ from a given point, which is the

$\boxed{}$ of the circle.

4. What is $m\overset{\frown}{ED}$ in the following diagram?

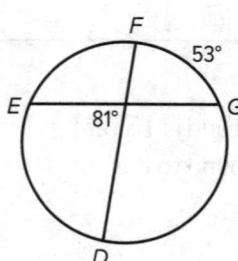

$m\overset{\frown}{ED} = \boxed{}$

5. A parabola has its focus at $(0, 5)$ and directrix $y = 1$. What is an equation for the parabola?

$\boxed{}$

6. In the diagram below, $m\overset{\frown}{BD} = 100$ and $m\angle ABC = 26$.

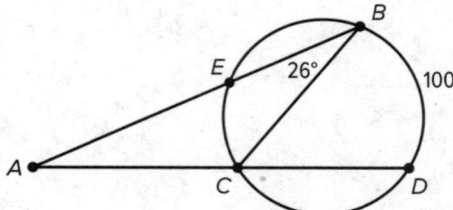

What is $m\angle BAC$?

$m\angle BAC = \boxed{}$

7. Circle *A* has a circumference that is 3 times the circumference of circle *B*. If circle *B* has a radius of 7, the diameter of circle *A* is $\boxed{}$.

8. In the diagram below, quadrilateral *DEFG* is circumscribed about circle *Q*.

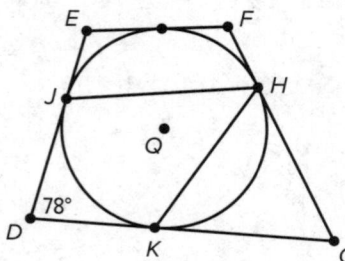

If $m\angle EDG = 78$, what is $m\angle JHK$?

$m\angle JHK = \boxed{}$

9. The points $(4, -2)$, $(5, 5)$, and $(-3, 5)$ lie on a circle. What is the equation of the circle?

10. Consider the diagram of circle *T*.

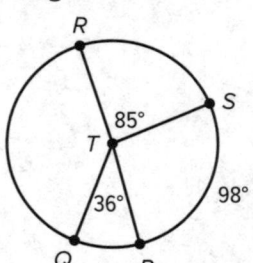

What is $m\overset{\frown}{QRS}$?
$m\overset{\frown}{QRS} = \boxed{}$

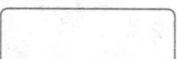

11. Consider whether each statement is *sometimes*, *always*, or *never* true and check the appropriate choice for each row.

Statement	Sometimes	Always	Never
If the radius of circle A is twice the radius of circle B, then the circumference of circle A is twice the circumference of circle B.			
In a circle, a chord bisects the radius.			
If the x-axis is tangent to a circle, then the y-axis is also tangent to the circle.			
A parabola is the set of points equidistant from the focus and the line of symmetry.			
If a triangle inscribed in a circle is a right triangle, then its hypotenuse is a diameter of the circle.			

12. What is the circumference of circle *P* in the diagram if $BD = 5$ and $AD = DC = 9$? Round to the nearest tenth. Explain.

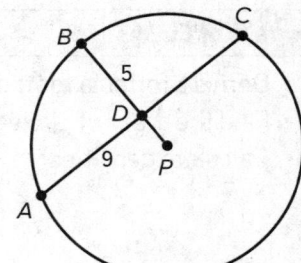

_____

_____

_____

13. Consider the diagram shown below. If $JL = 16$, what is the perimeter of $\triangle NKL$? Explain how you found your answer.

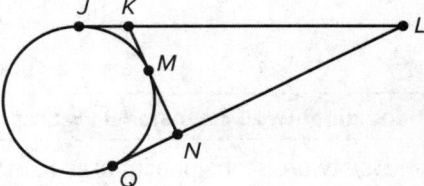

_____

_____

_____

14. A circle circumscribes an equilateral triangle that circumscribes a circle.

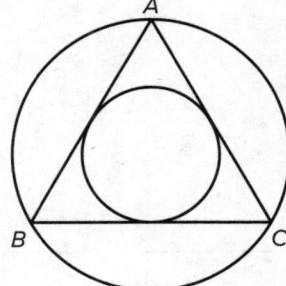

a. If the perimeter of the triangle is 24 inches, find the areas of each of the circles. Round to the nearest tenth.

_____

_____

_____

_____

b. If the coordinates of the vertices of the triangle are $A(4, 4\sqrt{3})$, $B(0, 0)$, and $C(8, 0)$, write the equation of the smaller circle.

_____

_____

CHAPTER FOCUS Learn about some of the objectives that you will explore in this chapter. Answer the preview questions. As you complete each lesson, return to these pages to check your work.

What You Will Learn	Preview Question
Areas of Parallelograms and Triangles	
• Use the distance formula to find areas of parallelograms and triangles on the coordinate plane. • Use area to solve real-world problems.	**SPM 2** A right triangle has coordinates $(4, 0)$, $(2, 4)$, and $(-2, -3)$. What is the area of the triangle? **15 unit²**
Areas of Circles	
• Derive a formula for the area of a sector of a circle • Find the areas of circles and sectors of circles • Calculate densities	**SMP 4** A pizza with a diameter of 12 inches is cut into 24 equal pieces. Arrange the pieces to show how to find the approximate area of the pizza. What is the approximate area using this method? 6π in. 6 in. $A \approx 6(6\pi) = 36\pi \text{ in}^2$
Modeling: Two-Dimensional Figures	
• Use two-dimensional figures to model real-world objects and situations on and off the coordinate plane. • Solve problems involving perimeter and area.	**SMP 4** The end of a shaft in a machine is a square with side length 1 inch. The end of the shaft fits into a hole in a circular disc. The disc has a radius of 3 inches. Draw a model of the disc and then determine the area of the circular face of the disc. $A = 9\pi - 1$, or about 27.3 in²

Objectives

- Use the distance formula to find areas of parallelograms and triangles on the coordinate plane.
- Use area to solve real-world problems.

Mathematical Practices

1, 2, 3, 5, 6, 7

The formula for finding the area of a parallelogram is $A = bh$. The formula for finding the area of a triangle is $A = \frac{1}{2}bh$. In both formulas, b represents the base and h represents the height. The base can be any side of either figure. The height is always perpendicular to the base.

EXAMPLE 1 Investigate the Areas of Parallelograms and Triangles

EXPLORE Use The Geometer's Sketchpad to explore the areas of parallelograms and triangles.

a. **USE TOOLS** Draw parallelogram $ABCD$. Next, draw altitudes from B to $\overline{AD}$ and from B to $\overline{CD}$. Label the points of intersection H and G, respectively. Use a measuring tool to find the length of each base and height. Then find the area of parallelogram $ABCD$.

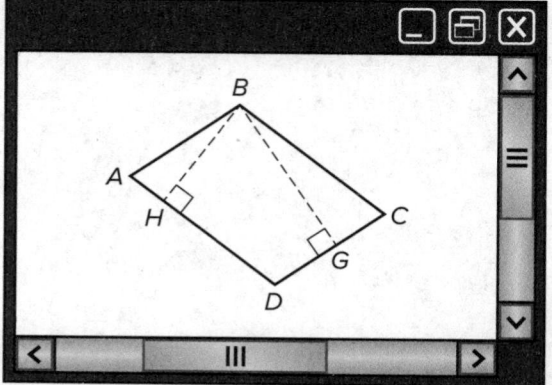

AD _____ BH _____ area of $ABCD$ _____

DC _____ BG _____ area of $ABCD$ _____

b. **MAKE A CONJECTURE** What do you notice about the two areas?

_____

c. **USE TOOLS** Draw $\triangle TRS$ with altitudes $\overline{RW}$, $\overline{TV}$, and $\overline{SU}$. Use a measuring tool to find the measure of each side and the corresponding height. Then find the area of $\triangle TRS$.

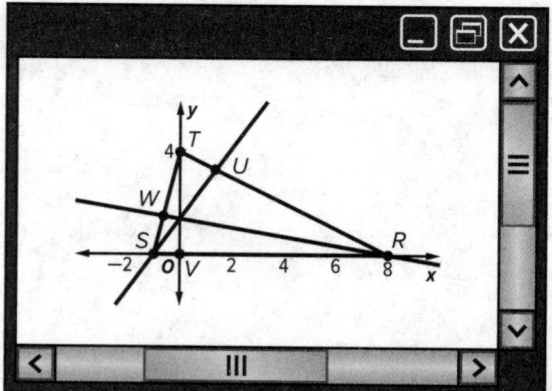

TR _____ SU _____ Area of $\triangle TRS$ _____

RS _____ TV _____ Area of $\triangle TRS$ _____

TS _____ RW _____ Area of $\triangle TRS$ _____

d. INTERPRET PROBLEMS Why does it not matter which side of a parallelogram or triangle is designated as the base?

_____

_____

e. PLAN A SOLUTION Suppose you use graph paper to draw a parallelogram or a triangle on a coordinate plane. How could you find the lengths of the sides if they were neither horizontal nor vertical?

_____

f. USE TOOLS On a new screen create parallelogram *JKLM* and add a coordinate plane to the sketch. Next, drag the points so that they have the following positions:

$J: (-3, 1)$ $K: (0, 3)$ $L: (5, 0)$ $M: (2, -2)$

Construct a line that is perpendicular to *JM* and that passes through point *K*. Label the point of intersection of *JM* and the new line *N*. Find the coordinates of point *N*.

N: _____

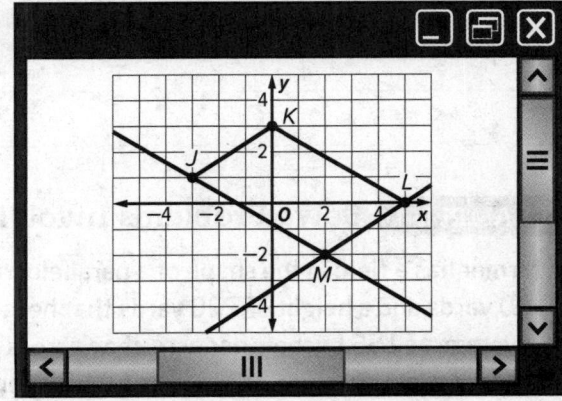

g. FIND A PATTERN Use the coordinates and the distance formulate to calculate *JM* and *KN*. Then find the area of *JKLM*.

JM _____

KN _____

_____

Area of JKLM _____

EXAMPLE 2 **Find the Area of a Parallelogram**

Find the area of parallelogram *ABCD*.

a. PLAN A SOLUTION What do you need to know to find the area of *ABCD*?

_____

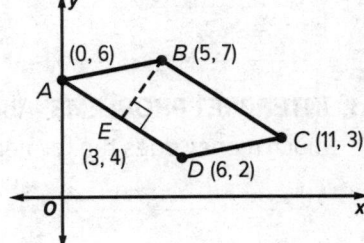

b. USE STRUCTURE Use slope to prove that $\overline{BE}$ is an altitude of *ABCD*.

_____

_____

c. USE STRUCTURE Use the distance formula to find *BE* and *AD*. Then find the area of *ABCD*.

_____

_____

EXAMPLE 3 Find the Area of a Triangle

Find the area of △RST.

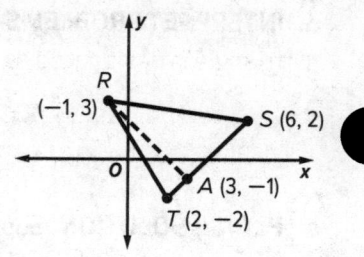

a. **PLAN A SOLUTION** What do you need to know to find the area of △RST?

_____

b. **USE STRUCTURE** Use slope to prove that $\overline{RA}$ is an altitude of △RST.

_____

c. **USE STRUCTURE** Use the distance formula to find ST and AR. Then find the area of △RST.

_____

_____

_____

EXAMPLE 4 Solve Problems Involving Density

A farmer has a field in the shape of a parallelogram that has a base of length 1100 yards and a height of 220 yards that he is planting with corn that yields an average of 155 bushels per acre. If an acre is equal to 4840 square yards and the average projected price for corn is $4.50 per bushel, approximately how much does he expect to make when he harvests his corn?

a. **INTERPRET PROBLEMS** What is the density of the corn?

_____

b. **PLAN A SOLUTION** Explain how you would solve this problem.

_____

_____

_____

c. **CALCULATE ACCURATELY** What is the area of the farmer's field in square yards? About how many acres of corn will he have planted?

_____

_____

d. **INTERPRET PROBLEMS** About how many bushels of corn can the field produce? How much money does the farmer expect to make from the sale of his corn?

_____

_____

e. **REASON QUANTITATIVELY** How does knowing the density of the crop aid the farmer?

_____

_____

1. **CALCULATE ACCURATELY** Find the area of each parallelogram. Show your work.

   a.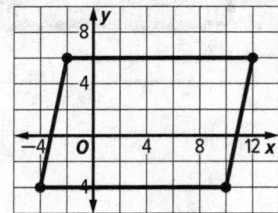

   b.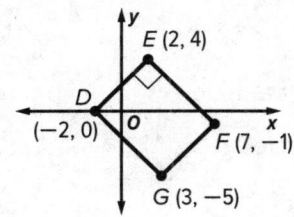

   c.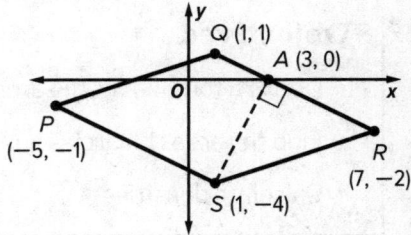

2. **CALCULATE ACCURATELY** Find the area of each triangle. Show your work.

   a.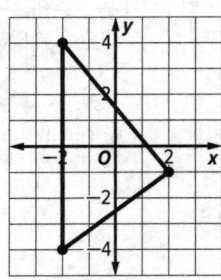

   b.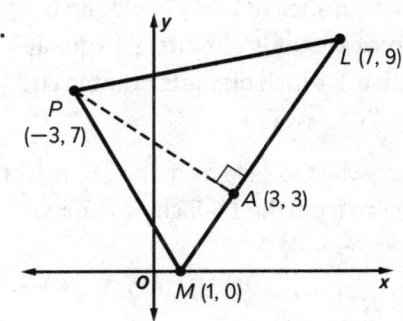

   c.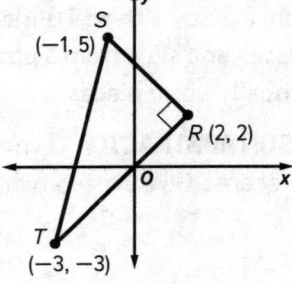

3. **INTERPRET PROBLEMS** A research team studied the prevalence of a type of wild mushroom to determine the best type of area for its growth. In a moist, heavily shaded region of the forest they found 84 specimens in an area in the shape of a right triangle with legs of 35 yards and 30 yards. Next to a stream they found 63 specimens in a sunny area shaped liked a parallelogram with a base of 45 yards and a height of 10 yards. Which area had the greater density of the mushrooms? Explain your reasoning.

   _____

   _____

   _____

4. **REASON QUANTITATIVELY** Gavin is the owner of a new fast food restaurant. Local regulations require a dining room density of no more than 1 customer per 12 square feet. He wants to be able to seat a maximum of 80 customers. What must be the area of his dining room?

   _____

   _____

Objectives

- Derive a formula for the area of a sector of a circle
- Find the areas of circles and sectors of circles
- Calculate densities

A **sector** of a circle is a region bounded by two radii and their intercepted arc. The angle formed by the radii is called a **central angle**.

EXAMPLE 1 **Investigate the Area of Sectors**

EXPLORE Alma wants to determine which slice of pizza is bigger: a slice from a pizza with an 18-inch diameter that is cut into 12 equally-sized pieces or a slice from a pizza with a 14-inch diameter that is cut into 8 equally-sized pieces.

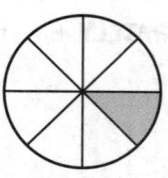

14-inch diameter   18-inch diameter

a. **REASON ABSTRACTLY** Give a reason why the slice from the 18-inch pizza might have a greater area. Give a reason why the slice from the 14-inch pizza might have a greater area.

b. **CALCULATE ACCURATELY** Find the area of each pizza and the area of each slice. Which slice has a greater area? Round your answers to the nearest tenth.

c. **FIND A PATTERN** What two dimensions affect the area of a sector?

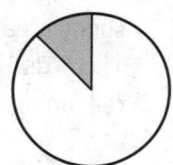

d. **REASON QUANTITATIVELY** How can you find the measure of the central angle of the sector in a pizza with a given number of equal slices? What is the degree measure of the central angle of an 8-slice pizza? What is the degree measure of the central angle of a 12-slice pizza?

e. **FIND A PATTERN** What is the measure of the central angle in the 14-inch pizza? Divide this measure by the full circle (360°). What is the ratio of the area of the slice to the area of the whole pizza? Make a conjecture about these two ratios, and confirm your conjecture by looking at the 18-inch pizza.

EXAMPLE 2 **Calculate Density**

Joshua is trying to determine which slice of pizza has the greatest density of olives.

a. **CALCULATE ACCURATELY** Complete the table. Round each density to the nearest hundredth.

Diameter of Pizza	Number of Equal Slices	Area of Slice	Average Number of Olives per Slice	Density
10 inches	6		18	
12 inches	8		20	
16 inches	10		24	

b. **REASON QUANTITATIVELY** Does the slice with the most olives have the greatest density of olives? Explain.

_____

_____

_____

EXAMPLE 3 **Develop a Formula for the Area of a Sector**

Gabriel and Tamika calculate the area of the sector of the circle A_s in different ways.

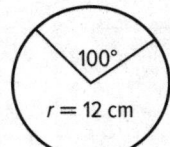

Gabriel:
$$\frac{100}{360} = \frac{A_s}{\pi(12^2)}$$

Tamika:
$$A_s = \frac{100}{360} \cdot \pi \cdot 12^2$$

a. **DESCRIBE A METHOD** Explain the reasoning behind each solution.

_____

_____

_____

b. **CALCULATE ACCURATELY** Do they get the same answer? Explain.

_____

c. **USE STRUCTURE** Using the equal ratios, write a formula for the area of a sector A_s in terms of the radius r and the degree measure of the central angle m. Show your work.

_____

d. **USE STRUCTURE** You have learned that angles can be measured in units known as radians. The entire circle has a measure of 2π radians. Suppose the central angle has a measure of x radians. Write the equal ratios using radian measures and solve to find the formula for the area of the sector.

_____

EXAMPLE 4 Solve a Real World Problem

A landscape architect designs a group of 8 flowerbeds. Each flowerbed is
shaped like a sector of a circle with a 45° central angle and a radius of
20 feet. She wants to find the total area of the flowerbeds. Her colleagues
each have a different approach for finding the area.

a. **USE STRUCTURE** Richard thinks you can find the area of one sector and multiply the
result by 8. Use his plan to find the total area, to the nearest tenth of a foot.

b. **USE STRUCTURE** Cora thinks the design looks like a parallelogram with a height of 20
feet and a base that is 4 times the length of one arc. Use her plan to find the total area,
to the nearest tenth of a foot.

c. **CONSTRUCT ARGUMENTS** Explain why both approaches have the same result.

PRACTICE

1. **REASON QUANTITATIVELY** A sector of a circle has an area A and a central angle that
measures $x°$. Explain how you can find the area of the whole circle.

2. **CALCULATE ACCURATELY** Find the area of each sector. Round to the nearest tenth,
if necessary.

   a. radius: 3.9 cm; central angle: 35° _____

   b. diameter: 42 in.; central angle: 72° _____

   c. radius: 20 cm; central angle: 2.5 radians _____

   d. diameter: 21.4 ft; central angle: $\frac{7\pi}{12}$ radians _____

CHAPTER 10 Extending Area

3. **DESCRIBE A METHOD** We know that the ratio of the central angle (x) to the full circle $(2\pi$ radians) is equal to the ratio of the area of the corresponding sector (A_s) to the area of the full circle (πr^2). In other words, $\frac{x}{2\pi} = \frac{A_s}{\pi r^2}$.

   a. Solve the proportion for the measure of the central angle (x).

   _____

   b. If the area of a sector is 9 square inches in a circle of radius 6, find the measure of the central angle in radians.

   _____

4. Lorenzo wants to use sectors of different circles as part of a mural he is painting. The specifications for each sector are shown.

     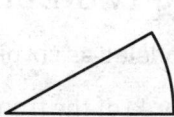

   Red:
   radius: 14 ft
   central angle: 60°

   Purple:
   radius: 12 ft
   central angle: 75°

   Green:
   radius: 18 ft
   central angle: 30°

   a. **CALCULATE ACCURATELY** Find the area of each sector to the nearest tenth.

   _____

   b. **REASON ABSTRACTLY** Was the sector with the largest area the one with the largest radius? Was it the one with the largest central angle? What can you conclude from this?

   _____

   _____

   _____

   c. **CALCULATE ACCURATELY** Lorenzo plans to paint 235 stars on each of the purple sectors and 153 stars on each of the green sectors. To the nearest tenth, what is the density of stars for each color of sector?

   _____

   d. **REASON QUANTITATIVELY** He also plans to paint stars on the red sectors. He wants the density to be twice the density of the stars on the green sector. How many stars should he paint on the red sectors? Round to the nearest whole star.

   _____

5. **CALCULATE ACCURATELY** Solve the proportion $\frac{m}{360} = \frac{A_s}{\pi r^2}$ to find a formula for the radius of a circle. Then use it to find the radius of a circle that contains a sector with a central angle of 30° and has an area of 100 square inches. Round to the nearest tenth of an inch.

   _____

Objectives

- Use two-dimensional figures to model real-world objects and situations on and off the coordinate plane.
- Solve problems involving perimeter and area.

You can use two-dimensional figures and their properties to model real-world objects and to solve problems.

EXAMPLE 1 Model Area Using Two-Dimensional Figures

A design for a medium t-shirt can be modeled as six pieces of material.

a. CALCULATE ACCURATELY Find the area of the front and back of the t-shirt. Round to the nearest square centimeter.

Rectangle area = (_____)(_____) = _____ cm²;

semicircle area = $\frac{1}{2}\pi($_____$)^2$ = _____ π cm²;

area of front = _____ − _____ $\pi \approx$ _____ cm².

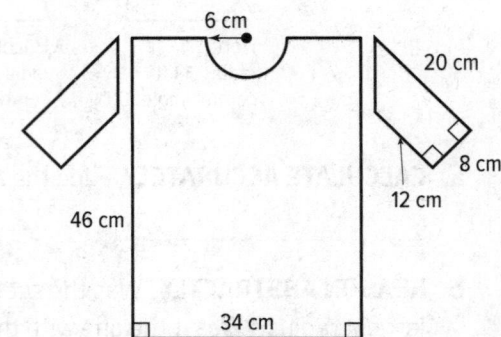

6 cm

20 cm

8 cm

12 cm

46 cm

34 cm

b. CALCULATE ACCURATELY Find the area of each of the four sleeve pieces as the sum of the areas of a triangle and a rectangle.

_____

_____

c. CALCULATE ACCURATELY Find the area of material for each t-shirt.

_____

d. COMMUNICATE PRECISELY Recall that the *scale factor* is the ratio of the lengths of the corresponding sides of two similar polygons. For a 5 cm square, a scale factor increases the perimeter to 60 cm. How did the area of this square change? Describe in terms of the scale factor.

_____

_____

e. PLAN A SOLUTION This design of t-shirt comes in small (S), medium (M), large (L), extra large (XL), and extra-extra large (XXL) sizes. Multiplying or dividing each dimension by 1.1 creates larger or smaller sizes. Explain how to find the amount of material needed for S and XXL sizes of t-shirt.

_____

_____

EXAMPLE 2 **Model with Two-Dimensional Figures on a Coordinate Plane**

Lake Superior has been superimposed on a coordinate grid. Each grid unit represents 27 miles.

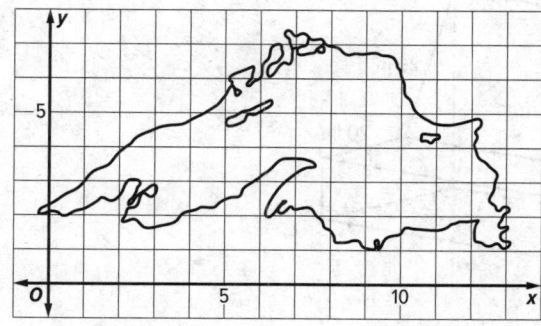

a. **USE A MODEL** Use points with whole-number coordinates to create a polygon that approximates the outline of Lake Superior.

b. **CALCULATE ACCURATELY** Divide the polygon into shapes whose area you can calculate. Find the approximate area of Lake Superior to the nearest thousand.

_____

_____

c. **CALCULATE ACCURATELY** Using the distance formula where necessary, find the approximate length of the shoreline of Lake Superior. Explain your method.

_____

_____

_____

_____

d. **USE A MODEL** Atma has a motorboat with a 27 gallon fuel tank. His boat gets 15 miles per gallon. Can he travel the perimeter of Lake Superior on a full tank of gas? Show your work or justify your answer.

_____

_____

e. **EVALUATE REASONABLENESS** Which approximation do you think is more reliable, the area or the shoreline length? Explain.

_____

_____

_____

_____

1. **a. USE A MODEL** The model shows the dimensions of a sofa. Draw a diagram to show how to calculate the total surface area of the sofa that would be covered by a fitted cover. Explain your technique.

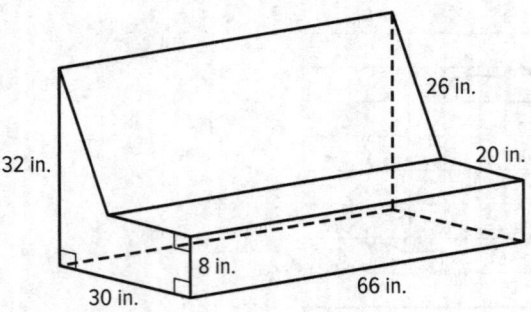

_____

_____

_____

b. REASON QUANTITATIVELY How much material is needed for a fitted sofa cover?

_____

_____

2. **a. REASON QUANTITATIVELY** Miguel is planning to renovate his living room. How much finish will he need for the hardwood floor? Assume 1 L of finish covers 4.5 m^2 and round to the nearest tenth.

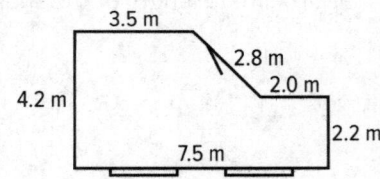

_____

_____

_____

b. CALCULATE ACCURATELY The height of the room is 2.6 m. Approximate how much paint is needed for the walls. Assume that 1 L of paint covers 7.5 m^2 and round to the nearest tenth.

_____

_____

_____

c. EVALUATE REASONABLENESS Why might Miguel adapt your answers in practice?

_____

_____

_____

3. **USE A MODEL** A two-lane running track is made by connecting two parallel straightaways with semicircular curves on each end. Each lane of the track has width 1.1 meters and the length of each straightaway is 100 meters.

   a. If the radius of the semicircle made by the inside of the first lane is $\frac{100}{\pi}$ meters, draw a model of the track labeling the information provided. Starting from the same spot on the track, find the distance that a runner must travel to complete a full lap in each of the two lanes. Measure from the inside of each lane.

   _____

   _____

   b. Are the distances to run a full lap equal for each lane? If not, how could you make a lap for each runner the same distance without changing their lanes or the finish line?

   _____

   _____

4. Main Street and 1st Street are perpendicular and intersect at a traffic light. The library is on 1st street and the community center is on Main Street. The distance from the traffic light to the library is 9 miles. The length of a direct path between the library and the community center is 17 miles. A city planner wants to put a bike path alongside the streets from the library to the traffic light to the community center and back to the library.

   a. **USE A MODEL** What shape best represents the bike path? Draw a model of the bike path.

   _____

   b. **CALCULATE ACCURATELY** The bike path is 8 feet wide along the entire route. Using the perimeter of your model, find the number of square feet, to the nearest square foot, of blacktop that the city planner will have to pour to cover the entire bike path. Explain your reasoning.

   _____

   _____

   _____

   _____

   c. **CALCULATE ACCURATELY** Based on the perimeter of your model, find the area of the city, to the nearest tenth of a mile, that will be enclosed by the bike path. Describe your solution process.

   _____

   _____

Paving a Surface

Provide a clear solution to the problem. Be sure to show all of your work, include all relevant drawings, and justify your answers.

Mr. Gleason has decided to pave the lot in front of his store. This diagram shows the lot.

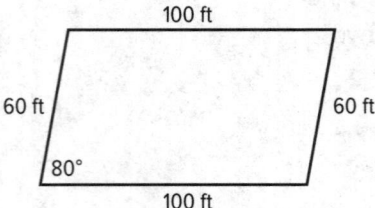

He is considering gravel, asphalt, or concrete as paving material.

Part A
Determine the area of the lot. Round to the nearest whole number. Explain your answer.

Part B

How much material, according to each option, would be required to cover the lot? Round calculations to the nearest hundredth.

Here is the needed depth information.

- **Gravel:** 2 inch depth
- **Asphalt:** 2.5 inch depth
- **Concrete:** 4 inch depth

Part C

Mr. Gleason has $8000 for paving materials. Here are cost estimates for the paving materials.

- **Gravel:** $8.00 per cubic foot
- **Asphalt:** $6.64 per cubic foot
- **Concrete:** $75.00 per cubic yard

Which of the three options can he choose?

Part D

Mr. Gleason is also considering the use of paving stones such as the one shown here.

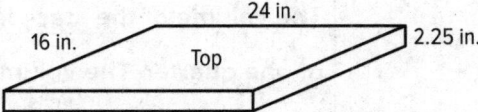

Estimate the number of paving stones he will need to cover the lot.

If each paving stone costs $6.36, will the cost of using paving stones be within his budget?

CHAPTER FOCUS Learn about some of the objectives that you will explore in this chapter. Answer the preview questions. As you complete each lesson, return to these pages to check your work.

What You Will Learn	Preview Question
Representations of Three-Dimensional Figures	
• Identify the shape of the two-dimensional cross section of a three-dimensional object. • Identify the shape of a three-dimensional object generated by the rotation of a two-dimensional object.	**SMP 6** A plane intersects a cylinder. Describe the shapes of the three possible two-dimensional cross sections and describe how they are formed. **Horizontal cross sections form circles. Vertical** **cross sections form rectangles. Angled cross** **sections form ellipses.**
Volumes of Prisms and Cylinders	
• Find the volumes of right and oblique prisms and cylinders. • Apply concepts of density.	**SMP 4** A manufacturer of ice cream cones is exploring new cone sizes. The cone the company currently sells is shown below. If the height of the cone is doubled and the diameter is halved, how does the volume of the cone change? 3.5 in. 4 in. **The volume of the new cone is one-half the** **volume of the original cone.**
Volumes of Pyramids and Cones	
• Understand volume formulas for pyramids and cones. • Use volume formulas for pyramids and cones to solve problems.	**SMP 2** A U.S. quarter has a diameter of 24.26 mm and is 1.75 mm thick. Explain how to use Cavalieri's principle to find the volume of a stack of 28 quarters. **The volume of the stack is 28 times the volume** **of one quarter. The volume of one quarter is** $V = \pi(12.13)^2(1.75) \approx 257.5\pi$ **mm³. So the volume** **of the stack is about** $28(257.5\pi) = 7210\pi$ **mm³.**

What You Will Learn	Preview Question
Volumes of Spheres	
• Find volumes of spheres. • Use the volume formula for spheres to solve problems.	**SMP 1** A cylindrical can holds three tennis balls. Each ball has a diameter of 2.5 inches. What are the minimum dimensions of a cylindrical can that can hold the balls? How much of the can is not occupied by the balls?
Spherical Geometry	
• Understand properties of lines and triangles in spherical geometry. • Compare and contrast spherical and Euclidean geometries. • Calculate areas of polygons in spherical geometry.	A manufacturer of ice cream cones is exploring new cone sizes. The cone the company currently sells is shown below. If the height of the cone is doubled and the diameter is halved, how does the volume of the cone change?
Modeling: Three-Dimensional Figures	
• Use three-dimensional figures to describe objects. • Model and solve problems using figures and their measurements.	**SMP 4** A plastic beach ball has a diameter of 15 inches. About how much plastic is used to make the beach ball? **SMP 1** A candy factory makes peanut butter chocolate balls with a 1-inch diameter using a chocolate shell and a peanut butter ball center. If the diameter of the peanut butter ball is 0.5 inch, find the volume of chocolate used to make each ball. What percentage of each ball is peanut butter?

Objectives

- Identify the shape of the two-dimensional cross section of a three-dimensional object.
- Identify the shape of a three-dimensional object generated by the rotation of a two-dimensional object.

A **cross-section** is the intersection of a three-dimensional figure and a plane. The shape of the cross-section is determined by the type of solid and the angle of the plane.

EXAMPLE 1   Investigate Cross Sections

a. REASON ABSTRACTLY Julio has a piece of modeling clay that is molded into a perfect sphere. Julio cuts the sphere so that he has two shapes with flat surfaces remaining. Identify the cross sections formed by horizontal cuts, vertical cuts, and diagonal cuts.

_____

_____

_____

b. REASON ABSTRACTLY Are any other cross sections possible?

_____

_____

_____

c. REASON ABSTRACTLY Julio also had a right cylinder made of clay and set it on one of its circular bases. Name the shapes for two possible cross sections and describe how you would slice the cylinder into two pieces to obtain each shape.

_____

_____

_____

d. CRITIQUE REASONING Julio believes that he can slice the cylinder in a way that would result in the shape to the right. Do you agree? Explain.

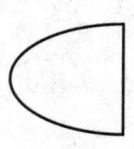

_____

_____

_____

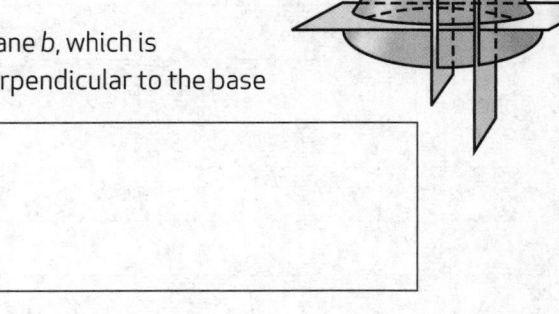

EXAMPLE 2 **Identify Cross Sections of a Cone**

REASON ABSTRACTLY Sketch the intersection of each of the following planes with the cone at right. Describe each cross section.

a. Plane *a*, which intersects the vertex, and is perpendicular to the base

b. Plane *b*, which is perpendicular to the base

c. Plane *c*, which is parallel to the base

d. Plane *d*, which does not intersect the base and is not parallel to the base

EXAMPLE 3 **Identify How to Obtain a Triangular Cross Section of a Solid**

REASON ABSTRACTLY Draw a plane that intersects each solid to form a triangular cross section. Use dashed lines to outline the triangle that is formed by the intersection of the plane with the solid. Describe how the plane intersects the solid. If it is not possible, justify your answer.

a. triangular prism

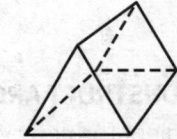

b. rectangular prism

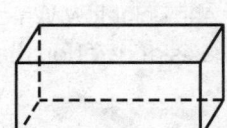

c. cylinder

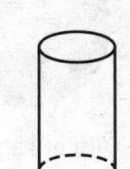

d. square pyramid

e. cone

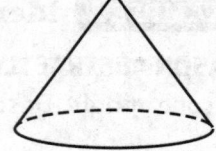

_____

_____

_____

f. sphere

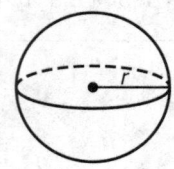

_____

_____

_____

Rotating a two-dimensional shape about a line known as an axis of rotation will form a three-dimensional object.

EXAMPLE 4 **Rotate Two-Dimensional Objects to Create Three-Dimensional Objects**

a. REASON ABSTRACTLY Dale and Ebony each rotated an index card around an axis of rotation containing one of the edges. What object did they each create? How are they different from each other?

Axes of Rotation

DALE   EBONY

_____

_____

b. REASON ABSTRACTLY If you rotate a right triangle about an axis of rotation containing one of the legs, what object would you create? Name another two-dimensional figure and the axis of rotation that can be used to generate the same type of object.

_____

_____

_____

c. CONSTRUCT ARGUMENTS Four students were asked to sketch a three-dimensional object formed by rotating a circle about an axis of rotation. Each of their figures are shown below. Which students successfully completed the assignment? Describe the axis of rotation that they used.

      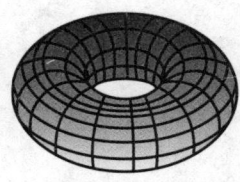   

Angel's answer   Isabella's answer   Nguyen's answer   Pam's answer

_____

_____

_____

1. **REASON ABSTRACTLY** Identify and sketch the cross section of an object made by each cut described.

a. Cylinder cut perpendicular to both bases

b. Square pyramid cut perpendicular to base but not through the vertex

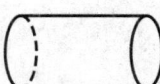

c. Rectangular prism cut diagonally from a top edge to a bottom edge on the opposite side

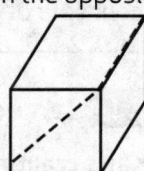

2. **CONSTRUCT ARGUMENTS** You want to cut a geometric object so that the cross section is a circle. There are three objects given below. Give the name for each object. Then describe the cut that results in a circle.

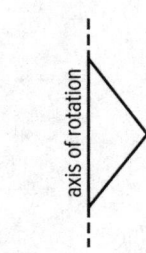

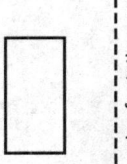

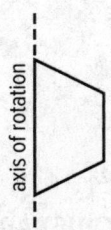

3. **USE TOOLS** Sketch and describe the object that is created by rotating each shape around the indicated axis of rotation.

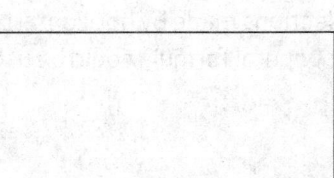

Objectives

- Find the volumes of right and oblique prisms and cylinders.
- Apply concepts of density.

EXAMPLE 1 Investigate Volumes of Cylinders

EXPLORE Kai is creating cylindrical shapes by stacking coins on top of each other.

a. **USE A MODEL** His first stack is made of 8 coins. Each coin is 1 unit high and 20 units across. What is the area of the face of each coin in terms of π?

_____

_____

b. **FIND A PATTERN** Kai remembers that the formula for finding the volume of a rectangular prism is $V = \ell \cdot w \cdot h$. He thinks finding the volume of his stack of coins should be similar. Explain how he can use this idea to find the volume of his stack of coins.

_____

_____

_____

_____

_____

c. **DESCRIBE A METHOD** Suppose Kai wants to write a formula for the volume of a stack of any number of coins. Explain how he could adapt the process he used in **parts a** and **b** to write a formula for the volume of a stack of coins that has a radius of r units and a height of h units. Write the formula in terms of π.

_____

_____

_____

_____

d. **CONSTRUCT ARGUMENTS** Explain how Kai could find the volume of any other tower of congruent solids if all the cross sections made by horizontal planes have the same area as the base of the solids. What general formula could be used to find the volume?

_____

_____

_____

EXAMPLE 2 **Compare Volumes**

One container is a rectangular prism with a base that is 6 inches by 6 inches and a height that is 8 inches. Another container is a cylinder with a radius of 3 inches and a height of 8 inches.

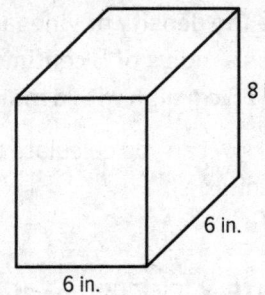
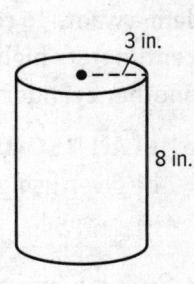

a. **MAKE A CONJECTURE** How do you think the volumes of the solids compare?

_____

_____

_____

_____

b. **CALCULATE ACCURATELY** Find the difference in their volumes to the nearest whole number.

_____

_____

_____

One part of Cavalieri's Principle states that if two solids have the same height, h, and the same cross-sectional area, B, at every level, then they have the same volume.

EXAMPLE 3 **Use Cavalieri's Principle to Find Volume**

a. **CALCULATE ACCURATELY** What right solid would have the same volume as the one shown at right? What is the volume of the solid shown?

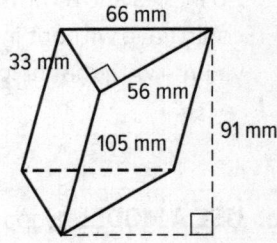

_____

_____

b. **CONSTRUCT ARGUMENTS** How can you use Cavalieri's Principle to show that the two solid figures have the same volume?

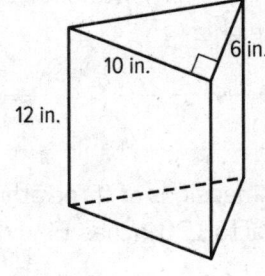
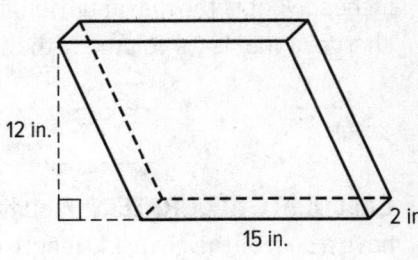

_____

_____

_____

_____

_____

_____

Volumes of Prisms and Cylinders **355**

Copyright © McGraw-Hill Education

EXAMPLE 4 | Find Density Based on Volume

Jenna wants to compare the density of vinegar and olive oil. She fills a cylinder that is 15 centimeters high and has a radius of 5 centimeters with 1200 grams of vinegar. Then she fills another cylinder that is 13 cm high with a radius of 6 cm with 1350 grams of olive oil.

a. **PLAN A SOLUTION** How can you calculate the density of each liquid? The density will be given using what units?

_____

b. **CALCULATE ACCURATELY** Find the density of each liquid to the nearest hundredths. Show your work.

_____

_____

c. **CONSTRUCT ARGUMENTS** Jenna knows that oil floats on vinegar. How can she use this knowledge to determine if her answers are reasonable?

_____

EXAMPLE 5 | Explore the Effect of Dimensions on Volume

In order for Lydia to avoid an extra shipping charge, the sum of the length and girth of her package must be no more than 130 inches. The length is defined as the greatest of the three dimensions. She wonders if all packages that have the maximum length and girth also have the same volume.

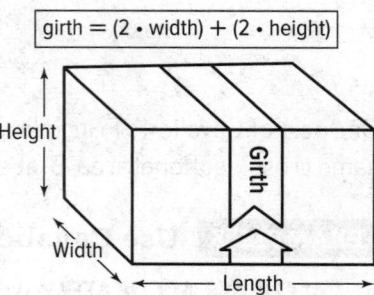

girth = (2 · width) + (2 · height)

a. **REASON QUANTITATIVELY** Suppose a package has a height of 20 inches and a width of 25 inches. What is the maximum length it could have without incurring an extra shipping charge? What is its volume with that height?

_____

b. **USE A MODEL** Suppose a package has a height of 10.5 inches and a width of 15.25 inches. What is the maximum length it could have without incurring an extra shipping charge? What is its volume with that height?

_____

c. **CALCULATE ACCURATELY** Find the dimensions of three other packages that each have a sum of the girth and length equal to 130 inches. Find the volume of each box.

_____

d. **FIND A PATTERN** List the dimensions of your three packages and those in **parts a** and **b** in order from least volume to greatest volume. What do you notice about the dimensions that produce the greatest volume?

_____

1. **CALCULATE ACCURATELY** Find the volume of each solid figure.

a.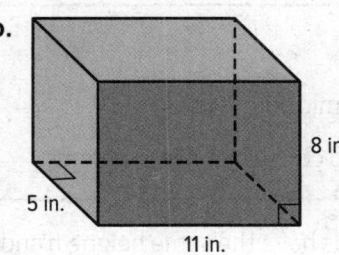

b.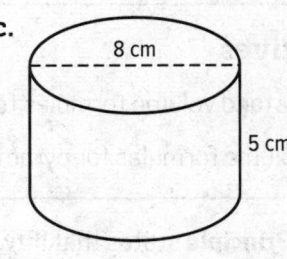

c.

_____   _____   _____

d. Discuss how the formulas that you used to find the volume for each of the above figures are similar.

_____

_____

_____

2. **PLAN A SOLUTION** A container company manufactures cylindrical containers with a radius of 3 inches and a height of 10 inches. They decided to produce a different cylindrical container with the same volume, but with an 8-inch height. What radius must the new container have for the volumes to be equal? What steps would you use to find the radius of the new cylinder? What is the radius of the new cylinder to the nearest tenth?

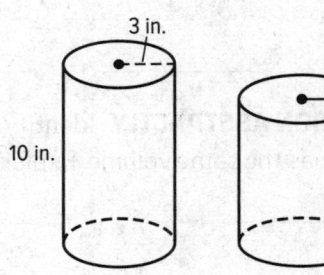

_____

_____

3. Benjamin finds that the density of baking soda is 2.2 grams/cm³ and the density of corn flakes is 0.12 grams/cm³. A box of baking soda has dimensions of 8 cm long by 4 cm wide by 12 cm high. The dimensions for a box of corn flakes are 30 cm long by 6 cm wide by 35 cm high. Benjamin wants to find the weight of the contents of each box if it is filled to within 2 cm of the top.

a. **USE A MODEL** How can Benjamin determine the weight of the contents of each box?

_____

_____

b. **CALCULATE ACCURATELY** Find the weight of the contents of each box.

_____

_____

Objectives

- Understand volume formulas for pyramids and cones.
- Use volume formulas for pyramids and cones to solve problems.

Mathematical Practices
1, 2, 3, 4, 6, 7, 8

Cavalieri's Principle states that if two solids have the same height, h and the same cross-sectional area, B at every level, then they have the same volume. This principle may be applied to find the areas of both cones and pyramids.

EXAMPLE 1   Investigate Volume of Pyramids

EXPLORE When a wooden puzzle is complete, it is a solid in the shape of a triangular prism as shown at the right. When taken apart, the puzzle consists of three solid pieces formed by making straight cuts from T to S, S to U and Q to U.

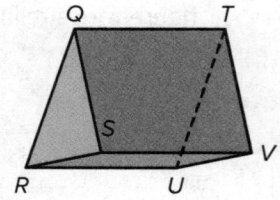

a. **USE STRUCTURE** Draw solid and dashed lines to represent the cuts of the puzzle pieces. What type of solid is each piece? Explain.

_____

_____

b. **REASON ABSTRACTLY** Identify two pairs of puzzle pieces such that each piece in the pair has the same volume. Explain your reasoning.

_____

_____

_____

c. **MAKE A CONJECTURE** What conjecture can you make about the volumes of the three puzzle pieces? Justify your reasoning.

_____

_____

_____

d. **INTERPRET PROBLEMS** Based on your conjecture, write a formula for the volume of a pyramid. Explain your reasoning.

_____

_____

_____

The formula for the volume of a pyramid can be used to find the volume of both right pyramids and oblique pyramids.

KEY CONCEPT

Complete the table by writing the missing information and labeling the models.

Volume of a Pyramid	
Words	**Models**
The volume of a pyramid is $V =$ _____ , where B is the _____ and h is the _____ of the pyramid.	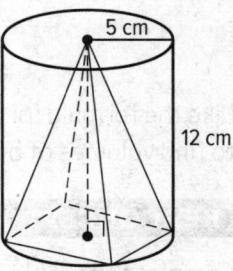
Symbols $V =$ _____	

The formula for the volume of a pyramid can be used to find the volume of a pyramid with any shape base.

EXAMPLE 2   Solve Problems Involving Volumes of Pyramids

Stacy makes and sells pyramid-shaped candles. She packages her candles in cylindrical boxes as shown at the right.

a. **INTERPRET PROBLEMS** Stacy makes a candle in the shape of a square pyramid. The candle's volume is the greatest possible volume given the dimensions of the box. What is the volume of the candle? Explain your answer.

_____

_____

_____

b. **USE STRUCTURE** In addition to the square pyramid candle, Stacy also makes pyramid candles with bases that are regular hexagons and regular octagons. For each type of candle, Stacy wants to have the greatest volume possible given the dimensions of the cylindrical box. Each circle at the right represents the base of a box. Sketch the base of each pyramid. For each base, determine and label the length of a side s and the apothem a. What is the area of each base rounded to the nearest tenth? Explain your answer.

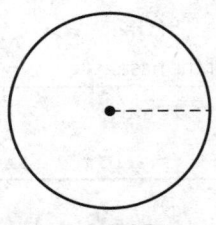 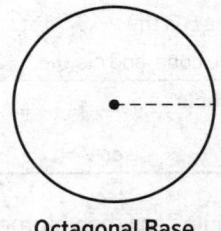

Hexagonal Base   Octagonal Base

_____

_____

_____

_____

_____

_____

c. FIND A PATTERN Complete the table below so that each type of candle has the greatest volume possible given the dimensions of the cylindrical box. Round your answer to the nearest tenth. The row for the square pyramid has been completed for you.

Base Number of Sides (n)	Base Area (B)	Height (h)	Volume ($V = \frac{1}{3}Bh$)
4	50 cm²	12 cm	200 cm³
6			
8			

d. MAKE A CONJECTURE What do you notice as the number of sides *n* increases? Make a conjecture about the shape of the base and the shape of the pyramid as *n* approaches infinity. How is the formula for the volume affected?

_____

_____

_____

Like the formula for volume of a pyramid, the formula for the volume of a cone can be used to find volumes of both right cones and oblique cones.

KEY CONCEPT

Complete the table by writing in the missing information and labeling the models.

Volume of a Cone	
Words	**Models**
The volume of a cone is $V = $ _____ , or $V = $ _____ , where *B* is the _____ , *h* is the _____ of the cone, and *r* is the _____ of the base.	
Symbols	
$V = $ _____ or $V = $ _____	

Formulas for the volumes of solids can be used in real-world situations, such as solving design problems involving constraints.

EXAMPLE 3 Solve Problems Involving Volume of Cones

The management team at a ballpark is thinking of selling peanuts and popcorn in cone-shaped containers that, when empty, can be turned into souvenir megaphones. Containers will come in two sizes, large and small, and will be filled to the brim with no overflow.

a. REASON QUANTITATIVELY What is the volume of each container? Round to the nearest tenth. Show your work.

_____

_____

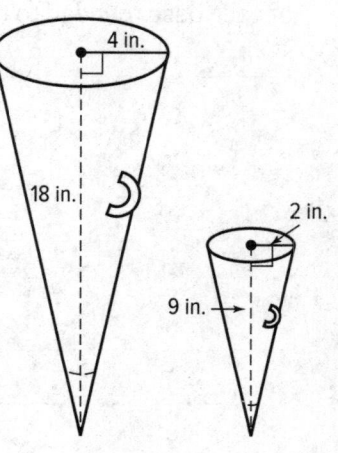

b. CRITIQUE REASONING One member of the team says that since the height and radius measurements of the large container are twice those of the small container, then the volume of the large container will be twice the volume of the small container. Do you agree? Justify your answer.

_____

_____

_____

c. USE A MODEL Use the information in the table at the right to determine the weight of each size container when it is filled with peanuts and when it is filled with popcorn. Do not include the weight of the container itself in your calculations.

Item	Density (ounce per cubic inch)
Peanuts	0.44
Popcorn	0.02

_____

_____

_____

d. CRITIQUE REASONING Suppose the management team asked you to evaluate their ideas for the new containers. What suggestions would you make? Explain your reasoning.

_____

_____

_____

PRACTICE

1. A garden shop sells pyramid-shaped lawn ornaments that each have a base area of 900 square centimeters and a height of 40 centimeters. The lawn ornaments are made of concrete, granite, or marble.

Material	Density (kg/m³)
Concrete	2371
Granite	2691
Marble	2711

a. REASON QUANTITATIVELY What is the volume of a lawn ornament in cubic meters? Explain.

_____

_____

b. USE A MODEL Find the weights of three pyramid lawn ornaments each made from a different material. Describe your calculations.

_____

_____

c. MAKE A CONJECTURE What generalization can you make about the relationships among the volume of a pyramid lawn ornament, the weight of the lawn ornament, and the density of the material used to make it?

_____

_____

d. USE STRUCTURE The garden shop also sells lawn ornaments in the shape of cones. The cones have the same height and volume as the pyramids. What can you say about the base of the cones?

_____

_____

2. CRITIQUE REASONING Jade says that the volume of the cone is about 1139 in³. Tiana says that the volume of the cone is about 1005 in³. Who is correct? Explain.

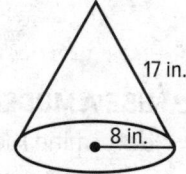

17 in.

8 in.

_____

_____

3. USE STRUCTURE A model pyramid has a volume of 270 ft³ and a base area of 90 ft². What is the height if the pyramid is a right pyramid? What is the height if the pyramid is an oblique pyramid? Explain your reasoning.

_____

_____

4. PLAN A SOLUTION Which solid has a greater volume: a cone with a base radius of 7 centimeters and a height of 28 centimeters or a pyramid with base area of 154 square centimeters and height of 28 centimeters? Explain your reasoning.

_____

_____

_____

5. Tristan makes and sells sugar-free candies. She packages them in pyramid-shaped boxes with a 2-inch by 2-inch base and a height of 3 inches. She sells each box for $2.00.

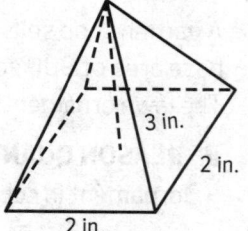

3 in.

2 in.

2 in.

a. CALCULATE ACCURATELY What is the volume of sugar-free candies in each box. What is the price per cubic inch?

_____

b. PLAN A SOLUTION Tristan wants to make a bigger package by doubling the lengths of the sides of the square base. How can she figure out how much to charge if she wants to keep the price per cubic inch the same?

_____

_____

c. REASON QUANTITATIVELY Tristan wants to design a box in the shape of a square-based pyramid that holds between 7 and 8 cubic inches of sugar-free candies. She wants the height to be within $1\frac{1}{4}$ inches of the length of each side of the square. What is one possible set of dimensions that she can use?

_____

_____

6. USE A MODEL Alonzo is building a box in the shape of a right triangular prism for his magic act. Inside the box, he is making a secret compartment. The compartment will be a pyramid with base △ABE and vertex at point C. After the secret compartment has been made, how is the volume of the space remaining inside the box related to the volume of the secret compartment? Explain your reasoning.

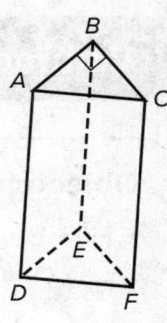

_____

_____

_____

_____

7. John wanted to verify that his calculations for the volume of a cone were reasonable. He modeled it using a series of stacked cylinders.

a. CALCULATE ACCURATELY Find the volume each figure to the nearest cubic meter.

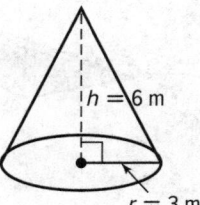

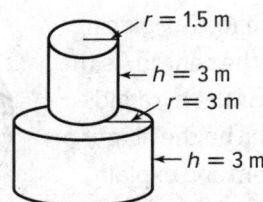

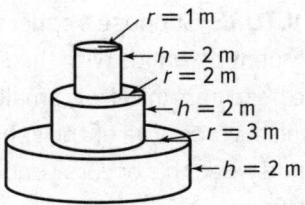

_____   _____   _____

_____   _____   _____

b. INTERPRET PROBLEMS How do the volumes of the figure compare?

_____

_____

c. USE STRUCTURE What would the stack of cylinders look like if there were 24 cylinders that were each $\frac{1}{4}$ inch high and their diameters gradually decreased?

_____

d. MAKE A CONJECTURE What can you predict about the volume of a stack of $24\frac{1}{4}$-inch-high cylinders with gradually decreasing diameters?

_____

e. MAKE A CONJECTURE How could we use the same procedure to estimate the volume of a square pyramid?

_____

_____

_____

Objectives

- Find volumes of spheres.
- Use the volume formula for spheres to solve problems.

A **sphere** is the set of all points in space that are a given distance from a given point, called the **center** of the sphere. A **radius** of a sphere is a segment from the center to a point on the sphere.

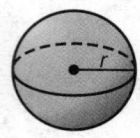

EXAMPLE 1   Investigate Volumes of Spheres

EXPLORE Paola is deriving the formula for the volume of a sphere with radius r from the formula for its surface area, $S = 4\pi r^2$. She starts by drawing the diagram of a sphere below.

a. **USE STRUCTURE** Suppose a sphere is approximated by placing many congruent pyramids with their vertices at the center. As the area of the base gets infinitely small, the bases of the pyramids will form the surface area of the sphere. What will be the height of each pyramid when this occurs? Label the diagram and explain your reasoning.

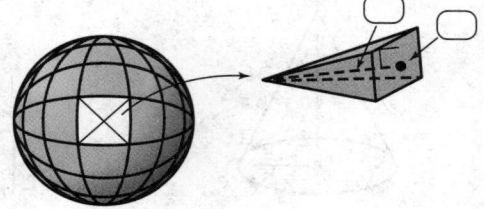

_____

_____

_____

_____

b. **INTERPRET PROBLEMS** Suppose each pyramid has a base area of B. Label the diagram. What is the formula for the volume of one pyramid? How are the volumes of the pyramids and the sphere related?

_____

_____

c. **CONSTRUCT ARGUMENTS** Complete the steps and reasons to show how Paola derived the formula for the volume of a sphere by using the pyramids.

$V = \frac{1}{3}B_1 r + \frac{1}{3}B_2 r + \cdots + \frac{1}{3}B_n r$ Sum of volumes of pyramids

$V = \frac{1}{3}r(B_1 + B_2 + \cdots + B_n)$ _____

$V = \frac{1}{3}r(\underline{\hspace{2cm}})$ As the area of the base gets smaller, the sum of the areas of the bases approaches the surface area of the sphere.

$V = \underline{\hspace{2cm}}$ Simplify.

KEY CONCEPT   Volume of a Sphere

Complete the table by writing the missing information and labeling the models.

Words	Model
The volume of a sphere is $V =$ _____ , where _____ is the radius of the sphere.	
Symbols	
$V =$ _____	

You can use the formula for the volume of a sphere to find the volume, diameter, or radius of a sphere.

EXAMPLE 2   Solve Problems Involving Volumes of Spheres

The Unisphere in Flushing Meadows, New York, was built as the centerpiece for the 1964–1965 New York World's Fair. The sphere has a diameter of 120 feet.

a. **REASON QUANTITATIVELY** What is the volume of the Unisphere? Explain how you found your answer.

_____

_____

_____

_____

120 ft

b. **FIND A PATTERN** The Daily Planet in Raleigh, North Carolina, is also a large spherical representation of Earth. With a volume of about 195,432 cubic feet, the Daily Planet is not quite as large as the Unisphere. What is the diameter of the Daily Planet? Explain.

_____

_____

c. **CALCULATE ACCURATELY** The Mapparium in Boston, Massachusetts, is a 20-story-tall hollow sphere of stained glass that, when viewed from the inside, is a giant globe showing a map of the world. Visitors can go inside the Mapparium by walking across a bridge that spans a diameter of the sphere. If the volume of the Mapparium is about 14,137 cubic feet, about how far is a visitor from the surface of the sphere if he or she stands exactly in its center?

_____

_____

d. EVALUATE REASONABLENESS With a diameter of about 28 feet, the Babson Globe in Wellesley, Massachusetts, was once the world's largest rotating globe. As of 1998, that distinction has belonged to Eartha, a revolving globe in Yarmouth, Maine, with a diameter of about 41 feet. About how many times greater is the volume of Eartha than the volume of the Babson Globe? Explain how you can find the answer without finding the volume of the globes.

_____

_____

_____

PRACTICE

At a pet store, toy tennis balls for pets are sold in 3 different sizes. Use the table at the right for Exercises 1–2.

1. USE STRUCTURE Complete the table by calculating the volume for each size ball. Record the volume of each tennis ball in terms of pi. What pattern do you notice as the diameter increases?

Size	Diameter (cm)	Volume (cm³)
Small	3	
Medium	4.5	
Large	6.75	

_____

_____

2. REASON QUANTITATIVELY The volume of a new Extra Large toy tennis ball for pets is about 221 cubic centimeters. If 3 Extra Large toy tennis balls are packaged and sold in a cylindrical package as shown, what is the approximate volume of the cylindrical package? Explain.

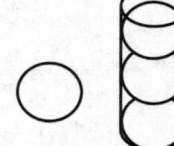

_____

_____

_____

_____

3. FIND A PATTERN A wooden sphere is carved from a solid cube of wood so that the least amount of wood is carved away.

a. If the block of wood had a volume of 729 in³, what is the volume of the sphere? Explain.

_____

_____

b. CRITIQUE REASONING Devon notes that $729 \cdot \frac{\pi}{6}$ also equals about 382. He says that he can multiply the volume of any cube by $\frac{\pi}{6}$ to find the volume of the sphere that shares the same diameter as the cube's side. Is he correct?

_____

_____

_____

4. **FIND A PATTERN** Jeremy collects glass fishing floats, which are spheres of glass that were once used to keep fishing nets afloat. There are three types of fishing floats in Jeremy's collection. The first set is blue, and the spheres have a diameter of 2 inches. The second set is red, and these spheres have a diameter of 3 inches. The largest set is amber with a diameter of 4 inches. Find the volume of the blue floats using the formula for the volume of a sphere. For the red and amber floats, find the volume, but use the observation you made about increasing radii and volume in **Exercise 1**.

_____

_____

_____

5. **CALCULATE ACCURATELY** Jenna is mailing a green gazing ball that is 10 inches in diameter to a friend. She packs it in a box shaped like a cube that allows a minimum of 2 inches of padding between the sphere and the box. To the nearest tenth, how many cubic inches of packing material will she need? Explain.

_____

_____

_____

6. **INTERPRET PROBLEMS** Reginald is creating a scale model of a building using a scale of 4 feet = 3 inches. The building is in the shape of a cube topped with a hemisphere whose circular base is inscribed in the square base of the cube. At its highest point, the building has a height of 30 feet and the radius of the hemisphere is shown. Find the volume of his scale model to the nearest cubic inch. Explain.

10 ft

_____

_____

_____

7. **USE STRUCTURE** The allium plant has a spherical bloom that is formed by tiny florets sprouting from a globe-shaped flower head in the center, as shown. Trinh is designing an abstract sculpture inspired by the allium plant. In her sculpture, the diameter of the small sphere in the center will be 1.5 centimeters and the volume of the entire spherical sculpture will be about 381.7 cubic centimeters. What will be the distance x? Explain.

_____

_____

_____

Objectives

- Understand properties of lines and triangles in spherical geometry.
- Compare and contrast spherical and Euclidean geometries.
- Calculate areas of polygons in spherical geometry.

The study of geometry on a plane is known as **Euclidean geometry.** There are numerous non-Euclidean geometries in which some of the most familiar concepts of Euclidean geometry are no longer true including **spherical geometry**, in which a plane is the surface of a sphere.

A line in spherical geometry is a great circle. A **great circle** is a circle formed when a plane intersects a sphere with its center at the center of the sphere.

EXAMPLE 1   Investigate Spherical Geometry

There are three great circles shown. They form a closed three-sided polygon called a *spherical triangle*.

a. **MAKE A CONJECTURE** In Euclidean geometry, two lines are defined to be parallel if they do not intersect. Draw several great circles (lines) on a sphere and make a conjecture about parallel lines.

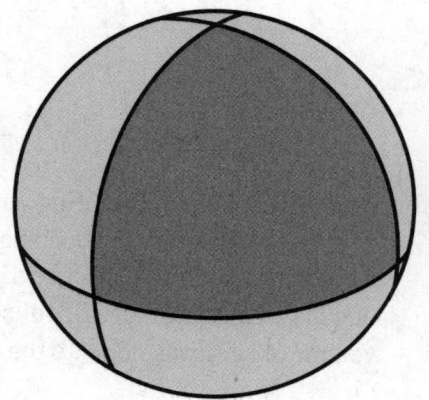

_____

_____

_____

_____

_____

b. **CONSTRUCT ARGUMENTS** The three great circles intersect at right angles, forming eight congruent spherical triangles as shown. What is the sum of the angles for the shaded triangle?

_____

_____

c. **COMMUNICATE PRECISELY** Contrast the answer of **part a** with what you know about triangles in Euclidean geometry.

_____

_____

d. REASON ABSTRACTLY Given any two distinct points on a sphere, the shortest distance between them along the surface of the sphere is along the great circle connecting them. A triangle in spherical geometry is formed by three noncollinear points and the three great circles containing them. Make a conjecture about the sum of the angles in a spherical triangle. Support your conjecture by imagining looking at the triangle formed by the three points in a plane that contains the points.

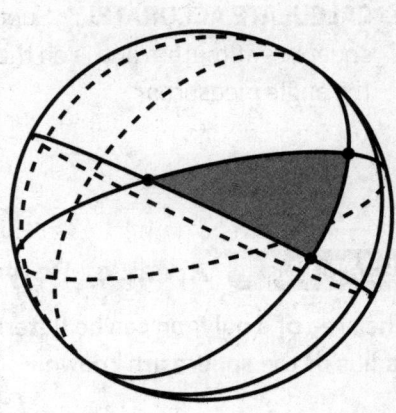

_____

_____

_____

_____

_____

EXAMPLE 2 **Area of a Spherical Triangle**

The area of a spherical triangle can be determined if the radius of the sphere and the angle measures are known. The area of a triangle with angles α, β, and γ on a sphere of radius r is

$$A = \tfrac{1}{180}\,\pi r^2(\alpha + \beta + \gamma - 180).$$

a. CALCULATE ACCURATELY Find the area of one of the eight congruent equilateral triangles if the radius of the sphere is 1 inch.

_____

b. CALCULATE ACCURATELY Use proportional reasoning and the formula for the surface area of a sphere to find the answer to **part a** in a different way.

_____

_____

c. REASON ABSTRACTLY Since all of the angles are congruent, the triangle is an equilateral triangle. What happens to the sum of the angle measures if the radius of the sphere remains equal to 1 but the triangle reduces in size while remaining an equilateral triangle?

_____

_____

d. CALCULATE ACCURATELY Find the area of a spherical triangle with angles that measure 80°, 100°, and 110° if the triangle is on a sphere of radius 4 inches.

_____

_____

e. CALCULATE ACCURATELY Suppose an equilateral spherical triangle has an area of 6π square centimeters and is on the surface of a sphere of radius 3 centimeters. What are the angle measures?

_____

_____

EXAMPLE 3 **Areas of Polygons**

The area of a polygon can be determined if the measures of the interior angles and radius of the sphere are known.

a. PLAN A SOLUTION The figure shown suggests a possible way to find the area of a spherical quadrilateral. Describe the method.

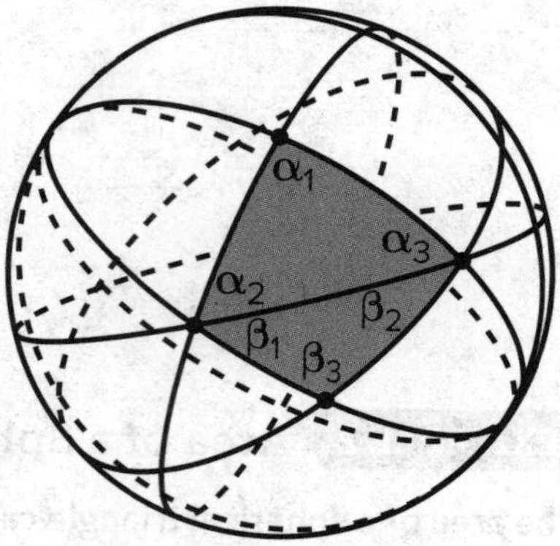

_____

_____

b. CONSTRUCT ARGUMENTS Suppose the sum of the interior angles of a spherical quadrilateral is S. Find a formula for the area of the spherical quadrilateral in terms of S and r. Use the figure as a reference.

_____

_____

_____

c. MAKE A CONJECTURE Make a conjecture about a formula for the area of a spherical polygon in terms of the sum of the interior angles S, the number of sides n, and the radius of the sphere r. Justify your answer.

_____

_____

_____

d. CALCULATE ACCURATELY Find the area of a regular hexagon on a sphere of radius 2 feet if each angle has measure 135°.

_____

_____

EVALUATE REASONABLENESS Determine whether each statement is *always,* *sometimes,* or *never* true. Explain.

1. A circle on the sphere is a line in spherical geometry.

2. Two lines intersect in exactly one point.

3. The sum of the interior angles of a spherical triangle is at most 180°.

4. The sum of the interior angles of a spherical octagon is at least 1000°.

5. **CALCULATE ACCURATELY** Determine the area of the figure with the given properties.

   a. A spherical triangle on a sphere of radius 1 with two right angles and a third angle whose measure is 120°.

   b. A spherical triangle on a sphere of radius 10 cm with angle measures of 100°, 120°, and 60°.

   c. A pentagon on a sphere of radius 4 in with angles that measure 100°, 110°, 120°, 130° and 140°.

6. **CONSTRUCT ARGUMENTS** The sum of the angles in a spherical triangle is always greater than 180°. Must the sum be less than 540°? Explain.

Use with Lesson 11-7

Objectives

- Use three-dimensional figures to describe objects.
- Model and solve problems using figures and their measurements.

You have learned how to find the volume of spheres, cylinders, prisms, and pyramids. The combinations of these four solids can be used to construct other solids that are more interesting and that can be used to describe real-world objects and their dimensions. The volume formulas you have learned will help you find the volumes of these more complicated solids.

EXAMPLE 1 **Describe an Everyday Object**

EXPLORE Consider the dinner glass shown at the right.

a. **USE A MODEL** Name two solid figures you could use to estimate the volume of the glass. Explain how you might use each to obtain the volume.

_____

_____

_____

_____

_____

_____

_____

b. **DESCRIBE A METHOD** A frustum is a portion of a solid that lies between a plane containing the base and a parallel plane that cuts through the solid. Describe how to create a frustum that will model the shape of the glass.

_____

_____

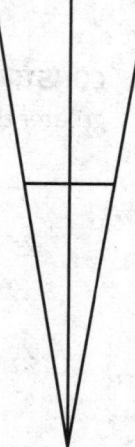

c. **USE A MODEL** Sketch the side view of the glass and continue the sides until they intersect to form a cone. Label the radius of the top of the glass R, the height of the glass h, and the radius of the bottom of the glass r. Let x be the height from the small base of the glass to the vertex of the cone. Express x in terms of the other three variables.

_____

_____

_____

d. PLAN A SOLUTION How could you find the volume of the frustum that represents the glass in **part c**? Express your solution in terms of the variables.

_____

_____

_____

e. USE TOOLS Find a frustum-shaped glass. Trace each base in the space below or on a separate sheet of paper. Measure the radii, as well as the height of the glass. Place your measurements on the figure from **part c**. Find any missing measurements and then calculate the volume of the glass.

_____

_____

| EXAMPLE 2 | Model a Volume Constraint Problem |

A water cooler bottle contains 5 gallons of water and each gallon of water has a volume of 231 cubic inches. A manufacturer is asked to create conical paper drinking cups, such that a pack of 100 cups would contain all of the water.

a. CALCULATE ACCURATELY If the radius of the cup is 1.5 inches, what should be the height of the cup? Justify your answer.

_____

b. CALCULATE ACCURATELY If the height of the cup is 6 inches, what should be the radius of the cup? Justify your answer.

_____

The manufacturer decides to use dimensions for the cup that will minimize the amount of paper used.

c. USE STRUCTURE The equation $L = \pi r \sqrt{h^2 + r^2}$ gives the lateral area of the conical cup. Use the volume constraint to solve for h and to rewrite the equation in terms of r.

_____

d. USE TOOLS Use a graphing calculator to graph the equation from **part c**. Use a window such as $0 < r < 10$ and $0 < L < 100$. Use the minimum feature of your calculator to find the minimum point on the graph. What does the value of each coordinate represent?

_____

e. CALCULATE ACCURATELY Find the dimensions of the cup that minimize the amount of paper used. Show your work.

_____

_____

f. COMMUNICATE PRECISELY Are there any reasons why the manufacturer may not want to use these exact dimensions? Explain your reasoning.

_____

_____

PRACTICE

USE A MODEL Describe how you could use three-dimensional figures to describe the given object.

1. A traffic cone with a base.

2. One pound weight.

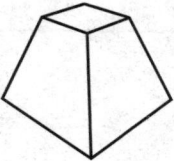

_____   _____

3. PLAN A SOLUTION Maly wants to compare the thickness of two brands of paper towel that she found at the store. The paper towels come in cylindrical rolls. She visits the website of the company and finds the following information:

Brand A

SHEET SIZE	SHEETS/ ROLL	ROLL DIAMETER	CORE DIAMETER
11" × 6"	77	4.5"	1.75"

Brand B

SHEET SIZE	SHEETS/ ROLL	ROLL DIAMETER	CORE DIAMETER
11" × 12"	35	5"	2"

Determine the thickness of each roll of paper towels for Maly. If she prefers a thicker paper towel, which brand should she buy?

_____

_____

_____

_____

_____

_____

4. A liquid propane tank contains 5 gallons of propane when full. The diagram shows approximate dimensions of the tank in inches.

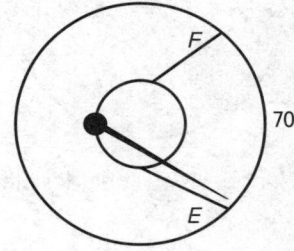

a. **USE A MODEL** Choose a solid to model the tank. Use the formula of the volume of this solid to approximate the volume of the tank.

_____

_____

b. **REASON ABSTRACTLY** Let h represent the height of propane left in the tank at any time and g represent the gallons of propane remaining. If the ratio of h to 14 is always equal to the ratio of g to 5, express h in terms of g.

_____

c. **REASON ABSTRACTLY** The fuel gauge shown indicates the approximate amount of propane remaining in the tank. Express the angle θ that the pointer makes from the E (empty) marker in terms of g, the volume of fuel in the tank. Express θ in terms of h, the height of fuel in the tank. Is your answer reasonable?

_____

_____

_____

d. **CONSTRUCT ARGUMENTS** Alex's grill uses 0.5 gallons of propane per hour. He has invited friends over for a cookout, and he has estimated that it will take one hour of grilling time to cook all of the food. If θ from **part c** is 3.5°, does he have enough propane for the cookout?

_____

_____

5. **REASON ABSTRACTLY** In **Example 1** you found the volume of a frustum that was created by cutting off the top of a cone. In this exercise you will find the volume of the frustum when a square pyramid is cut by a plane parallel to its base.

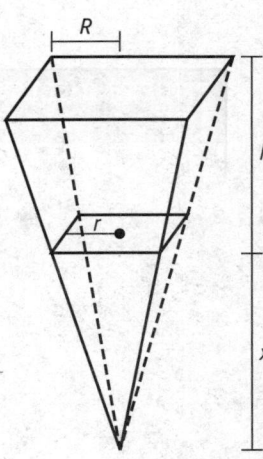

a. Let R be half the length of one side of the base. Let h be the height of the solid. Let r be half the length of the side of the smaller base, and let x be the length needed to "complete" the pyramid. Does the side view of the solid look similar to the sketch from **Example 1**? Find x in terms of the other variables.

_____

b. Use the formula for the volume of a square pyramid and your result from **part a** to derive a formula for the volume of this solid.

_____

_____

_____

Estimating Volumes of Planters

Provide a clear solution to the problem. Be sure to show all of your work, include all relevant drawings, and justify your answers.

Nancy designs and makes planters. This diagram shows the top, bottom, and side views of a planter design she is considering.

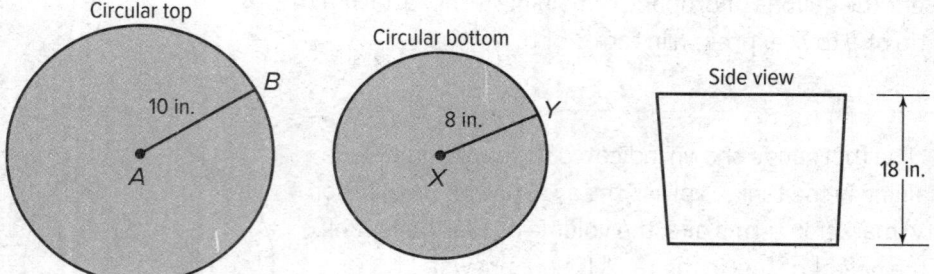

Circular top

10 in.

B

A

Circular bottom

8 in.

Y

X

Side view

18 in.

Her goal is to consider planter shapes, volume estimates, and costs of potting soil.

Part A

Using cylinders, determine a lower estimate and an upper estimate for the volume of the planter.

Part B

From online research, Nancy has gathered this information.

        1 cubic foot = 25.71404638 quarts (Dry U.S.)

        1 quart (Dry U.S.) = 0.0388892508 cubic feet

        1 cubic inch = 0.0148808138 quarts (Dry U.S.)

        1 quart (Dry U.S.) = 67.2006254 cubic inches

She has also found that a 20-pound bag of potting soil contains 16.1 dry quarts and costs $4.99. How much can she expect to spend on potting soil using the lower and upper estimates of volume from **Part A**?

Part C

Suppose that Nancy decides to make some changes in her design. She replaces the circular bases with regular octagons as shown. Using regular octagonal prisms, determine a lower estimate and an upper estimate for the volume of this planter.

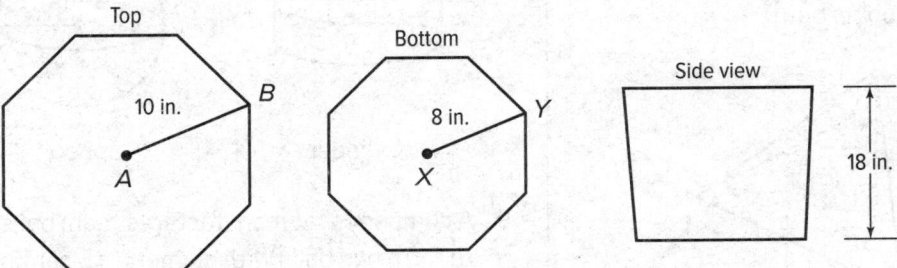

Top   10 in.   B   A

Bottom   8 in.   Y   X

Side view   18 in.

Show how she can estimate the cost of potting soil for this planter by using regular octagonal prisms. What cost estimates can she expect?

1. Determine the area of △*ABC*.

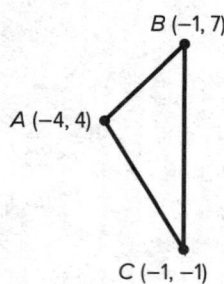

area: ☐ units²

2. The grain silo shown below is shaped like a cylinder with a hemisphere on top.

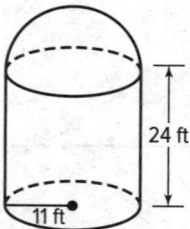

Rounded to the nearest cubic foot, what is the volume of the grain silo?

☐ ft³

3. The gift shop at a history museum sells miniature pyramids filled with bubble bath.

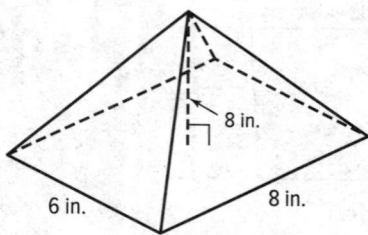

What is the volume of the rectangular pyramid?

area: ☐ in³

4. A solid steel sphere has a radius of 4 mm. Find the radius of a second solid steel sphere that has twice the volume of the original sphere. Round to the nearest tenth of a millimeter.

radius: ☐ mm

5. Parallelogram *ABCD* has coordinates *A*(1, −2), *B*(−3, −2), *C*(−1, 3), and *D*(3, 3). What are the coordinates of point *E* if △*ABE* has the same area as parallelogram *ABCD*, ∠*ABE* is a right angle, and *E* is in the second quadrant?

☐

6. The edges of a square pyramid are all the same length and have a sum of 120 centimeters. What is the volume of the pyramid? Round your answer to the nearest tenth.

☐

7. A candy company makes hard candies by pouring a syrup mixture into spherical molds and letting it harden. They make large candies with a radius of 11 mm, and small candies with a radius of 6 mm. How many small candies can be made with the same amount of syrup needed to make 25 large candies?

about ☐ candies

8. Which of the solids has the smaller volume?

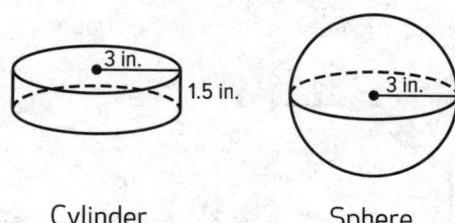

Cylinder   Sphere

9. A company that manufactures foam balls wants to form two balls with diameters 5 centimeters and 6 centimeters, each having density 0.95 grams per cubic centimeter. How many total grams of foam do they need to make the balls? Round to the nearest tenth of a gram.

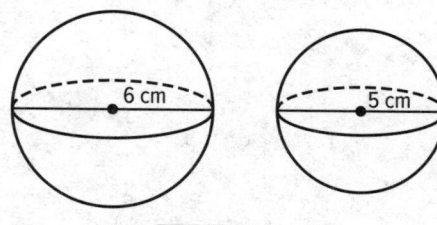

Approximately ☐ grams

10. Rosalinda designed the quilt block shown at right. The block is 12 inches by 12 inches. Complete the table to show the area of each color. Round to the nearest whole number.

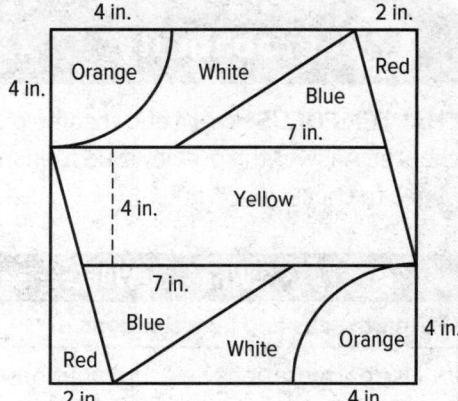

Color	Area (in²)
Orange	
White	
Blue	
Red	
Yellow	

11. A set of steps has the dimensions shown.

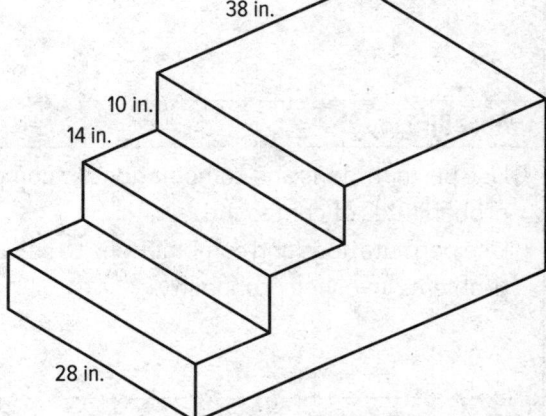

   a. Describe the set of steps as a composite of three-dimensional figures.

   _____

   _____

   _____

   b. The steps will be made out of concrete. If concrete costs $4.50 per cubic foot, how much will the steps cost? Show your work.

   _____

   _____

   _____

   _____

12. The base of a rectangular prism is twice as long as it is wide. A cone has a radius that is equal to the width of the base of the prism. If the two solids have the same volume and the height of the prism is 15 centimeters, what is the exact height of the cone? Show your work.

   _____

13. A gardener has created a flower bed in the shape of a parallelogram as shown in the figure. The units of vertices are feet.

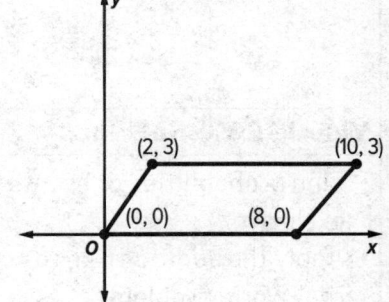

   a. The gardener wants to spread fertilizer over the flower bed to a thickness of 2.5 inches. How many cubic feet of fertilizer should the gardener use? Show your work.

   _____

   _____

   b. If the fertilizer for the flower bed is sold by the pound and the density of the fertilizer is 10 pounds per cubic foot, how many pounds should the gardener buy? Show your work.

   _____

12 Probability

CHAPTER FOCUS Learn about some of the objectives that you will explore in this chapter. Answer the preview questions. As you complete each lesson, return to these pages to check your work.

What You Will Learn	Preview Question
Permutations and Combinations	
• Use permutations and combinations to solve problems. • Use permutations and combinations with probabilities.	**SMP 2** Eight friends are going to ride a roller coaster. Only 4 people can sit in each car. How many ways are there to select a group of 4 of the friends to sit together on the roller coaster? **70**
Probability	
• Use permutations and combinations to compute probabilities of compound events. • Use permutations and combinations to solve problems involving probability.	**SMP 4** Five movies are randomly selected from a set of 6 action movies and 9 comedies. What is the probability that 3 action movies and 2 comedies are chosen? **0.2397**
Independent and Dependent Events	
• Identify whether two events are independent or dependent. • Find probabilities of independent and dependent events.	**SMP 6** Two number cubes are rolled. How does the event "the sum of the numbers is even and greater than 12" compare to the event "the sum of the numbers is not even or not greater than 12"? **The first event includes all even sums greater** **than 12. The second event is all sums between** **2 and 12 and all odd sums.**
Mutually Exclusive Events	
• Find probabilities of events that are mutually exclusive. • Apply the addition rule to solve mathematical and real-world problems.	**SMP 3** A six-sided number cube is rolled. Find the probability of drawing an even number or a prime number. Explain why the event is not mutually exclusive. **0.8333; The number 2 is both even and prime.**

What You Will Learn	Preview Question
Conditional Probability	
• Find the conditional probability of one event given another. • Use conditional probability to identify independent events.	**SMP 3** Two cards are drawn from a standard deck of 52 cards. The first card is not replaced before the second is drawn. Jason says the probability of drawing two kings is 0.0059. Lisa says it is 0.0045. Who is correct? Explain. _____ _____
Two-Way Frequency Tables	
• Construct and interpret two-way frequency tables of data. • Use two-way frequency tables to find conditional probabilities.	**SMP 4** The table shows the results of a sample survey asking whether people are for or against a new traffic light. Use the table to find the probability that a surveyed man does not want the traffic light.

	For	Against
Men	54	32
Women	86	28

Objectives

- Use permutations and combinations to solve problems.
- Use permutations and combinations with probabilities.

A **permutation** is an arrangement of objects in which order is important. The expression $n!$, which is read as *n factorial*, can be used to calculate the number of permutations of n objects. The **factorial** of a positive integer n, written $n!$, is the product of the positive integers less than or equal to n.

KEY CONCEPT

Permutations	Permutations with Repetition	Circular Permutations
The number of permutations of n distinct objects taken r at a time is denoted by $_nP_r$ and given by $$_nP_r = \frac{n!}{(n-r)!}.$$	The number of distinguishable permutations of n objects in which one object is repeated r_1 times, another is repeated r_2 times, and so on, is $$\frac{n!}{r_1! \cdot r_2! \cdot \ldots \cdot rk!}.$$	The number of distinguishable permutations of n objects arranged in a circle with no fixed reference point is $$\frac{n!}{n} \text{ or } (n-1)!.$$

EXAMPLE 1   Investigate Permutations

EXPLORE Hayley, Aja, and Arturo need to create 6-letter passwords from the letter tiles below.

Hayley's Letters	Aja's Letters	Arturo's Letters
F R Z L K Q G M	F R Z Q G M	F R Z F G Z

a. **INTERPRET PROBLEMS** How many possible passwords can Hayley create with her letters? Explain.

b. **CALCULATE ACCURATELY** How many possible passwords can Aja create with her letters? Explain.

c. **CALCULATE ACCURATELY** How many possible passwords can Arturo create with his letters? Explain.

d. MAKE A CONJECTURE Whose password has the least probability of someone randomly guessing it? Which password has the greatest probability? Explain your reasoning.

_____

_____

_____

A **combination** is an arrangement of objects in which order is *not* important.

KEY CONCEPT **Combinations**

> The number of combinations of n distinct objects taken r at a time is denoted by $_nC_r$ and given by
>
> $$_nC_r = \frac{n!}{(n-r)!r!}$$
>
> When order is important in a problem, use permutations.
> When order is not important in a problem, use combinations.

EXAMPLE 2 **Use Permutations and Combinations to Solve Problems**

At the annual Chess Club banquet, all 10 members will pose for a photograph. Six members will stand in the front row.

a. REASON QUANTITATIVELY Enter names in the chart below to show one possible arrangement for the front row. Should a combination or permutation be used to find the total number of possible groups of six members to stand in the front row? Is this different than finding the total number of possible arrangements for the front row? Explain your reasoning.

Chess Club Members			
Name	Year	Name	Year
Zoe	JR	Luisa	FR
Feng	SR	Nancy	SR
Jamal	SO	Imani	SR
Pedro	JR	Dan	JR
Bob	SR	Naomi	SO

◯ ◯ ◯ ◯ ◯ ◯

_____

_____

_____

b. INTERPRET PROBLEMS Find the total number of possible groups of six members to stand in the front row. What is the probability that the group of six members you chose in **part a** will be in the front row? Explain your reasoning.

_____

_____

_____

_____

c. CALCULATE ACCURATELY If there is an added requirement that a senior (SR) must stand on each end, find the total number of possible arrangements for the front row. If you choose a group of six members that satisfy the requirements and list them in order, what is the probability that the group of six members you chose will be standing in the exact order you arranged them? Explain your reasoning.

_____

_____

_____

d. USE A MODEL At the banquet, some circular tables have 7 seats and some have 10 seats, as shown. How many different groups of the chess club members can be seated at a 7-seat table? How many different seating arrangements are possible for one group of 7?

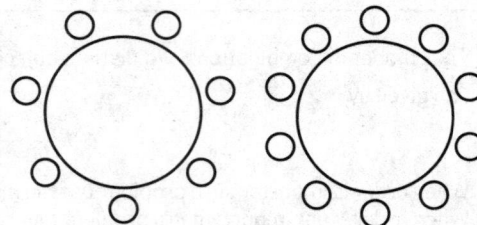

_____

_____

_____

e. INTERPRET PROBLEMS How many different seating arrangements are possible if all 10 members sit at the 10-seat circular table? Explain.

_____

_____

PRACTICE

1. In a computer game, locations that must be visited to complete a mission are represented on a map in the shape of a pentagon, as shown. Dotted lines represent additional pathways between the locations.

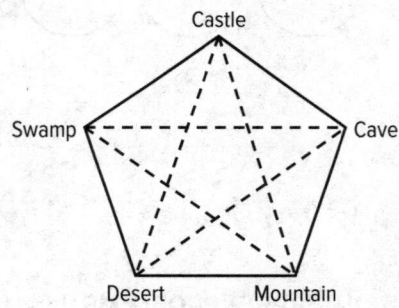

a. REASON QUANTITATIVELY To complete the mission at the advanced level of the game, all 5 locations must be visited in any order by choosing a starting location and then visiting each of the others. If each location is visited only once, in how many possible orders can all the locations be visited? Explain.

_____

b. USE STRUCTURE Trace one possible order on the map above. If an order that visits all 5 locations is chosen at random, what is the probability that it is your order? Explain.

_____

c. **REASON QUANTITATIVELY** At the beginner level of the game, any 3 locations can be visited in any order to complete the mission. How many possible combinations of three locations are there to complete the mission at the beginner level? What is the probability that the three locations will be the swamp, the desert, and the mountain? Explain.

_____

_____

d. **CONSTRUCT ARGUMENTS** At the intermediate level, players must visit 4 of the 5 locations. If each location is visited only once, will there be more possible ways to complete the mission if order matters or if order does not matter? Justify your answer.

_____

2. **FIND A PATTERN** At a party store, Latoya's mother bought the cut-out letters shown. What is the probability that if she randomly arranges all the letters, they will spell "WAY TO GO LATOYA" disregarding the spaces between words? Explain your reasoning.

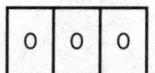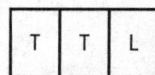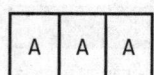

| O | O | O | | T | T | L | | A | A | A | | G | | Y | Y | | W |

_____

_____

3. **CRITIQUE REASONING** Charles claims that the number of ways n objects can be arranged if order matters is equal to the number of permutations of n objects taken $n - 1$ at a time. Do you agree with Charles? Justify your answer.

_____

_____

4. Jenna makes circular necklaces out of the 12 birthstones. Some necklaces include a gold heart pendant while others do not, as shown.

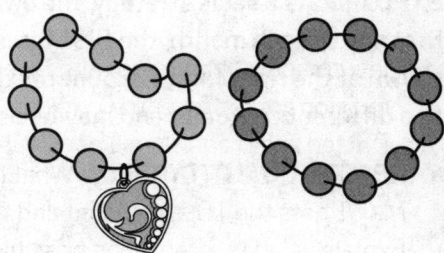

a. **CALCULATE ACCURATELY** If Jenna makes a necklace with 12 different birthstones arranged at random, what is the probability that they will be arranged in order chronologically by month? Explain.

_____

_____

b. **REASON QUANTITATIVELY** How many necklaces can Jenna make with the birthstones for January and February next to each other?

_____

_____

_____

Probability

Objectives

- Use permutations and combinations to compute probabilities of compound events.
- Use permutations and combinations to solve problems involving probability.

Mathematical Practices

1, 2, 3, 4, 5, 7, 8

The strategies you used for calculating the number of outcomes with permutations and combinations can be used to determine **probability**.

KEY CONCEPT

Complete the table by writing in the remaining information.

Probability
Probability is the ratio of the number of favorable equally likely outcomes to the number of possible equally likely outcomes. $P(\text{event}) =$ ______________

You can solve real-world problems involving probability by using permutations and combinations to calculate the numbers of both favorable and possible outcomes.

EXAMPLE 1 Investigate Probability with Permutations and Combinations

EXPLORE As a sales strategy, the owner of a hair salon advertises that on the first day of each month, the first 6 customers will receive one of the coupons shown at the right for a discount off their total bill. Each coupon is given at random to a different customer and may be used only once.

5% OFF	10% OFF
15% OFF	20% OFF
25% OFF	50% OFF

a. **REASON QUANTITATIVELY** What is the probability that the first customer on May 1 gets the 10% discount and the second customer gets the 25% discount? Explain using favorable and possible outcomes.

b. INTERPRET PROBLEMS How many different groups of two coupons can the first two customers on August 1 receive regardless of order? Given a specific set of two coupons, how many different ways can these two coupons be distributed to the first two customers? How can this be used to compute the probability in **part a**?

_____

_____

_____

_____

c. INTERPRET PROBLEMS How is finding the probability that the first 3 customers get discounts of 50%, 25%, and 20%, in that order, different from finding the probability that 3 customers get those discounts in any order?

_____

_____

_____

_____

d. FIND A PATTERN The salon owner adds a second coupon special: on the 15th day of every month, each of the 15 coupons shown is given at random to a different customer. What is the probability that all the 10% coupons are given out first, all the 20% coupons are given out second, all the 25% coupons are given out third, and the 50% coupon is given out last? Explain.

10% OFF	20% OFF
7 coupons	4 coupons

25% OFF	50% OFF
3 coupons	1 coupon

_____

_____

_____

You can use permutations and combinations to find the probabilities of **compound events** which consist of two or more simple events.

EXAMPLE 2   Solve Problems with Permutations and Combinations

As part of a school beautification project, 12 alumni each donated a tree to be planted on the school grounds. The types of trees are shown in the table. There will be a sign next to each tree with the donor's name.

Donated Trees	
Type	Number of Trees
Cherry	5
Dogwood	4
Crabapple	2
Redbud	1

a. USE TOOLS If the trees are planted in a row at random, what is the probability that they will be in alphabetical order by donor name? Explain.

_____

_____

_____

b. USE REASONING If 4 trees are randomly selected and planted near the school entrance, what is the probability that they will all be dogwood trees? Explain.

_____

_____

_____

c. USE STRUCTURE Of the five cherry trees, Bonita and Terrance each donated one. If all nine cherry and dogwood trees are randomly planted in a row, what is the probability that Bonita's tree will be the first tree on the left? If the first tree on the left is already chosen, what is the probability that Terrance's is the second? Use these probabilities to find the probability that Bonita's tree is the first on the left and Terrance's is the second. Explain.

_____

_____

_____

_____

d. INTERPRET PROBLEMS If all 12 trees are planted in a row, what is the probability that the two crabapple trees will be planted in the first and last positions?

_____

_____

_____

_____

_____

e. USE STRUCTURE If the cherry and the crabapple trees are randomly planted in a circle as shown, what is the probability that the trees will be in alphabetical order by donor name? Explain.

_____

_____

_____

f. CRITIQUE REASONING Suppose the cherry, crabapple, and redbud trees are planted randomly in a row. Luis says that the probability that the redbud is the first followed by all the crabapples and then all the cherry trees is $\frac{1}{40,320}$. Is Luis correct? If not, find and correct his error.

_____

_____

_____

1. **USE STRUCTURE** If the numbers 1, 2, 3, 4, 5, and 6 are listed in random order, what is the probability that the sum of the first two numbers listed is 7? Explain.

_____

_____

_____

2. **USE TOOLS** The scores on a math midterm exam are shown at the right. If the test papers are arranged in a stack, what is the probability that the 5 tests on top will be those with the 5 highest scores, in descending order? Explain.

Midterm Exam Scores				
95	62	71	93	86
88	100	90	75	51
55	82	72	71	86
86	95	74	75	60

_____

_____

_____

_____

_____

_____

3. **USE A MODEL** A discount bin at the supermarket contains 15 cans of a generic tomato soup. Of these 15 cans, 7 are dented. If Martha selects 4 cans at random from the bin, what is the probability that all 4 are dented? Explain.

_____

_____

_____

4. **CRITIQUE REASONING** The 10 letter tiles below, when unscrambled, spell a word that is unusual because it has three consecutive pairs of double letters. Enrique says that it is not necessary to know the word in order to calculate the probability that a random arrangement of all 10 letters will spell the word. Is Enrique correct? Justify your answer.

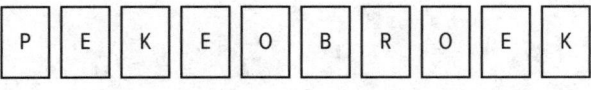

| P | E | K | E | O | B | R | O | E | K |

_____

_____

_____

Objectives

- Identify whether two events are independent or dependent.
- Find probabilities of independent and dependent events.

Compound events consist of two or more simple events.
Suppose that a compound event consists of events A and B. If the outcome of event A does not affect the outcome of event B, then A and B are **independent events**. If the outcome of event A affects the outcome of event B, or vice versa, then A and B are **dependent events**. The probability that event B occurs given that event A has already occurred is represented by $P(B|A)$.

KEY CONCEPT

Probability of Two Independent Events	Probability of Two Dependent Events
The probability that two independent events both occur is the product of the probabilities of each individual event.	The probability that two dependent events both occur is the product of the probability that the first event occurs and the probability that the second event occurs given the first event has already occurred.
If two events A and B are independent, then $P(A \text{ and } B) = P(A) \cdot P(B)$.	If two events A and B are dependent, then $P(A \text{ and } B) = P(A) \cdot P(B|A)$.

Before finding the probability of a compound event, decide if it consists of independent or dependent events.

EXAMPLE 1 Investigate Independent and Dependent Events

EXPLORE Keesha, Raoul, Janelle, and Jacob are designing a game for the school carnival. Their game consists of a small wading pool filled with rubber ducks, some of which have prize codes on the bottom. Players can buy a chance to select a duck. To get started, each student created a plan as shown in the table.

a. **REASON QUANTITATIVELY** Suppose the first player selects 2 ducks. Is the probability of choosing a winning duck the same for the first and second duck under Keesha's plan? Under Raoul's plan? Is the probability of winning both times greater under Keesha's plan or Raoul's plan? Explain.

Plans for the Duck Game			
Name	Total Number of Ducks	Number of Prize Ducks	Replace the Ducks?
Keesha	200	30	Y
Raoul	300	30	N
Janelle	300		Y
Jacob	200	20	

b. INTERPRET PROBLEMS The probability of selecting 2 winning ducks is $\frac{1}{144}$ under Janelle's plan. Complete Janelle's row in the table. Explain your reasoning.

_____

_____

c. FIND A PATTERN Under Jacob's plan, if the first player selects 2 ducks, the probability of winning both times is $\frac{38}{3980}$. Complete Jacob's row in the table. Explain.

_____

_____

A **uniform probability model** is a model in which each outcome has an equal probability of occurring. However, each event may not be equally likely.

EXAMPLE 2 **Apply the Multiplication Rule to Uniform Probability Models**

Apply the general Multiplication Rule in each situation.

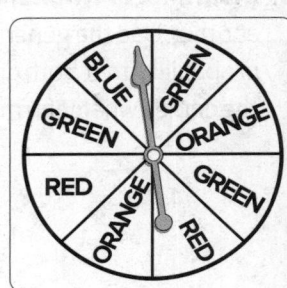

a. CONSTRUCT ARGUMENTS If Fred spins the spinner twice, determine the probability that he lands on sections labeled "orange" and "green." Show that $P(orange) \cdot P(green|orange) = P(green) \cdot P(orange|green)$ in this situation and explain why this is true in terms of the model. Justify your answer.

_____

_____

_____

_____

_____

b. USE STRUCTURE Yolanda takes one tile out of the sack at random and does not replace it. Then she takes out another tile at random. Why is this an example of a uniform probability model? If Yolanda does the same process but is only concerned about the letter on the tile she takes out, is it still a uniform probability model? If Yolanda chooses two tiles, determine the probability that she selected an R and then an E. Explain.

_____

_____

_____

_____

You can analyze the probability of two events occurring together and determine whether or not the events are independent. For two events A and B, if the general Multiplication Rule can be used to show that $P(B|A) = P(B)$ or $P(A|B) = P(A)$ then the two events are independent.

EXAMPLE 3 **Identify Independent Events**

At a picnic, there is a cooler with some bottles in it. The table contains some information about the contents, but other information is missing.

Drinks in the Cooler	
Type of Juice	Number of Bottles
Orange	
Cranberry	1
Grape	1
Apple	
TOTAL	

a. **REASON QUANTITATIVELY** Suppose Joe randomly takes out one bottle and replaces it. Then he randomly takes out another bottle and replaces it. If the probability that the first bottle was cranberry juice and that the second bottle was grape juice is $\frac{1}{196}$, what information in the table can you fill in? Are the two events independent? Explain.

_____

_____

_____

_____

b. **INTERPRET PROBLEMS** Jodi randomly takes 1 bottle of juice, keeps it, and then takes another. Use the general Multiplication Rule to determine if the events are independent. The probability that both drinks are apple juice is $\frac{12}{182}$ or $\frac{6}{91}$. If Jodi then puts both bottles back into the cooler, what information in the table can you fill in based on this knowledge? Explain.

_____

_____

_____

_____

c. **FIND A PATTERN** A second cooler contains the sandwiches shown in the table at the right. Tyrone selects one drink at random from the first cooler and one sandwich at random from the second cooler. What is the probability that he selects a bottle of orange juice and a tuna sandwich? Are the events independent? Explain.

Sandwiches in Another Cooler	
Type of Sandwich	Number of Sandwiches
Tuna	5
Ham	1
Egg Salad	2
TOTAL	8

_____

_____

_____

d. **REASON QUANTITATIVELY** Suppose Tyrone chooses one sandwich at random from the cooler but then decides to exchange his current sandwich for another one at random. What is the probability that he will choose an egg salad sandwich and exchange it for another egg salad sandwich? Are the events independent? Explain your reasoning.

_____

_____

_____

A magician is doing a trick with a standard deck of 52 cards. Each new card trick starts with a fresh deck of cards. Use this information for Exercises 1 and 2.

1. **REASON QUANTITATIVELY** A volunteer randomly selects a card, looks at it, and puts it back in the deck. Then the magician randomly selects the same card. Are the events independent? What is the probability that they both pick the queen of spades? Explain.

_____

_____

2. **FIND A PATTERN** The magician shuffles a new deck of cards and randomly selects 4 of them without replacing each card. Are the events independent? What is the probability that all 4 cards are aces? Explain.

_____

_____

Two boxes each contain computer chips, some of which are defective, as shown in the table. Use this information for Exercises 3 and 4.

3. **INTERPRET PROBLEMS** Derek randomly selects one chip from Box B, puts it in his pocket, and then randomly selects another chip from Box B. What is the probability that both chips are defective? Explain.

	Number of Chips	Defective
Box A	100	4%
Box B	150	2%

_____

_____

4. **USE STRUCTURE** Sunita randomly selects one chip from Box A, and then she randomly selects another chip from Box A. The probability that both chips are defective is $\frac{1}{625}$. Did Sunita replace the first chip before selecting the second one? Explain.

_____

_____

5. **USE A MODEL** Each square of the area model represents 1 square foot of a garden. For each plant, a random square is to be chosen as a planting site. Is this a uniform probability model? A square is to be chosen for a sunflower and then another square chosen for a tomato plant. What is the probability that shaded squares are chosen for both plants? Explain.

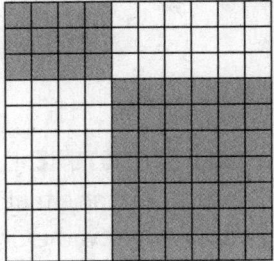

_____

_____

_____

_____

Mutually Exclusive Events

Mathematical Practices
1, 2, 3, 6, 7, 8

Objectives

- Find probabilities of events that are mutually exclusive.
- Apply the addition rule to solve mathematical and real-world problems.

If two events are **mutually exclusive**, then they cannot occur at the same time. Another way to say this is that they have no outcomes in common.

KEY CONCEPT

Complete the table by writing the missing information.

Probability of Mutually Exclusive Events
Words
If two events, A and B, are mutually exclusive, then the probability that A or B occurs is the sum of the probabilities of each individual event.
Symbols
If two events, A and B, are mutually exclusive, then $P(A \text{ or } B) = P(A) + P(B)$.

EXAMPLE 1 Investigate Mutually Exclusive Events

EXPLORE Each year, one student attending the annual high school picnic is randomly selected to win a free copy of the yearbook. Numbers of students attending the picnic are given in the table.

a. **REASON QUANTITATIVELY** Clarissa wants to know the probability that a junior or senior will win. Are these events mutually exclusive? Find the probability and explain your reasoning.

_____

_____

_____

b. **USE STRUCTURE** Draw a Venn diagram in the space provided to represent the situation. How did you decide how many circles to draw and how the circles are related?

_____

_____

_____

Number of Students Attending the Annual School Picnic	
Class	Number of Students
Freshman	112
Sophomore	78
Junior	124
Senior	226
Total	540

c. CRITIQUE REASONING Nelida is a junior and her sister Rosa is a freshman. Nelida claims that the probability that either she or her sister will win the yearbook is about 44%. Do you agree with Nelida's claim? If not, identify Nelida's error and correct it.

_____

_____

_____

If you want to find the probability that event A occurs or event B occurs if A and B are not mutually exclusive, you can apply the Addition Rule for events that are not mutually exclusive.

KEY CONCEPT

Complete the table by writing in the missing information.

Probability of Events That Are Not Mutually Exclusive
Words
If two events, A and B, are not mutually exclusive, then the probability that A or B occurs is the sum of their individual probabilities minus the probability that both A and B occur.
Symbols
If two events, A and B, are not mutually exclusive, then $P(A \text{ or } B) = P(A) + P(B) - P(A \text{ and } B)$.

EXAMPLE 2 Find Probability of Events That Are Not Mutually Exclusive

Dan has won many awards in various sports. The table shows how many blue, white, and red ribbons he has won in four different types of sports.

a. CALCULATE ACCURATELY If Dan randomly selects one of his ribbons, what is the probability that he selects a blue ribbon or a ribbon for track and field? Explain.

Sport	Blue	White	Red
Swimming	1	7	3
Track & Field	6	2	4
Equestrian	2	4	8
Soccer	1	5	5

_____

_____

_____

_____

b. INTERPRET PROBLEMS Draw a Venn diagram in the space provided to represent the situation in **part a**. Label the diagram and use the numbers from the table to write the appropriate quantities in each area.

c. USE STRUCTURE Use the Venn diagram you drew in **part b** to help you explain why the probability that both events occur must be subtracted when the two events are not mutually exclusive.

_____

_____

_____

d. DESCRIBE A METHOD Find the probability that if Dan selects a ribbon at random, it will be red or white. How does your strategy for solving this problem compare to your strategy for solving the problem in **part a**?

_____

_____

_____

There are instances in which the probability of an event not occurring is needed. The **complement** of event A, referred to as *not A*, is the event that includes all outcomes in which A does not occur. Because the events A and *not A* are mutually exclusive and account for the entire sample space, $P(A) + P(not\ A) = 1$. In general, the probability of the complement of A is $P(not\ A) = 1 - P(A)$.

EXAMPLE 3 **Find the Probability of a Complement**

a. INTERPRET PROBLEMS A card game using a standard deck gives greater rewards to cards drawn with higher face values. If two cards are to be drawn with replacement, find the probability of drawing a 2 each time. Then find the probability of not drawing a 2 each time.

_____

_____

b. USE STRUCTURE If a standard die is rolled two times, find the probability that the sum of the two rolls is not 2.

_____

_____

PRACTICE

A polling company conducted a survey of 1200 registered voters from 4 states. Use the table for Exercises 1 and 2.

1. CRITIQUE REASONING An analyst at the company says that the probability that a randomly selected voter in the study will be from Maine or Florida is 4.6%. Do you agree? If not, identify the analyst's error and correct it.

_____

_____

_____

_____

Survey of Registered Voters	
State	**Number of Voters**
California	456
Florida	320
Maine	207
Washington	217
Total	1200

2. **USE STRUCTURE** The analyst drew a Venn diagram to show that 176 of voters surveyed were under age 26 and that 28 of those were also voters from Maine. If a voter from the survey is randomly selected, what is the probability that the voter is under age 26 or from Maine? Explain.

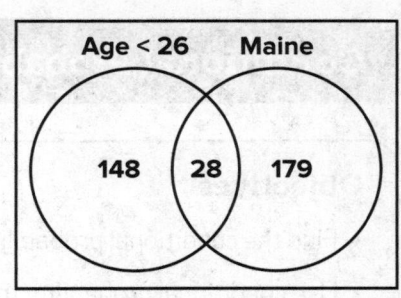

_____

_____

_____

3. **REASON QUANTITATIVELY** A standard die is rolled twice. Find the probability that the sum of the two rolls is not 4.

_____

_____

Barb is planning her summer garden. The table shows the number of bulbs she has according to type and color of flower. Use the table for Exercises 4–5.

Flower	Orange	Yellow	White
Dahlia	5	4	3
Lily	3	1	2
Gladiolus	2	5	6
Iris	0	1	4

4. **USE STRUCTURE** If Barb randomly selects one of the bulbs, what is the probability that she selects a bulb for a yellow flower or a dahlia? Are the events mutually exclusive? Explain.

_____

_____

_____

5. **INTERPRET PROBLEMS** What is the probability that a randomly selected bulb will be for an iris or an orange flower? Are the events mutually exclusive? Explain your reasoning. Draw a Venn diagram in the space provided to support your explanation.

_____

_____

6. **CRITIQUE REASONING** Rita says that if the word _or_ is used to describe the relationship between two events, then the events must be mutually exclusive. Wayne says that the word _or_ sometimes indicates that the events are mutually exclusive, but not always. Do you agree with either student? Justify your answer.

_____

_____

_____

Objectives

- Find the conditional probability of one event given another.
- Use conditional probability to identify independent events.

Sometimes additional information is known about an event that affects its probability. The probability that is affected is called **conditional probability**. The notation $P(A \mid B)$ represents the probability of event A given that event B has already occurred. To find the conditional probability of event A given B, first find the probability of B and the probability of A and B.

KEY CONCEPT

Conditional Probability

The conditional probability of event A given event B is

$P(A \mid B) = \dfrac{P(A \text{ and } B)}{P(B)}$, where $P(B) \neq 0$.

EXAMPLE 1 Investigate Conditional Probability

EXPLORE Fifteen cards numbered 1–15 are placed in a hat.

a. **USE STRUCTURE** What is the probability of picking a card with an odd number? What is the probability of picking a card that has a multiple of 3 on it? Explain.

_____

b. **REASON QUANTITATIVELY** A card is picked with an odd number on it. What is the probability that the card has a multiple of 3 on it? Compare this probability with the probability of picking a card with a multiple of 3 on it from **part a**. What does this say about the events? Explain.

_____
_____
_____
_____

EXAMPLE 2 Conditional Probability with Percentages

A school picnic offers students hamburgers, hot dogs, fries, and a drink.

a. **INTERPRET PROBLEMS** At the picnic, 60% of the students order a hamburger and 48% of the students order a hamburger and fries. What is the conditional probability that a student who orders a hamburger also orders fries? Explain.

_____
_____

b. REASON QUANTITATIVELY If 50% of the students ordered fries, are the events ordering a hamburger and ordering fries independent? Explain.

_____

_____

c. USE STRUCTURE If 80% of the students who ordered a hot dog also ordered a drink and 35% of all the students ordered a hotdog, find the probability that a student at the picnic orders a hot dog and drink. Explain.

_____

_____

EXAMPLE 3 Conditional Probability with Cards

A game offers prizes depending on the cards drawn from a standard deck of 52 cards.

a. USE STRUCTURE What is the probability that a player randomly draws an ace from the deck of cards? If the player returns the card to the deck, what is the probability that the next randomly drawn card is a club? Explain.

_____

_____

b. INTERPRET PROBLEMS A player draws a card, returns it to the deck, and draws another card. What is the probability that the first card is an ace and the second card is a club? Use the Key Concept box to find the probability of drawing a club for the second card after drawing an ace for the first card. How does this compare to the probability of drawing a club for the first card? Explain this in terms of the context.

_____

_____

_____

_____

PRACTICE

1. **PLAN A SOLUTION** Twenty balls numbered 1–20 are placed in a box.

   a. If a ball is randomly chosen from the box and has an odd number on it, find the conditional probability that the ball has a prime number on it. Explain.

   _____

   _____

   b. If a ball is randomly chosen from the box and has a multiple of 5 on it, find the conditional probability that the ball has a multiple of 3 on it. Explain.

   _____

   _____

c. Two balls are randomly chosen from the box without replacing the first one. If the first ball chosen has an even number on it, find the probability that the second ball has an odd number on it. How does this compare to the probability of choosing the first ball with an odd number on it? Explain the probabilities in terms of independent and dependent events.

_____

_____

_____

_____

_____

2. REASON QUANTITATIVELY Answer the following questions about the courses taken by students at a high school.

a. Of all the students, 25% are enrolled in Algebra and 20% are enrolled in Algebra and Health. If a student is enrolled in Algebra, find the probability that the student is enrolled in Health as well.

_____

_____

b. If 50% of the students are enrolled in Health, are being enrolled in Algebra and being enrolled in Health independent events? Explain.

_____

_____

c. Of all the students, 20% are enrolled in Accounting and 5% are enrolled in Accounting and Spanish. If being enrolled in Accounting and being enrolled in Spanish are independent events, what percent of students are enrolled in Spanish? Explain.

_____

_____

_____

3. COMMUNICATE PRECISELY Let event A be owning a house and event B be owning a car. Do you expect the two events to be independent or dependent? How do you think $P(A \mid B)$ compares to $P(B \mid A)$? Explain your reasoning.

_____

_____

_____

_____

_____

The Venn diagram represents the results of a random survey about where students study for final exams. Let *L* represent the library and *H* represent at home. Use this information for Exercises 4–6.

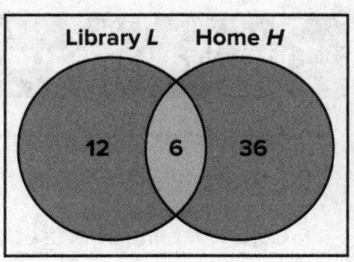

Library *L*   Home *H*

12   6   36

4. **USE A MODEL** A total of 60 students responded to the survey. Determine the number of students who replied that they study neither at the library nor at home.

_____

_____

5. **USE STRUCTURE** What is the probability that if a student selected the library, he or she selected the library and at home? Explain.

_____

6. **CRITIQUE REASONING** A student says that selecting the library and selecting at home are independent events. Do you agree? Explain.

_____

_____

7. There are 52 cards in a standard deck. A face card is a jack, queen, or king.

   a. **INTERPRET PROBLEMS** Find the probability of randomly drawing a face card. If a face card is drawn, what is the probability that it is a jack?

   _____

   _____

   b. **CALCULATE ACCURATELY** Face-value cards are the cards numbered 2–10. Two cards are to be randomly drawn without replacing the first card. Find the probability of drawing two face-value cards and the conditional probability that exactly one of those cards is a 4. Explain.

   _____

   _____

   _____

   _____

8. **REASON QUANTITATIVELY** A high school calculus class has 40 students, 24 of whom are males. Of the 40 students, 9 are males wearing blue shirts.

   a. Find the conditional probability that a randomly selected male is wearing a blue shirt.

   _____

   _____

   b. If there are 14 students total wearing blue shirts, find the conditional probability that a randomly selected student wearing a blue shirt is a female.

   _____

   _____

Objectives

- Construct and interpret two-way frequency tables of data.
- Use two-way frequency tables to find conditional probabilities.

Mathematical Practices
1, 2, 3, 4, 6, 7, 8

A **two-way frequency table** is used to show the frequencies of data that pertain to two variables. The rows represent one variable and the columns represent the other. **Joint frequencies** are reported in the interior of the table, while **marginal frequencies** are reported in the *Totals* row and column along the outside of the table. A **relative frequency table** displays the ratio of the number of observations in each category to the total number of observations expressed as a fraction or percent.

EXAMPLE 1   Investigate Two-way Frequency Tables

EXPLORE A market research firm asks a random sample of 240 adults and children at a movie theater whether they would rather see a new summer blockbuster in 2-D or 3-D. The survey shows that 64 adults and 108 children prefer 3-D, while 42 adults and 26 children prefer 2-D.

a. USE A MODEL Organize the responses in the two-way frequency table at the right. How did you identify the variables and the categories for the variables?

Age	Prefers 2-D	Prefers 3-D	Totals
Adult			
Child			
Totals			

_____

_____

_____

b. REASON QUANTITATIVELY How many more people surveyed would prefer seeing the movie in 3-D rather than 2-D? Explain how you found the answer.

_____

_____

c. CALCULATE ACCURATELY Convert the table from **part a** into a two-way relative frequency table. Round to the nearest tenth of a percent. Out of every 10 people surveyed, about how many would prefer to see the movie in 3-D? Explain.

Age	Prefers 2-D	Prefers 3-D	Totals
Adult			
Child			
Totals			

_____

_____

d. USE STRUCTURE Find the probability that a person surveyed prefers seeing the movie in 3-D, given that he or she is an adult. Write the formula that you used to perform the calculation.

_____

_____

e. CRITIQUE REASONING An analyst at the firm claims that the probability that a person surveyed is a child given that he or she does not prefer to see the movie in 3-D is 10.8%. Do you agree? Justify your answer.

_____

_____

f. INTERPRET PROBLEMS Is a preference for 2-D or 3-D movies independent of age? Explain your reasoning.

_____

_____

_____

_____

Two-way frequency tables can also be used to analyze the results of experiments.

EXAMPLE 2 **Determine Independence of Events**

In an experiment, Marlena rolled a pair of dice 75 times and organized the results in the frequency table at the right. Unfortunately, some of the data in the table was mistakenly erased.

Die	1 or 6	2, 3, 4, or 5	Totals
Die 1	25		75
Die 2		50	
Totals			150

a. EVALUATE REASONABLENESS Complete the table by writing in the values that were erased. What patterns, if any, do you notice?

_____

b. CALCULATE ACCURATELY Convert the table from **part a** to a two-way relative frequency table. Round to the nearest tenth of a percent. What fraction of the rolls that were a 1 or a 6 were rolled on die 1? Explain how the fraction is related to the conditional probability that a roll was on die 1 given that it was a 1 or a 6.

Die	1 or 6	2, 3, 4, or 5	Totals
Die 1			
Die 2			
Totals			

_____

_____

c. CONSTRUCT ARGUMENTS Is the event of rolling a 1 or a 6 independent of whether die 1 is rolled or die 2 is rolled? Explain your reasoning.

_____

_____

d. REASON QUANTITATIVELY Find the conditional probability that the second die rolled will be a 2, 3, 4, or 5 given that the first die was a 1 or 6. Are the two events independent? Explain your reasoning.

_____

_____

Amanda asked a random sample of seniors at her high school whether they own a car and whether they have a job. The results of her survey are shown in the two-way relative frequency table. Use the table for Exercises 1–3.

1. **CRITIQUE REASONING** Amanda says that the conditional probability that a student has a job given that they have a car is 46.7%. Do you agree?

Car	Has a Job	Does Not Have a Job	Totals
Has a Car	21.9%	12.5%	34.4%
Does Not Have a Car	25%	40.6%	65.6%
Totals	46.9%	53.1%	100%

2. **USE STRUCTURE** Find the probability that a student has a job given that he or she does not have a car. Explain how you found the answer.

3. **REASON QUANTITATIVELY** If there are 270 seniors at Amanda's high school, about how many would you predict have a car if they say they have a job? Explain.

For a business report on technology use, Darnell asks a random sample of 72 shoppers whether they own a smart phone and whether they own a tablet computer. His survey shows that out of 51 shoppers who own smart phones, 9 of them also own a tablet, while out of 21 shoppers who do not own smart phones, 15 of them don't own tablets either.

4. **USE A MODEL** Organize Darnell's data into the two-way frequency table. Explain how you decided on the variables.

Device	Tablet	No Tablet	Totals
Smart Phone			
No Smart Phone			
Totals			

5. **CALCULATE ACCURATELY** Convert the table from **Exercise 4** to a two-way relative frequency table. Round to the nearest tenth of a percent. What is the probability that a shopper has a smart phone and a tablet computer? Explain.

Device	Tablet	No Tablet	Totals
Smart Phone			
No Smart Phone			
Totals			

_____

_____

6. **REASON QUANTITATIVELY** Find the conditional probability that a shopper has a tablet computer, given that he or she has a smart phone. Explain.

_____

_____

7. **EVALUATE REASONABLENESS** What is a possible trend Darnell could report based on the data? Explain.

_____

_____

8. **CONSTRUCT ARGUMENTS** Are the variables in this two-way frequency table independent? Explain how you can tell.

_____

_____

_____

9. **USE STRUCTURE** A random sample of 500 citizens in a town with population 20,000 was selected to complete a survey on upcoming changes. They were to answer whether or not they wanted a new highway and whether or not they wanted to have a mobile home park built. The results are shown in the two-way relative frequency table.

Changes	For M.H. Park	Against M.H. Park	Totals
For Highway	21.8%	37.6%	59.4%
Against Highway	15.4%	25.2%	40.6%
Totals	37.2%	62.8%	100%

a. Find the probability that a citizen is for the mobile home park if they are against the highway.

_____

b. Find the probability that a citizen is for the highway if he or she is for the mobile home park.

_____

c. Of the 20,000 individuals in the town, about how many would you expect to be against the mobile home park and against the highway?

_____

_____

Copyright © McGraw-Hill Education

Two-Way Frequency Tables **405**

Ball-Toss Game

Provide a clear solution to the problem. Be sure to show all of your work, include all relevant drawings, and justify your answers.

A ball-toss game is played at a carnival. You toss a ball and earn a prize based on where it lands. If the ball lands in a blue cup, you win the large prize. If the ball lands in a green cup, you win a small prize. If the ball lands in a white cup, you win a sticker. There are 5 blue cups, 20 green cups, and 75 white cups. The diameter of each cylindrical cup is 3.5 inches. The square table holding the cups has a side length of 35 inches. Assume that a ball thrown always lands on the table or in a cup.

Part A
Find the area of the table and the area of a cup opening.

Part B
It is possible for the ball to land on the table between the cups. In that case, you do not win a prize. What is the probability that you do not win a prize?

Part C

What is the probability that you win a large prize? What is the probability that you win a small prize? What is the probability that you win a sticker?

Part D

Takala, the carnival owner, decides to set up the table such that the players cannot see the colors of the cups and thus throw the ball onto the table at random. She sees that Aleta, the carnival worker running the table, has arranged the cups such that all of the blue and green cups are placed together in the center. Takala tells Aleta to arrange the cups such that all blue and green cups are surrounded by white cups, as this will decrease the probability that a player wins a small or large prize. The carnival worker says that the arrangement does not matter. Who is correct? Explain your reasoning.

Step Right Up and Win a Car

Provide a clear solution to the problem. Be sure to show all of your work, include all relevant drawings, and justify your answers.

You pay $50 for one of 100 raffle tickets sold for a fundraising campaign. A person can only buy one raffle ticket. The person whose raffle ticket is selected in a random drawing gets to play a game to win a new car. In the game, the contestant selects one of three identical closed boxes and wins the contents of the box. One box contains the keys to a new car. Each of the other boxes contains a toy car. The new car is worth $18,000. Each of the toy cars is worth $1.29.

Part A
Is the decision as to which person gets to play the game a fair decision? Explain.

Part B
Is the event of winning the new car independent of the raffle outcome? Explain your answer in terms of the probabilities of these events.

Part C

A local car dealership donated the new car and the toy cars to the fundraiser in exchange for free advertising. Assume that, instead, the fundraiser had to purchase all three cars. Calculate the cost per ticket that would make the raffle a fair contest in such a scenario. Explain whether the raffle as part of a fundraising event *should* be a fair contest.

Part D

A raffle ticket is selected and the contestant selected wins a toy car. The host decides to randomly select another raffle ticket and the contestant can choose between the two remaining boxes. Given that you were not selected as the first contestant, find the probability of winning the new car.

1. Emily, Rashaad, Nezra, Greg, and Enrique are working the late shift. The boss will randomly choose two of them to unload a truck. What is the probability that Nezra and Enrique will be selected?

2. The Venn diagram shows the results of survey of 200 students about pet ownership.

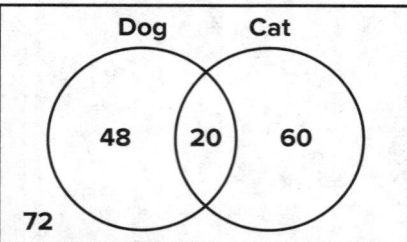

Complete the two-way frequency table with this information.

	Dog	No Dog	Total
Cat			
No Cat			
Total			

If a dog owner is selected at random, what is the probability that the person will not also have a cat?

If a person who doesn't own a cat is selected at random, what is the probability that the person will have a dog?

3. Two dice are rolled. The sum of the two dice is 8. What is the probability that one die is a 5?

The dice are rolled again. The sum of the two dice is 8. What is the probability that one die is a 1?

4. Bianca, Albert, Cara, Darren, and Edwin have tickets to see a play. Their seats are together in the same row. If they sit in the five seats in random order, what is the probability that they will sit in alphabetical order, from left to right, by their first names?

5. A bag contains marbles of various colors. The table below shows how many marbles of each color are in the bag.

Red	Green	Blue	Yellow	Black
12	8	11	13	6

Three marbles are chosen without replacement. What is the probability that the first marble will be green, the second yellow, and the third green?

Two marbles are chosen with replacement. What is the probability that both will be black?

6. Determine whether the events listed in the table are independent or dependent. Check Independent or Dependent in each row.

Events	Independent	Dependent
You roll two number cubes, and they land on a 3 and a 4.		
A bag contains 26 letter tiles corresponding to the letters of the alphabet. Two letter tiles are randomly selected and placed on a game board. The tiles spell the word "up."		
A spinner lands on red on the first spin. Then it lands on yellow on the second spin.		
Ms. Jones randomly selects two students to pass out papers. She selects the two oldest students.		

7. Ana and Molly are going to see a movie together, but they cannot agree on what movie to see. They decide to roll two number cubes. If the product is even, Ana will choose the movie. If the product is odd, Molly will choose the movie. Is this a fair way to make the decision? Explain your answer.

_____

_____

8. Seth needs to select 2 beads from a box that contains 10 red, 15 yellow, 5 blue, and 6 white beads. If he randomly selects a bead from the box and then selects a second bead without replacing the first one, is it likely that he will not select 2 red beads? Explain your answer.

_____

_____

9. Consider the tree diagram at right.

   a. What is $P(E \mid B)$?

_____

   b. What is $P(C \text{ and } E)$? Show your work.

_____

_____

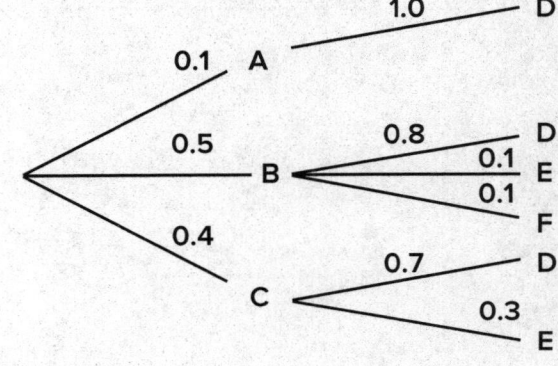

10. AnXuan is in a class of 36 students having 14 boys and 22 girls. The instructor wants to randomly select four students to represent the class in a school-wide awards ceremony.

   a. AnXuan is hoping to be selected along with one of her friends in the class. If the instructor randomly selects four students, what is the probability that AnXuan and her friend are selected for the group?

_____

_____

_____

   b. If the instructor decides to randomly choose two boys and two girls, is this a fair decision maker? Explain your reasoning.

_____

_____

_____

   c. If the group of students selected contains two randomly selected boys and two randomly selected girls, how many different groups of two boys and two girls can be selected to represent the class?

_____

_____

_____